D1055283

Frontiers in
Biochip Technology

Wan-Li Xing
Jing Cheng

Editors

Frontiers in Biochip Technology

With 158 Illustrations

 Springer

Wan-Li Xing, PhD
Tsinghua University School
  of Medicine
National Engineering Research Center
  for Beijing Biochip Technology
  (NERCBBT)
CapitalBio Corporation
Beijing, China

Jing Cheng, PhD
Tsinghua University School
  of Medicine
National Engineering Research Center
  for Beijing Biochip Technology
  (NERCBBT)
CapitalBio Corporation
Beijing, China

Library of Congress Control Number: 2005923842

ISBN-10: 0-387-25568-0   e-ISBN-0-387-25585-0
ISBN-13: 978-0387-25568-2

Printed on acid-free paper.

© 2006 Springer Science+Business Media, Inc.
All rights reserved. This work may not be translated or copied in whole or in part without the written permission of the publisher (Springer Science+Business Media, Inc., 233 Spring Street, New York, NY 10013, USA), except for brief excerpts in connection with reviews or scholarly analysis. Use in connection with any form of information storage and retrieval, electronic adaptation, computer software, or by similar or dissimilar methodology now known or hereafter developed is forbidden.
The use in this publication of trade names, trademarks, service marks, and similar terms, even if they are not identified as such, is not to be taken as an expression of opinion as to whether or not they are subject to proprietary rights.

Printed in the United States of America.   (SPI/EB)

9 8 7 6 5 4 3 2 1

springeronline.com

F.W. Olin College Library

Contents

Preface

This book is a collection of manuscripts from twenty internationally renowned experts in the field of biochip technology. It covers a wide range of topics in the area of biochips, including microarray technology and its applications, microfluidics, drug discovery, surface chemistry, lab-on-chip technology, and bioinformatics. The idea for publishing this book came from International Forum on Biochip Technologies 2002 (IFBT 2002), the second time IFBT was held in Beijing. More than 300 scientists in the area of biochip technologies attended the forum to communicate their updated research results and discuss the development of biochip technologies. After the forum, twenty experts attending the forum from both academia and industry were invited to prepare manuscripts for this book. The authors not only describe their latest research, but they also discuss the current trends in biochip technology. We believe that both researchers in the field of life sciences and professionals in the field of biochips will benefit from this book. The collected papers from IFBT 2000 were published by Springer in a book entitled *Biochips: Technology and Applications*. Readers are encouraged to read these two books and witness the progress of biochip technologies.

Jing Cheng, PhD

Part I

Microarrays

1

A Homogenous Microarray for Enzymatic Functional Assays
Chemical Compounds Microarray

HAICHING MA*, YUAN WANG, AMY S. POMAYBO, AND CONNIE TSAI
Reaction Biology Corporation, Malvern, PA 19355, USA

Abstract: Microarrays as an emerging research tool promises to play a pivotal role in the post genomic era. However, in spite of the fast development of this technology special requirements, such as the immobilization and delivery of bio-reagents on the chip surface limit the utilization of microarrays, especially for small chemical compound libraries. We have developed a unique homogenous microarray system that overcomes these limitations and can be used to array most biofunctional molecules, such as small chemical compounds, peptides and proteins without pre-immobilization. A standard microscope slide containing up to 5000 microarray dots, with volumes less than 2 nanoliter each and acting as individual reaction centers, can be printed with standard DNA arrayer. An aerosol deposition technology was adapted to deliver extremely small volumes of biofluids uniformly into each reaction center. The fluorescence based reaction signals were then scanned and analyzed with standard DNA scanner and DNA array analyzing software. With this platform, we demonstrated that this chip format could be used for not only screening individual but also multiple enzymatic activities simultaneously with different fluorescent tagged small peptide libraries. We further demonstrated that this system could be a very powerful ultra high throughput screening tool for drug discovery, with which we identified potential "hits" after screening chips printed with small chemical compounds against caspases 1 and 3. This highly sensible chip is also able to monitor caspase protein expression profiles by activating the peptide chips with cell lysates undergoing apoptosis.

Key words: Small chemical compounds microarray, peptide array, enzymatic assay, aerosol deposition, ultra high-throughput screening.

Introduction

The large scale DNA sequencing of the human genome represents only a starting point in the future of biology. New demands are emerging in proteomics – the study of proteins, their expression, and their function. Proteins

3

are more complex, richer in information, and more relevant to biomedical research than genes. Genomes are essentially fixed for a lifetime; proteomes never stop changing. Protein profiles yield the most clues about the functions of cells and the activity of biological systems. Microarrays, such as peptide arrays, antibody arrays and small chemical compound arrays, have been developed in recent years to study protein function, such as, enzymatic activity, receptor ligand binding and protein-protein interaction (1-6). Traditionally, protein assays and small chemical compound screening are performed in solution-phase utilizing automatic liquid handling machines. Microarray technology has reduced reaction volumes more than a thousand fold from microliters to nanoliters, but the delivery of bioreagents robotically into these micron size reaction centers is problematic. The popularity of microarray technology has transformed traditional solution-phase screening into immobilized-format (or separation-format) screening (1-3). However, problems such as microarray surface modification, substrate binding specificity, uniformity and the orientation of molecular attachment have arisen (1-3). For example, Schreiber and co-workers immobilized newly synthesized small chemical compounds on a chip and used them for enzymatic assays, protein-protein interaction and high throughput screening (4,5). This immobilization technology is marginally useful for screening the millions of small compounds already on the shelves of pharmaceutical, chemical and biotechnology companies. It is cumbersome to immobilize libraries of this size on chips for high-throughput screening. The inconvenience and expensive cost of immobilization chemistry is also demonstrated in other microarray systems, such as the peptide array for kinase reaction (3). To solve this issue, a homogeneous microarray assay system mimicking the conventional protein assay scheme is needed. The major obstacle is how to deliver the biofluid onto the chip without cross-contaminating the individual reaction centers which are less than 2 nl in volume, and approximatly 300 microns apart, while simultaneously avoiding the problems of evaporation inherent to these nanoliter volumes.

Here we describe the first *H*omogeneous *M*ulti-functional *M*icroarray (HMM) system that can be used for multi-screening applications. With this platform, molecules, including small chemical compounds, peptides and proteins, can be arrayed without pre-immobilization. Chips containing up to 5000 dots, less than 2 nanoliter each, can be arrayed on a standard microscope slide with each dot acting as an individual reaction center. An aerosol deposition technology delivers a miniature volume of target material uniformly into each reaction center. By doing so, multi-aerosol mists with 2 picoliter average volume merge with each array dot to initiate the chemical reaction. In this study, we demonstrate that this platform is a versatile tool for multi-functional ultra High Throughput Screening, for enzyme substrate specificity, enzyme activity and antigen-antibody binding assays. We further demonstrate that by screening a small chemical compound library against caspases 1 and 3, in our microarray platform that potential 'hits' can be identified. By using a caspase substrate array created with this platform, we also show that our chip format is very sensitive for cell lysate screening and for monitoring caspase expression changes through a functional assay during apoptosis. This new functional proteomic assay system has great potential for studies including cell function, drug interaction assays and signaling pathway monitoring.

The HMM Platform and Applications

Overview of HMM system. Miniaturization is a key concept of current microarray technology for high throughput screening to meet the future needs of fast and cost-effective drug discovery, and for diagnostic screening to get more information with small amount available biomaterials. However, many physical problems associated with the producing homogeneous nano-liter biomolecule droplets on a solid surface, such as the microscope slides or chip. For example, nanoliter droplets of biomaterials will dry up within minutes on the surface of the glass slid, and are easily cross contaminated if additional solution added later on. With the size of the array dots, it requires a very prissily control arrayer to re-array each dot to delivery any new materials that needed. The strategy that HMM adapted is to use viscous inactive solvent to prevent the evaporation and cross contamination on the array, and then deliver any required biomaterials through an aerosol deposition system that we were licensed from the University of Pennsylvania. The general procedure is as follow: Compounds were mixed in a cocktail that includes 25-50% of a glycerol-like material for controlling evaporation, 1-10% of an organic solvent, such as DMSO, to enhance compound solubility, and buffer to maintain the biochemical reaction components. The compounds were arrayed on the surface of plain or polylysine coated slides with a conventional contact pin arrayer (Figure 1.1A). The chips were then activated by spraying the screening target material. We adapted an aerosol deposition technology that converts the biofluid into a fine mist for uniform spraying onto the surface of the chip (Figure 1.1B). After activation and incubation, the fluorescence signal was detected with a fluorescent microscope equipped with a cooled CCD camera, and interpreted using both imaging and data analysis software (Figure 1.1C).

Characterization of the Aerosol Deposition. Reaction cocktails (containing 40% glycerol, 10% DMSO and various small synthetic peptides or chemical compounds) were arrayed on the surface of glass slides with a GeneMachine OmniGrid (Figure 1.2A and insert). The arrayed chips were then sprayed with solutions such as water, DMSO, 1% glycerol in water, or caspase 3 enzyme reaction buffer (100 mM NaCl, 50mM HEPES 1mM EDTA and 100 mM DTT). Brightfield images (Figure 1.2B and insert) demonstrated that

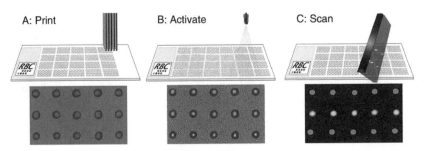

FIGURE 1.1. **HMM system.** (A) Peptides, proteins or small molecules are mixed with reaction cocktail and then arrayed onto glass slides as individual reaction centers. (B) The chip is then activated by a fine aerosol mist of biological sample; the mist droplets fuse with each array dot without causing cross-contamination between reaction centers. (C) Fluorescent signals were detected with imaging instruments such as a fluorescence microscope, and the data was analyzed with microarray software.

repeated spraying with reaction buffer did not significantly alter the reaction center morphology. The sprayed mist (5 to 20 mm in diameter) had a consistent distribution within each slide and throughout the entire slide tray that contained 20 slides. For instance, the actual counted number of droplets on slide areas 1, 2 and 3 (Figure 1.2B) had an average CV of 11% between slides and 7% within slides.

The array design, spot to spot spacing, and array size are critical for generating a protein chip that can be sprayed later with multiple solutions and stored for longer periods of time. The characteristics of this array were: Array spot spacing: 500 um center to center; Array spot diameter: 180 ± 14 um; Array dot volume: 1.6 ± 0.3 nl; Spray droplet diameter on chip: 18.1 ± 6.3 um; Spray droplet volume: 2.2 ± 0.2 pL. Successful aerosol delivery to the slide was determined by the optimal operational parameters of the spray systems. Determined by repeated testing, we set our parameters as follows in most situations: biological sample flow rate 800 nL/s; slide deck velocity: 2.54 cm/s. Distance of nozzle to the slide: 2.54 cm; Nozzle orifice diameter: 0.09 inches.

Another critical requirement for this homogenous array environment is that no adjacent array spots on the slide mix or cross-contaminate other spots during the spray activation of the chip. To illustrate the absence of cross-contamination, we arrayed alternating rows of FITC and rhodamine on chips with a spacing of 500 μm. Multiple sprays (up to 8 mist applications of caspase 3 reaction buffer) did not cause a single example of cross-contamination between the rows or columns for over one thousand spots. The fluorescence signal changes after spraying were minimal, for rhodamine the

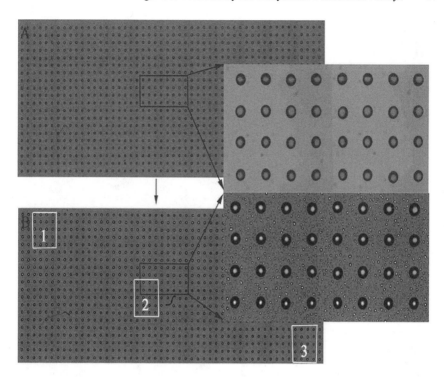

FIGURE 1.2. **Demonstration of operational parameters for the Morewood enzyme chip platform.** Small synthetic peptides were dissolved in 10% DMSO and 40% glycerol and arrayed on the surface of glass slides. Pictures were taken before (A) and after (B) aerosol deposition. The whole array pictures show that array spot morphology was uniform across the slide after multiple sprays with caspase reaction buffer. The enlarged pictures show the close up views of pictures A and B. The mist is quite evenly distributed among the dots, and the array dots are well preserved after multiple sprays. The spray mist distribution (droplets per unit area) was quite uniform; the average mist droplets were very consistent through out the 3 regions (B). The inter-slide CVs of region 1, region 2 and region 3 are 16%, 14% and 3% respectively and the intra-slide CVs of slide 1, slide 2 and slide 3 are 1%, 4% and 16%.

change was <0.01%, and for FITC the change was <0.1% (Figure 1.3). This experiment was repeated with each of the spray solutions mentioned above with the same or similar results.

Validation of Enzymatic Activities in HMM System. The major difference between HMM reaction conditions from conventional solution phase reactions was that we used a higher concentration of glycerol-like material to reduce evaporation and enhance long term storage. Glycerol was a good protein stabilization material used in daily protein storage, but at a high concentration, it may also affect enzymatic activity or antibody-antigen binding characteristics. To investigate this, we carried out several enzyme kinetic

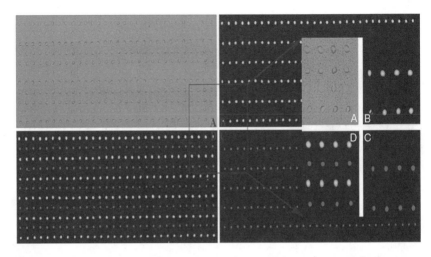

FIGURE 1.3. **Demonstration of no-cross reaction among reaction centers after spray.**
The FITC (green) and rhodamine (red) dyes were arrayed in alternating rows and the
chip was repeatedly sprayed (8 times) using a caspase 3 reaction buffer. Panel A
showed the morphology of dots after spray, the fine mist of spray could be observed
very clearly in the close up bright field view. Panel B and C showed each fluorescence
channel separately and panel D are the combined pictures. The fluorescence value of
FITC dots is roughly equal to the background value when the rhodamine signal was
collected, and vice versa. No red-green co-mixing, indicative of cross-contamination
was detected after this exposure to 8 separate sprayings of the chip.

studies and antibody-antigen binding assays under 40% glycerol reaction
conditions and compared the results with standard reaction conditions. For
enzymatic reactions, we used Caspases 1, 3 and 6 with their specific substrates, S1 (Ac-YVAD-AMC); S3 (Ac-DEVD-AMC) and S6 (Ac-VEID-
AMC), respectively. The experimental K_m values for caspases 1, 3 and 6
under the 40% glycerol conditions were 13.2 mM, 7.5 mM and 30.1 mM for
S1, S3 and S6, respectively, compared to published K_m values from the substrate manufacturer (BioMol) of 14 mM, 9.7 mM and 30 mM in 10% glycerol. The specific activity of these enzymes is 20 to 30% lower in 40%
glycerol than in 10% glycerol, but with a longer reaction time, the total substrate conversion could reach the same level (data not shown)

Applications in Single Enzymatic Assay. Caspase cross activity on other caspases' substrates was the first application tested in the HMM system. We
arrayed a number of different caspase substrates and activated them with an
aerosol of purified caspases. Figure 1.4A showed that caspase 1 not only had
enzymatic activity on S1 (subarray 1), but also had cross-reactivity with both
S3 (subarray 2), S6 (subarray 3), which was confirmed during a conventional
384-well format experiment (Figure 1.4B).

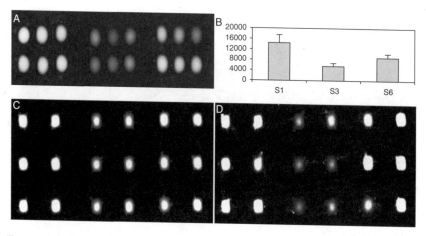

FIGURE 1.4. **Single enzymatic detection.** Substrates for caspases 1, 3 and 6 were arrayed and activated by caspase 1 (A). The cross activities between caspase 1 with substrates of caspase 3 and 6 were well reflected in 384-well reaction, showed in (B). The fluorescently tagged PKA peptide substrate from the IQ™ assay kit was arrayed (C). The left subarray has substrate, inhibitor (50 μM) and PKA. The middle subarray has no inhibitor and the right subarray has no PKA. ATP was sprayed to activate the kinase reaction, and quencher solution was sprayed several hours later to detect the reaction results (D), and only the central subarray showed the desired quench effect.

The HMM system can also be adapted for other commercial available homogenous assays, such as the Kinase reaction illustrated in Figure 1.4C. We arrayed the IQ™ PKA Assay reagents with or without PKA inhibitors on chips, and activated the reaction by spraying ATP. The PKA substrate in IQ™ PKA Assay Kit is fluorescence labeled (Figure 1.4C), and the fluorescence signal will be quenched upon the addition of a phosphate during kinase reaction (Figure 1.4D). Our experiment showed that only the middle subarray had a fluorescence quenching effect caused by the phosphorylation reaction followed by quencher binding. The data clearly demonstrated that the HMM system was capable of differentiating this kinase reaction with or without inhibitors. This application without surface immobilization or radioisotopes is clearly more favorable than other similar microarray approaches (3).

These data demonstrated that utilizing such a peptide array, HMM can be used for screening substrate libraries of single or multiplexed enzymes to search for the best specific substrate, even for enzymes belonging to the same family and having cross-reactivity with other family members. Similarly, we arrayed different enzymes on the chips and sprayed with a single substrate. We were able to detect the activity of each enzyme towards the sprayed substrate (data not shown). This application could be used by manufacturers that are looking for the best enzyme to convert certain substrates.

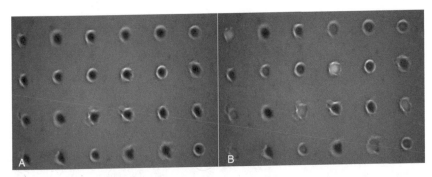

FIGURE 1.5. **Demonstration of multiplex detections.** Thrombin substrate with a blue fluorescent tag and chymotrypsin substrates with a red or green fluorescence tags were arrayed on the chip (A) and activated with thrombin and chymotrypsin simultaneously (B).

Application in Multi-Enzymatic Assay. We next tested the feasibility of multiplexing assays using the HMM system. We have randomly arrayed one thrombin peptide substrate (carbobenzoxy-VPR-MCA) two different chymotrypsin quenching substrates, BODIPY FL and BODIPY TR-X on a chip, then activated by spraying both thrombin and chymotrypsin simultaneously or sequentially (Figure 1.5). The results showed this peptide array could detect both enzymes' activities and the three specific substrates without mixing the signals. So that theoretically, a small chemical compound array in HMM system could be used for screening multi-enzymes activities, as long as the specific activities of enzymes could be differentiated with different fluorescence channels.

High-Throughput Screening of Small Chemical Compound Libraries. One major potential application of the HMM system is to use it as an ultra high-throughput screening tool for drug discovery. With an arrayer producing multiple sets of identical chips from the same small chemical compound library, we believe that each of these chip sets can be used for a single target screening. To establish this concept, we selected caspases as targets. Apoptosis is a genetically programmed, morphologically distinct form of cell death that can be triggered by a variety of physiological and pathological stimuli. The enzyme family of caspases plays a critical role in the initiation and execution of this process. Thus, various pathways in apoptosis are targets of pharmaceutical discovery.

We arrayed multiple identical sets of chips with a library of 380 small chemical compounds (Figure 1.6). On the same chip, a subarray of glycerol dots (A01, row A and column 1) without chemical compound were used as a negative control to show the uninhibited enzymatic activity and two subarrays with known peptide inhibitors of caspase 1 (B01) and 3 (C01) were

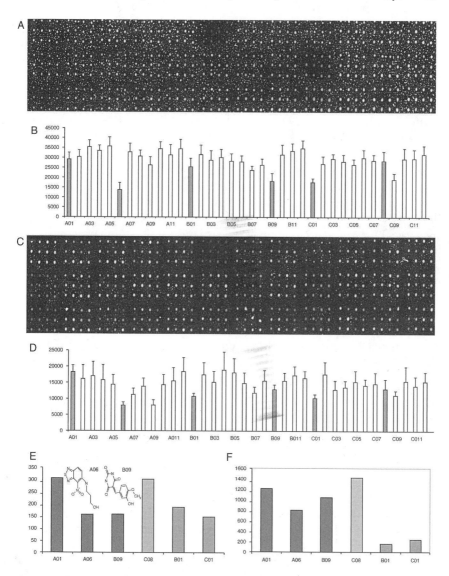

FIGURE 1.6. An ultra high-throughput screening assay with the HMM system. A library of small chemical compounds were arrayed on a chip and screened with caspase 3 (A and B) and caspase 1 (C and D) respectively. Subarray A01 (Blue bar) is a negative control and subarrays B01 and C01 (red bars) are positive controls in each chip (see text for details). Compounds A06 and B09 (Black bars) showed inhibitory effects on caspase 3 compared to the positive control based on image (A) and chip data analysis (B), but only A06 showed a similar effect on caspase 1 (C and D). The conventional 384-well format reactions confirmed these finding for caspase 3 (E) and caspase 1 (F). Compound C08 (Green bar) was included in both assays as an internal control (A-D).

used as positive controls to show inhibited activity (Figure 1.6A and 1.6C). The chips were then sprayed with caspase 1 and caspase 3 in two separate sets, and followed with a second spray of their specific substrates. After an overnight incubation, the slides were scanned on a fluorescent microscope-based scanner to detect potential 'hits' (Figure 1.6A and 1.6C). In this study, we found two potential inhibitors for caspase 3 (Figure 1.6A and 1.6B) and one for Caspase 1 (Figure 1.6C and 1.6D). We then repeated the same reactions in a 384-well format including the known inhibitors, identified inhibitors and a randomly selected compound (C08). This experiment confirmed our microarray findings with caspase 3 (Figure 1.6E) and caspase 1 (Figure 1.6F). The caspases have 4 binding subsites, S1 to S4, and previous researchers have indicated that binding at the S1 site confers the selectivity of caspases, while binding to the S3 and S4 subsites is believed to be critical for differentiating between caspases (7). By comparing the x-ray crystal structure of caspase 3 bound with an isatin sulfonamide inhibitor (8) with these two compounds, the benzothiadiazole ring of A06 has the potential to bind the S2 subsite of caspases, and the ketone carbonyl group of the general barbiturate ring in compound B09 may form a tetrahedral intermediate with the catalytic residue Cys163, and the benzene ring could occupy the S3 subsite. This experiment demonstrated the high-throughput screening capability of the HMM system and identified a compound with both S2 and S3 binding capability which could be further refined by screening a larger library containing additional structural variation (8,9).

HMM Used for Monitoring Apoptosis Process. The amplification techniques for RNA and DNA made the DNA array chip possible for evaluating the gene expression profiles of a given cell type. However, there is no equivalent technique for protein amplification, and it is also difficult to adapt a protein array to evaluate all protein expression profiles because of the complexity of the protein expression pattern and the protein levels at different stages. Antibody arrays have been used to screen protein expression changes (10), but the simple binding assay is not as useful as a functional bioassay. Based on the sensitivity of the fluorescent detection adapted for the HMM system, we believe that the HMM system could be used for detection of protein expression profiles by using a functional enzyme assay. Caspases afford an excellent opportunity for testing this hypothesis. By scanning a 60-compound fluorogenic, positional scanning library of Ac-X-X-X-Asp-AMC, Thornberry (11) has assigned each caspase a specific peptide substrate that has been used to evaluate caspase activity (12). To establish the concept that the HMM system is sensitive enough to be used for cell lysate screening, we compared 3 common caspase activities in Jurkat cells before (Figure 1.7A) and after (Figure 1.7B) camptothecin treatment. In this experiment, we arrayed 4 peptide substrates S1, S3, S6 and S6/8 [a substrate (Ac-IETC-AMC) that can be used for detecting both caspases 6 and 8] and then sprayed the arrays with Jurkat cell lysate. By comparing the fluorescent signals, we found that each substrate's turnover had increased significantly after camptothecin induction; with substrates 3 and 6 having the highest activity (Figure 1.7C). This assay indicated that the HMM chip is sensitive enough to develop a protein expression profile. Caspases, in general, have high cross-reactivity to the various substrates and inhibitors (11-13). Based on our initial screening assay (data not shown), building a substrate array for separating activities of caspases 1, 3 and 6, requires a minimum of these 4 substrates. To build an assay chip to cover the full spectrum of caspases to monitor the apoptosis process with cell lysates (other enzymes may also have activities on these substrates), an array may have to include all the caspase substrates and combinations of inhibitors (14). Such a chip will have a significant impact on basic apoptosis research, caspase drug development and drug-drug interaction studies.

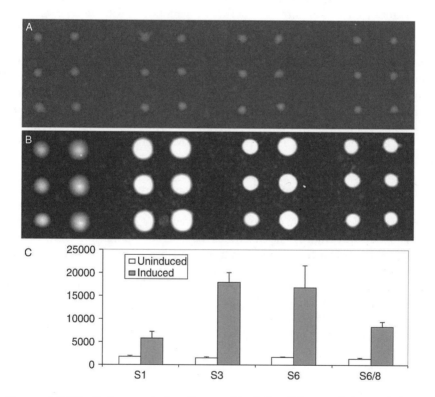

FIGURE 1.7. **Monitoring apoptosis pathways with whole cell lysates.** Substrates of caspases 1, 3, 6 and 6/8 were arrayed and activated with Jurkat cell lysate before (A) and after (B) camptothecin induction. Campothecin induction increased caspase activities 3-9 fold (C).

Conclusion

Despite the great advance in DNA microarray development, the protein chips and small chemical compound array chip are still a researcher or drug discoverer's fantasy (4,5,15). Existing protein array and small chemical compound array are two technologies that originated from DNA array and both of them adapted its immobilized-format (or separation-format). However, a homogenous format (or non-separation-format) microarray will be very attractive because it is simple, easy and fast. It is also very cost effective comparing to traditional assay. For example, the 1536-well with 5-10 µl reaction could save the cost dramatically comparing to the traditional 100-200 µl reaction in 96-well. However, further reducing these reactions to microarray format with nanoliter volume, many physical obstacles exist, and one of the most challenging one is how to precisely deliver sub-nanoliter volume of biomaterials into each reaction center. The aerosol deposition technology that HMM system adapted could be a perfect solution. It offers a versatile assay tool which can control the delivery of inhibitors, substrates, cofactors, or enzymes to each reaction center. With an average volume of <2 nl for each reaction center, a conventional 384-well plate with an average volume of 20 µl could be used to make a hundred sets of small chemical compound microchips, and then could be used for one hundred different target screenings. With an estimate of 3 µl of biological fluid to activate each slide containing up to 5000 reaction centers, the HMM ultra high-throughput screening system can save over 90% of the cost of drug discovery screening.

The functional proteomic activity of HMM could play an important role both in drug discovery and basic proteomic research, since enzymes represent about 28% of current drug targets. Many enzymes of interest, including tPA, kallikrein, plasmin, thrombin (16,17), activated protein C (18), factor Xa, factor XIa, factor VIIa (19), peptidases, matrix metalloproteinases, elastases, caspases, furin, cathepsins, trypsin, chymotrypsin (20), and viral proteases (21) already have fluorogenic substrates that can be optimized for analysis. In addition, quenched fluorescent peptides have been synthesized with phosphotyrosine which can be removed by a phosphatase to reveal a protease sensitive sequence whose cleavage results in a fluorogenic signal (22). Conversely, phosphorylation of a recognition site by a kinase would prevent cleavage by a protease, thus allowing a wide range of microarray-based studies of intracellular signaling enzymes. The drug discovery process is critically dependent upon the ability of screening efforts to identify "hits" with therapeutic potential and the screening efforts is one of the bottlenecks in the process of drug discovery. As an uHTS tool, HMM platform could save both time and money for drug discovery. HMM reduces the direct drug screening cost simply by requiring thousand fold less library compounds and drug targets. It also reduces the indirect cost, such as chemical synthesis and handling. In HMM system, a traditional small diversified library can be produced into hundreds sets and each of them for different target.

In summary, comparing to immobilized format array systems, the HMM system has several advantages. First, each arrayed dot may contain different chemicals and be used as a unique reaction center. The homogeneous reaction can generate data unobtainable from the immobilized format including enzyme kinetics. Second, HMM eliminates the need to immobilize the arrayed molecules, which allows the arraying of any existing library of peptides or small chemical compounds on chips for high throughput screening. Third, HMM conditions are very stable and yet flexible, and can also be used for arraying proteins and antibodies, which can be used for protein functional screening and antibody-antigen based assays. Fourth, HMM utilizes an aerosol deposition technology that converts a minimum amount of biological fluid into an extremely fine mist and then uniformly deposits it on the chip to activate each reaction center.

Materials and Methods

Materials

Purified caspases, peptide substrates and fluorescent dyes were purchased from BioMol (Plymouth Meeting, PA). The small chemical compound library was ordered from Nanosyn (Menlo Park, CA). Thrombin was from Enzyme Research Laboratories (South Bend, IN), and its fluorescent substrate was purchased from Peninsula Laboratories, Inc. (San Carlos, CA). The EnzChek Protease Assay Kits (both Green and Red) for the chymotrypsin quenching assay were purchased from Molecular Probes (Eugene, OR). The IQ™ Kinase Kit and PKA were purchased from Pierce Biotechnologies (Rockford, IL). The AlphaScreen™ Kit for antibody antigen detection was purchased from PerkinElmer (Meriden, CT). Caspase induced and un-induced cell lysates were purchased from Geno Technology (St. Louis, MO). The research grade nitrogen gas was purchased from BOC Gases (Baltimore, MD), gas flow regulators, tubing and syringes were purchased from Cole Parmer Instrument Company (Vernon Hills, IL). All glass slides, general chemicals and supplies were purchased from Fisher Scientific (Pittsburgh, PA).

Arraying and Aerosol Deposition

The small chemical compounds or peptides were dissolved in DMSO, then mixed with reaction buffer, including 25-40% glycerol as indicated in the text, in 96-well or 384-well plates, and then arrayed with a GeneMachine OmniGrid or stored at −20 to −80 °C for later use. The arrayed slides were then activated by aerosol mists of enzymes, cell lysates, or other fluids.

Fluorescence Detection and Data Analysis

The fluorescently tagged substrates were scanned with a fluorescence microscope (Nikon E600) equipped with a cooled CCD camera, X-Y automatic control stage and picture stitching program ImagePro (MediaCybernetics, Silver Spring, MD). The final array picture was then analyzed through a DNA array program, ArrayPro (MediaCybernetics, Silver Spring, MD), and fluorescence intensities were automatically assigned.

The Determination of K_m Values of Caspases at High Glycerol Concentration

Reactions were performed in a 384-well plate, and in each 10 μl reaction, 20 U of enzyme was added to activate the pro-fluorescence substrate. Reaction buffer contained 40% glycerol, and the substrate concentrations ranged from 10 μM to 160 μM. Reactions were carried out at 37 °C and were read with a Labsystems Fluoroskan Ascent FL every 2 to 5 minutes for 40 time points.

Peptide and Small Chemical Compound Chips

Peptide substrate concentrations were 200 μM for caspases 1, 3 and 6, 500 μM for thrombin and chymotrypsin. The concentration of small chemical compounds on the chips for caspase 1 and 3 screening was 300 μM. Purified caspases used for spraying were 10 Unit/μl with 3μl/spray/slide on the average. The induced and uninduced Jurkat cell lysates had a concentration of $2\text{-}2.4 \times 10^7$ cells/ml. Aerosol deposition parameters were: biofluid flow rate, 800 nl/s; slide deck velocity, 2.54 cm/s; distance of nozzle to the slide, 2.54 cm and nozzle orifice diameter: 0.09 inches.

Kinase Assay Chips

The IQ™ assay solutions were prepared as suggested by the manufacturer except that 40% glycerol was mixed into the reaction buffer. Fluorescently tagged substrate was arrayed with or without 50 μM inhibitor (TYADFIAS-GRTGRRNAI-MH2, Upstate Biotechnology, Lake Placid, NY), ATP and PKA were then sprayed to activate the kinase reaction. After 2-3 hours incubation, quencher solution was sprayed and the fluorescent signal change was recorded. When a no PKA subarray was used as a control, PKA was arrayed together with substrate and inhibitor in the remainder of the subarrays, and the reactions were then activated by spraying ATP only.

Acknowledgements. We thank Dr. Scott Diamond, the inventor of the aerosol deposition technology for delivering biofluid on microarrays, and for both technical assistance and invaluable scientific discussion.

References

1. Zhu, H., Bilgin, M., Bangham, R., Hall, D., Casamayor, A., Bertone, P., Lan, N., Jansen, R., Bidlingmaier, S, Houfek, T, et al. (2000). Analysis of yeast protein kinases using protein chips. *Nature Genetics* **26**, 283-289.
2. Zhu, H., Klemic, J.F., Chang, S., Bertone, P., Casamayor, A., Klemic, K.G., Smith, D., Gerstein, M., Reed, M.A. and Snyder, M. (2001). Global analysis of protein activities using proteome chips. *Science* **293**, 2101-2105.
3. Houseman, B.T., Huh, J.H., Kron, S.J. and Mrksich, M. (2002). Peptide chips for the quantitative evaluation of protein kinase activity. *Nat Biotechnol.* **20**, 270-274.
4. MacBeath, G. and Schreiber, S.L. (2000). Printing proteins as microarrays for high-throughput function determination. *Science* **289**, 1760-1763.
5. Kuruvilla, F.G., Shamji, A.F., Sternson, S.M., Hergenrother, P.J. and Schreiber, S.L. (2002). Dissecting glucose signalling with diversity-oriented synthesis and small-molecule microarrays. *Nature* **416**, 653-657.
6. De Wildt, R.M.T., Mundy, C. R., Gorick, B.D. and Tomlinson, I.M. (2000). Antibody arrays for high-throughput screening of antibody-antigen interactions. *Nat. Biotechnol.* **18**, 989-994.
7. Nicholson, D.W. and Thornberry, N.A. (1997). Caspases: killer proteases. *Trends Biochem. Sci.* **22**, 299-306.
8. Lee, D., Long, S.A., Adams, J.L., Chan, G., Vaidya, K.S., Francis, T.A., Kikly, K., Winkler, J.D, Sung, C.M, Debouck, C. et al. (2000). Potent and selective nonpeptide inhibitors of caspases 3 and 7 inhibit apoptosis and maintain cell functionality. *J. Biol. Chem.* **275**, 16007-16014.
9. Webber, S.E., Tikhe, J., Worland, S.T., Fuhrman, S.A., Hendrickson, T.F., Matthews, D.A., Love, R.A., Patick, A.K., Meador, J.W., Ferre, R.A. et al. (1996). Design, synthesis, and evaluation of nonpeptidic inhibitors of human rhinovirus 3C protease. *J. Med. Chem.* **39**, 5072-5082.
10. Knezevic, V., Leethanakul, C., Bichsel, V.E., Worth, J.M., Prabhu, V.V., Gutkind, J.S., Liotta, L.A., Munson, P.J., Petricoin, E.F. 3rd, Krizman, D.B. et al. (2001). Proteomic profiling of the cancer microenvironment by antibody arrays. *Proteomics* **10**, 1271-1278.
11. Thornberry, N.A., Rano, T.A., Peterson, E.P., Rasper, D.M., Timkey, T., Garcia-Calvo, M., Houtzager, V.M., Nordstrom, P.A., Roy, S., Vaillancourt, J.P. et al. (1997). A combinatorial approach defines specificities of members of the caspase family and Granzyme B. *J. Biol. Chem.* **272**, 17907-17911.
12. Amstad, P.A., Yu, G., Johnson, G.L., Lee, B.W., Dhawan, S. and Phelps, D.J. (2001). Detection of caspase activation in situ by fluorochrome-labeled caspase inhibitors. *Biotechniques* **3**, 608-610.
13. Talanian, R.V., Quinlan, C., Trautz, S., Hackett, M.C., Mankovich, J.A., Banach, D., Ghayur, T., Brady, K.D., Wong, W.W. et al. (1997). Substrate specificities of caspase family proteases. *J. Biol. Chem.* **272**, 9677-9682.
14. Garcia-Calvo, M., Peterson, E.P., Leiting, B., Ruel, R., Nicholson, D.W. and Thornberry, N.A. (1998). Inhibition of human caspases by peptide-based and macromolecule inhibitors. *J. Biol. Chem.* **273**, 32608-32613.
15. Mitchell, P. (2002). A prospective on protein microarrays. *Nature Biotech.* **20**, 225-229.
16. Morita, T., Kato, H., Iwanaga, S., Takada, K. and Kimura, T. (1977). New fluorogenic substrates for α-thrombin, factor Xa, kallikreins, and urokinase. *J. Biochem.* **82**, 1495-1498.

17. Backes, B.J., Harris, J.L., Leonetti, F., Craik, C.S. and Ellman, J.A. (2000). Synthesis of positional-scanning libraries of fluorogenic peptide substrates to define the extended substrate specificity of plasmin and thrombin. *Nat. Biotechnol.* **18**, 187-193.

18. Butenas, S., Drungilaite, V. and Mann, K.G. (1995). Fluorogenic substrates for activated protein C: substrate structure-efficiency correlation. *Anal. Biochem.* **225**, 231-241.

19. Butenas, S., Ribarik, N. and Mann, K.G. (1993). Synthetic substrates for human factor VIIa and factor VIIa-tissue factor. *Biochem.* **32**, 6531-6538.

20. Harris, J.L., Backes, B.J., Leonetti, F., Mahrus, S., Ellman, J.A, Craik. C.S. (2000). Rapid and general profiling of protease specificity by using combinatorial fluorogenic substrate libraries. *Proc. Natl. Acad. Sci. U S A.* **97**, 7754-7759.

21. Toth, M.V. and Marshall, G.R. (1990). A simple, continuous fluorometric assay for HIV protease. *Intl. J. Pept. Protein Res.* **36**, 544-50.

22. Nishikata, M., Suzuki, K., Yoshimura, Y., Deyama, Y. and Matsumoto, A. (1999). A phosphotyrosine-containing quenched fluorogenic peptide as a novel substrate for protein tyrosine phosphatases. *Biochem. J.* **343**, 385-391.

2

Improvement of Microarray Technologies for Detecting Single Nucleotide Mismatch

HONG WANG, ZUHONG LU*, JIONG LI, HEPING LIU,
AND QUANJUN LIU
*Chien-Shiung Wu Laboratory, Department of Biomedical Engineering,
Southeast University, Nanjing, China 210096*

Abstract: With the completion of the Human Genome Project and beginning of post genome era, there is an urgent need for a fast, specific, sensitive, reliable and cost-effective method for the genome-wide polymorphism analysis. Microarray technology is one of the most promising approaches to this need. There are two major concerns in the microarray technology; one is target labeling and the other is the reliability of single nucleotide mismatch discrimination. In this chapter, we reported our recent progress in molecular beacon arrays for detecting label-free targets and a microarray based melting-curve analysis method for improving the reliability of single nucleotide mismatch discrimination. Several successful practical applications of these improved microarray technologies have also been illustrated.

Key words: Microarray, SNPs, mutations, molecular beacon, melting-curve analysis.

1.   Introduction

The past few years have witnessed an extraordinary surge of interest in the microarray technology(1-3). A microarray is a collection of miniaturized test sites fabricated on a solid substrate either by robotically spotting or by *in situ* synthesizing with photo-deprotection method, inkjet spraying method or molecular stamping method(4-7). It permits many tests to be performed in a parallel and high-throughput way and offers the first great hope for 'global views' of DNA and RNA variation during biological processes instead of the traditional gene–by–gene approach.

   The application of the microarray technology in monitoring RNA expression levels of thousands of genes was well established and widely reported (8-9). But other applications, such as identification and genotyping of point mutations and single nucleotide polymorphisms (SNPs), are still in their infant stage and many issues remain to be worked out.

SNPs are most commonly occurred variant, and are estimated to be 1 out of every 1000 bases in the human genome (10). With the completion of the Human Genome Project and beginning of post genome era, more and more SNPs and point mutations are being uncovered and assembled into large SNP databases. The large number of SNPs provides a rich set of markers that can be used in a wide variety of genetic studies. The identification of a complex set of genes that cause a disease also requires both linkage and association analyses of thousands of SNPs across the human genome in thousands of individuals.

Since the important roles of SNPs and point mutations in molecular biology, many assay principles have been developed in the last 20 years (11-12). Even most of these principles were firstly illustrated in homogeneous solution; many efforts have been devoted to implement them in a microarray assay format to meet the urgent needs for a fast, specific, sensitive, reliable and cost-effective method for a genome-wide SNPs analysis. For example, the allele specific oligonucleotide hybridization is employed in most of the microarray based SNPs genotyping methods, such as the Genechip assay provided by Affymetrix. Other examples are microarray based single nucleotide primer extension reactions, oligonucleotide ligation reactions and enzymatic cleavage methods which provide a better power of discrimination between genotypes than allele specific oligonucleotides hybridizations.

Microarray technology is one of the most promising approaches for large scale and high throughput genotyping. There are two major concerns in the microarray technology; first is the target labeling. Labeling is an important step in most of the microarray based target preparing protocol. It is not only time consuming, and rather expensive, but can also change the levels of targets originally present in the sample. Some label free techniques used in biosensors, such as QCM, SPR and RIfS are not compatible with high throughput applications and lack sensitivity in low molecular weight DNA detection. Second is the reliability of single nucleotide mismatch discrimination. The melting curve of the immobilized duplex is greatly broadened and depressed which greatly reduces the fluorescence intensity difference between the perfect matched duplex and the single base mismatched one. Moreover, it is difficult to normalize the hybridization conditions for a microarray because of massive number of probes and the insufficient knowledge of hybridization reactions at the solid-liquid interface

In this chapter, we reported our recent progress in resolving the above problems. First, a molecular beacon array was constructed, which allowed one to work with unlabelled targets and to retain the high sensitivity of fluorescence techniques. Second, a microarray based melting curve analysis method was investigated. Several successful practical applications of these improved microarray technologies were also illustrated.

2. Molecular Beacon Arrays

2.1 Molecular Beacons

Molecular beacons are oligonucleotide probes that can report the presence of specific nucleic acids in homogeneous solutions (13). They are single-stranded oligonucleotides containing a loop sequence complementary to the target that is flanked by a self-complementary stem, which carries a fluorophore on one end and a quencher at the other end. In the absence of target, the self-complementary stem structure holds the fluorophore so close to the quencher that fluorescence does not occur. When binding to the target, the rigidity of the probe–target duplex forces the stem to unwind, causing the separation of the fluorophore and the quencher and the restoration of fluorescence (Figure 2.1).

Molecular beacons are useful in situations where it is either not possible or desirable to isolate the probe-target hybrids from an excess of the hybridization probes, such as in real-time monitoring of polymerase chain reactions in sealed tubes or for the detection of RNAs within living cells. Therefore, they have been widely used in detecting SNPs and mutations (14-15), virus and pathogens (16-17), amplicons generated in nucleic acid sequence based amplification (NASBA) (18), mRNA in *in vivo* applications (19), and single strand DNA binding protein (20).

Since the first report of molecular beacons in 1996, several improvements and new developments of the molecular beacon technology have been reported. Multicolored molecular beacons (21) were firstly used for allele genotyping in the same solution and they displayed excellent specificity for single nucleotide mismatch discrimination. They had also been used in detecting four different retroviruses. The high sensitivity and specificity of

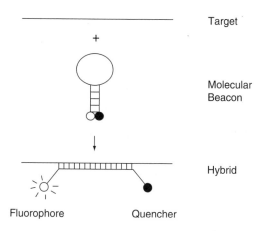

FIGURE 2.1. Scheme of molecular beacon.

each color reaction allowed the detection of fewer than ten copies of one virus amidst a background of unrelated viruses. To overcome the low excitation efficiency of multicolor molecular beacon by monochromatic light source provided by most of the commercialized instrument (such as PCR instrument, fluorescence microscopy, etc.), the wavelength-shifting molecular beacon was introduced (22). Using a combination of conventional molecular beacons and wavelength-shifting molecular beacons it will be possible to reliably perform six-plex or seven-plex PCRs simultaneously.

In another approach, the incorporation of the gold nanoparticles as a quencher instead of the commonly used (DABCYL) substantially increased the sensitivity and specificity of the assay because of the superior quenching ability of gold clusters (23). Moreover, the quenching ability of gold is not limited to its use in clusters. It is possible to construct a molecular beacon array on a gold surface and use the surface as a quencher of fluorescence.

Even many efforts were devoted to construct a serial of molecular beacons to resolve multiplex target simultaneously, the number of the targets was strictly limited due to the limitation of the suitable fluorescent labels. Recently, several groups intended to immobilize molecular beacons onto a solid surface and construct a molecular beacon array to resolve the target DNA sequences spatially.

Tan et al. firstly immobilized a set of molecular beacons on a silica surface through biotin-avidin binding to construct a micrometer DNA biosensor (24-25). They achieved the rapid response, stable, and reproducible results by such kind of DNA biosensors, which make it possible to detect a large number of targets simultaneously. Brown et al. synthesis molecular beacon attached to long chain alkyl amino-controlled pore glass (LCAA-CPG) (26). It can hybridize the target DNA or RNA and restore fluorescence that can be isolated and analyzed. Steemers et al. immobilized molecular beacons on a randomly ordered optic fiber to construct gene arrays (27). They greatly decreased the feature size to construct a miniaturized array capable of detecting unlabeled DNA targets at subnanomolar concentrations. Taking the advantages of low diffusion limitations and high local concentrations of sensing beads on the distal end of fiber, they improved the signal-to-background ratio. The above works opened up an important area of label-free large-scale and high-throughout detection of DNA sequence information.

However, there are still several bottlenecks unsolved for the molecular beacon array technology in practical applications. The electrostatic properties at the solid-liquid interface and the local ion strength of the immobilized molecular beacons are greatly different from that in the bulk solution. The stem structure of the molecular beacon is greatly destabilized which causes high fluorescent background and greatly decreases the signal-background ratio. Presently, the fluorescence increment ratio of the complementary immobilized molecular beacon probes to that of the noncomplementary ones is just about 1-2 after hybridization with targets, while this ratio of the same molecular beacons is tens to hundreds in homogeneous solutions.

Our group tried to improve the molecular beacon immobilization technique. We chose different modified substrates for immobilizing molecular beacons and investigated the annealing and hybridization properties of the molecular beacon arrays immobilized on different substrates (28).

2.2   Fabrication of Molecular Beacon Arrays

2.2.1   The Design of Molecular Beacons for Immobilization

To investigate the hybridization properties of immobilized molecular beacons and potential for constructing molecular beacon array for label-free target detection, we carefully designed amino-modified molecular beacon probes.

The molecular beacons designed for immobilization contain three parts (Figure 2.2): First, a single strand hairpin structure. Most studies have indicated that 15 to 25 base loop together with 5 to 7 base pair stem will provide an appropriate balance to form a hairpin structure, we choose a 16 base loop and 6 base pair stem in our design. Five molecular beacon probes were carefully

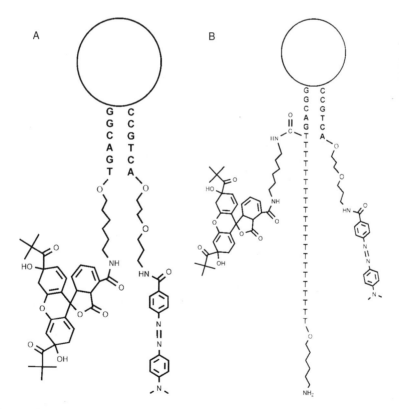

FIGURE 2.2. Structure of (A) molecular beacon and (B) amino modified molecular beacon for immobilization.

TABLE 2.1. Molecular beacon probe and target sequences

Symbol	Sequence
MB1	NH2-(T)20-FAM-<u>TGACGG</u> GAAGGTGGAATGGTTG <u>CCGTCA</u>-DABCYL
MB2	NH2-(T)20-FAM-<u>TGACGG</u> TGCAGAAG**C**GCCTGGC <u>CCGTCA</u>-DABCYL
MB3	NH2-(T)20-FAM-<u>TGACGG</u> TGCAGAAG**T**GCCTGGC <u>CCGTCA</u>-DABCYL
MB4	NH2-(T)20-FAM-<u>TGACGG</u> TGCAGAAG**G**GCCTGGC <u>CCGTCA</u>-DABCYL
MB5	NH2-(T)20-FAM-<u>TGACGG</u> TGCAGAAG**A**GCCTGGC <u>CCGTCA</u>-DABCYL
T2	GCCAGGC**G**CTTCTGCA
T3	GCCAGGC**A**CTTCTGCA

Note: Underline indicates the stem sequence of molecular beacon. Bold underline indicates the single nucleotide difference between sequences.

selected: four of them (MB2-MB5) targets an allele of human apoE gene and have one base difference at the central position of loop sequences. They are used to demonstrate the single nucleotide discrimination capability. MB1 is used as negative control and has a non-related loop sequence with MB2-MB5. All molecular beacon probes and corresponding target sequences are listed in Table 2.1. Second, a fluorescein in the internal location within the 5′ arm is used as the fluorophore and a 3′-end [4-(4-dimethylaminopherylazo) benzoic acid] (DABCYL) is used as the quencher. Fluorescein and DABCYL are most widely used fluorophore and quencher pair because DABCLY are nonfluorescent and can quench the fluorescein extremely well. Third, a 20 base thymine spacer with an amino group linked to 5′ end extends past the position of the fluorescein on the 5′ arm of our molecular beacons. The 20-base-thymine is used to increase the flexibility of molecular beacon and to minimize destabilization effects caused by 5′-end immobilization. The 5′ end amino-modified molecular beacon probes can be covalently immobilized on the activated substrates via the Schiff base aldehyde-amine chemistry.

The hybridization results in solution indicate the spacer structure of the molecular beacons for immobilization will decrease the quench efficiency, but the effects can be neglected and the probes still reserve stem-loop structure and high specificity for single nucleotide mismatch discrimination.

2.2.2   Substrates for Molecular Beacon Arrays

Several types of substrates are used in microarrays, such as membrane filters, glass slides and hydrogel films (29).

Glass slides have been a favored solid support for immobilization of probes because of their easy availability, low intrinsic fluorescence, high transparency, good thermal properties, excellent rigidity, and straightforward chemistries for surface modification. Due to the nonporous nature of glass, the labeled targets have direct access to immobilized probes without limitations of internal diffusion, enabling a high local concentration and rapid hybridization kinetics. The non porous surface also facilitates the rapid removal of excessive probes and fluorescence labeled targets. Even planar glass slides have many advantages in microarray applications; it is difficult for a structured biomolecule, such as an antibody or a structured oligonucleotide probe, to assume its native configuration at the surface of a planar glass slide. Moreover, the immobilization capacity of the planar glass slide is limited which results in a relatively low sensitivity.

To eliminate the disadvantages of planar glass slides and improve the performance of the microarray technology, three-dimensional functionalized hydrophilic microporous gel film substrates were introduced. Combining the advantages of the porous structure and the planar surface, these hydrogel microporous films provide a high binding capacity and a solution-like environment in which the hybridization and other processes resemble a homogeneous liquid phase reaction rather than a heterogeneous liquid-solid interface reaction. These films are compatible with the state-of-the-art microarray spotters or dispensers and detection instruments. The disadvantages of the hydrogel film are as follows, first, it will require a relatively long washing time to remove the unreacted targets and, second, that it is more cumbersome to prepare the slides or more expensive to buy these slides.

In this section, we introduced the modification and activation of three different substrates for immobilizing molecular beacons.

2.2.2.1 *Activation of Aminosilane Glass Slides*

The aminosilane derived glass slides (Cat No. S3003, Dako) were cleaned with deionized distilled water and incubated in 5% glutaraldehyde in 0.1 M PBS buffer (pH = 7.4) for 2 hours. Then the slides were thoroughly washed twice with methanol, acetone and deionized distilled water, and dried.

The aldehyde group of glutaraldehyde is attacked by primary amino group of aminosilane and forms a covalent bond, which can be stabilized by a dehydration reaction and leads to Schiff base formation.

2.2.2.2 *Preparation and Activation of Polyacrylamide Film (PAA film) Coated Glass Slides*

The preparation of PAA film coated glass slides were introduced elsewhere (30-32) and summarized as follows. The glass slides were cleaned in a piranha solution (7:3 v/v mixture of concentrated H_2SO_4 and 30% H_2O_2) at 80 °C for 2 hour and washed thoroughly with deionized distilled water. Polymerization solution contained 1 M acrylamide, 0.02 M N, N-methylene-bis-acrylamide, 0.1% TEMED and 1 mg/ml ammonium persulfate. It was injected into the small chamber formed by a Bind-Silane treated glass slide and a Repel-Silane treated glass slide separated by two 20 μ m thicker Teflon spacer strips. The PAA films were activated by immersion in 25% glutaraldehyde in phosphate buffer (pH = 7.5) at 40 °C overnight, then thoroughly rinsed with deionized distilled water for 2 hours and dried.

The PAA gel films are produced by polymerization of acrylamide into linear chains and cross-linking the acrylamide chains with bis-acrylamide. Polymerization is initiated by adding ammonium persulfate and the reaction is accelerated by TEMED which catalyzes the formation of free radicals from ammonium persulfate.

In aqueous solutions, two aldehyde groups of glutaraldehyde can be easily cross-linked and glutaraldehyde is present largely as polymers of variable sizes. The unsaturated $C=C$ bond can react with amide groups of PAA film. The free aldehyde groups sticking out of the side of each unit of glutaraldehyde polymer are readily combining with the primary amino group modified molecular beacons.

2.2.2.3   *Preparation and Activation of Agarose Film Coated Glass Slides*

Preparation and activation of agarose film coated glass slides were introduced elsewhere and summarized as follows (33): 1% agarose solution was prepared by adding 100 mg agarose to 10 ml deionized distilled water, mixing and boiling for 5 minutes. Then 2 ml of the agarose solution was poured over each of the aminosilane derived glass slides. After gelation of agarose, the slides were dried at 37 °C in a dryer over night. Before immobilization of the molecular beacon probes, the agarose films were activated by immersion in 20 mM $NaIO_4$ in 0.1 M PBS buffer (pH = 7.2) for 30 minutes at room temperature, then thoroughly rinsed twice with deionized distilled water and dried.

The vicinal hydroxyl groups of agarose can be oxidized by sodium periodate at a mild condition, forming aldehyde groups. The aldehyde groups can react with primary amino group via the Schiff base aldehyde-amine chemistry.

2.2.3   Fabrication of Molecular Beacon Arrays

2.2.3.1   *Spotting of Molecular Beacon Arrays*

All molecular beacon arrays were manufactured by the Cartesian Technologies PA Series of microarray spotting workstation, PixSys 5500. ChipMaker pin CPM3 was used to perform molecular beacon array spotting. The spotting diameter of CMP3 pin is 90 to 100 μm and the delivery volume of each spot is about 600 pL.

Spotting solutions were obtained by dissolving molecular beacon probes in sodium carbonate buffer (0.1 M, pH = 9.0) at the desired concentration.

After spotting, the agarose films and PAA films coated glass slides were incubated in a humid chamber at room temperature overnight and washed with 0.1% Tween, deionized distilled water and dried. Glutaraldehyde derived glass slides were incubated in a humid chamber at room temperature for 2 hours and at 37 °C for 2 hours. Then the slides were washed thoroughly in 0.1% Tween, distilled water and dried.

The Schiff base reaction is reversible at acid pH. For greater stability, the Schiff base was reduced with sodium borohydride. The sodium borohydride solution was prepared by dissolving 1.5 g NaBH4 in 450 ml phosphate buffered saline (PBS), then adding 133 ml 100% ethanol to reduce bubbling. Prepare the solution just before use and treat the molecular beacon arrays in the solution for 5 min at room temperature. This treatment also blocks unreacted free aldehyde groups by reducing them.

2.2.3.2   *Immobilization Capacities of Different Substrates*

To assess the immobilization capacity of different substrates, a serially diluted molecular beacon probes from 100μM to 1μM in spotting solutions were prepared and used to fabrication the molecular beacon arrays.

After immobilization and washing, the fluorescence images of the molecular beacon arrays are collected at the same laser power and PMT gain with a confocal microscope. The fluorescence intensities were extracted by ImageJ software and plotted in Figure 2.3(A).

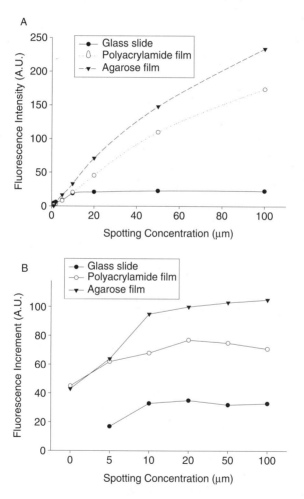

FIGURE 2.3. (A) Fluorescence intensity versus spotting concentration of molecular beacon array (B) Fluorescence intensity increment versus spotting concentration after hybridization with perfect matched targets in hybridization solution containing 10mM target in 20 mM Tris-HCl (pH = 8.0) and 10 mM MgCl$_2$.

The fluorescence intensities of molecular beacons immobilized on the glass slides increase with the increasing of spotting concentration. But the fluorescence intensities are steady when the spotting concentration is higher than 10μM, which indicate a saturated immobilization capacity of the slides. The results consist with the saturated concentration of common linear oligonucleotide probes on glass slides.

There is a nearly linear relationship between the spotting concentration and fluorescence intensity till 100μM for molecular beacons immobilized on the agarose films and the PAA films. But the agarose films show a liitle higher immobilization capacity than that of the PAA films. The tendency of the fluorescence intensity curves also indicates unsaturated immobilization capacity on both films. Further increase the spotting concentration will lead to increase of fluorescence intensities.

Further experiments were conducted to investigate the fluorescence recovery after the molecular beacon arrays hybridized with the perfectly matched targets. The results of hybridization reaction performed in hybridization solution containing 10mM target in 20 mM Tris-HCl (pH = 8.0) and 10 mM MgCl$_2$ were plotted in Figure 2.3(B). Experiments with other concentration of targets and MgCl$_2$ showed similar results.

The fluorescence intensity increments are low and showed no dependence on spotting concentration when it was higher than 10μM for molecular beacon arrays immobilized on the glass slides. It can be attributed to saturated immobilization capacity at 10μM.

The fluorescence intensity increment increase with the spotting concentration, but the increment is slow when the spotting concentration is higher than 10μM for molecular beacon arrays immobilized on the agarose films. We can see the same tendency in molecular beacon arrays immobilized on the PAA films at lower spotting concentration. But it is supervising to see the fluorescence intensity increment decrease when the spotting concentration is higher than 20μM. We contribute this decrease to the relatively lower immobilization capacity compared with the agarose films. The steric hindrance caused by the lower immobilization capacity and the higher immobilization density will retard the formation of stem structure of molecular beacons and would reduce the number of molecular beacons available for conformation change and fluorescence increment. The high density of molecular beacon probes also retards the diffusion of the target molecules.

We selection 10μM as spotting concentration based on the results and discussions above.

2.3   Hybridization of Molecular Beacon Arrays

2.3.1   Instrumentation and Software

All hybridization fluorescence images were collected by a laser scanning confocal microscope, Leica TCS SP. The 488 line of a Kr-Ar ion laser was employed as excitation source and 10X objective was used in all experiments. The standard FITC filter setting was used in fluorescence images collection. The accumulation of 4 times (in about 4 seconds) was used in image collection to reduce the random electronic noise. In most of time serial images collection, images were collected every minute for the first 20 minutes and every 5 minutes for the following time to reduce the potential photo bleaching of the fluorescein. If not specified, images were collected at the following conditions: laser power and PMT gain settings are carefully adjusted that the average fluorescence intensity of spots in an array is about 90% of the saturated value (255) before annealing. The setting should not be changed during the collecting process.

A fluidic sample cell made of anodized aluminum was fabricated, on which a molecular beacon array was mounted. The cell can be directly mounted on the microscope stage for real time fluorescence observation during the annealing and the hybridization process. Hybridization buffers and target solutions were pumped into the fluidic cell with a peristaltic pump.

Fluorescence images were analyzed with ImageJ version 1.27. ImageJ is a public domain Java image processing program inspired by NIH Image. It can be freely downloaded from http://rsb.info.nih.gov/nih-image/index.html.

In all images of this section, the molecular beacon probes were spotted on the substrates in triplet format. From the left to the right, the probes are MB1 (noncomplementary to T2), MB2 (perfectly matched with T2), and MB3, MB4, MB5 (single central base mismatched with T2).

2.3.2   Annealing of Molecular Beacon Arrays

When the molecular beacon arrays are allowed to be dry or incubated in buffers containing no cations, quench efficiency is low and high fluorescence background image of the molecular beacon arrays can be registered. It is because the negative charged phosphate backbone of the oligonucleotide would hinder the formation of the stem structure of the molecular beacon probes. The emission from the fluorescein can't be efficiently quenched without a stable stem structure.

In order to improve the signal-background-ratio, it is important to select the ion strength, especially divalent cations to counteract the negative charge of the phosphate backbone and to stabilize the stem structure.

2.3.2.1   Annealing Process of Molecular Beacon Arrays

The fluorescence images were collected for one hour after the hybridization buffer containing different concentration of $MgCl_2$ in 20 mM Tris-HCl (pH = 8.0) were pumped into the reaction cell. The fluorescence intensities of all the molecular beacons immobilized on the different substrates were averaged and normalized by the fluorescence intensity before annealing. Two typical plots were shown in Figure 2.4.

Figure 2.4 (A) show the annealing process of the molecular beacon arrays in hybridization buffer containing 10mM $MgCl_2$. The fluorescence intensities decrease with time for the molecular beacon arrays immobilized on all substrates. For the glass slides immobilized ones, the process is relatively faster and it complete in about 5 minutes. For the PAA and agarose film immobilized ones, the process is slower and complete in about 20 minutes.

Figure 2.4(B) show the annealing process of the molecular beacon arrays in hybridization buffer containing 500mM $MgCl_2$. Compared with the results in Figure 2.4 (A), the annealing process is much faster for the PAA film and the agarose film immobilized arrays.

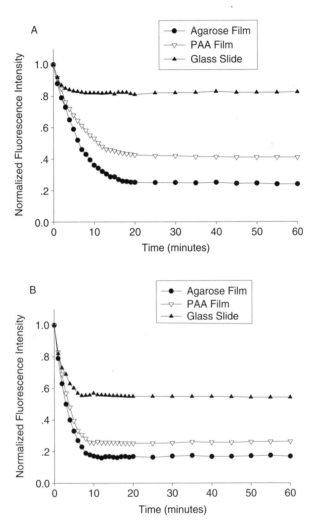

FIGURE 2.4. Normalized fluorescence intensity versus time of the molecular beacon arrays immobilized on the different substrates. The fluorescence intensity was the average of all molecular beacons on the substrates. The hybridization buffer is 20 mM Tris-HCl (pH = 8.0) and (A) 10 mM, (B) 500 mM $MgCl_2$.

The annealing time course in Figure 2.4 indicates that the annealing process complete in 20 minutes for arrays in lower concentration $MgCl_2$ buffer. So we choose 30 minutes annealing time for all the experiments in the following sections of this thesis.

2.3.2.2 Annealing Properties in Different Ion Strength

The relationship between the normalized fluorescence intensity and $MgCl_2$ concentration is illustrated in Figure 2.5. In the case of the glass slide immobilized molecular beacon arrays, the fluorescence background decrease with the increase of the $MgCl_2$ concentration. When the concentration of $MgCl_2$ reaches 100 mM, no significant changes can be observed after incubation. The fluorescence intensity of the molecular beacon arrays in hybridization buffer containing 500 mM $MgCl_2$ is about 60% of the initial intensity. In the case of the agarose film immobilized arrays; low concentration of $MgCl_2$ can effectively maintain the stem structure and quench the fluorescence. The fluorescence intensity is just about 20% of the initial intensity in hybridization buffer containing 10 mM $MgCl_2$. Further increasing of $MgCl_2$ concentration has little effect on improving the quench efficiency. The fluorescence intensity curve of PAA film immobilized arrays is similar to that of agarose film immobilized ones. This indicated similar hydrophilic microenvironments provided by the PAA films and the agarose films.

The quench efficiency is higher than 99% in bulk solutions containing 1 mM to 5 mM $MgCl_2$.

There are two effects contributed to the decrease in quenching efficiency of the agarose film and the PAA film immobilized molecular beacon arrays: first, there are still some surface effects which destabilize the stem

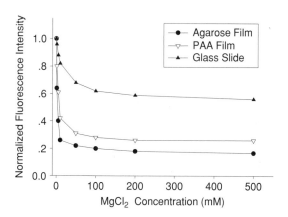

FIGURE 2.5. Normalized fluorescence intensity versus $MgCl_2$ concentration of the molecular beacon arrays immobilized on the different substrates. The fluorescence intensity is the average of all molecular beacons on the substrates. The hybridization buffer is 20 mM Tris-HCl (pH=8.0) and 500 mM $MgCl_2$.

structure of molecular beacons. Second, steric effects caused by high immobilization density retards the formation of stem structure and lower the quench efficiency.

2.3.3 Hybridization of Molecular Beacon Arrays

To investigate the specificity of the immobilized molecular beacons, hybridization solutions containing different ion strength were applied to the molecular beacon arrays on the different substrates. In all images of this section, the molecular beacon probes were spotted on the substrates in triplet format. From the left to the right, the probes are MB1 (noncomplementary to T2), MB2 (perfectly matched with T2), and MB3, MB4, MB5 (one central base mismatched with T2), respectively.

2.3.3.1   Hybridization in Different Ion Strength Solutions

Figure 2.6 displayed the hybridization results of the molecular beacon arrays immobilized on the glass slides in hybridization buffers containing 5 mM, 50 mM and 500 mM $MgCl_2$, respectively. The concentration of target T2 was 10mM. The background and hybridization fluorescence intensities were extracted with ImageJ software and the fluorescence intensity increments are calculated.

When the hybridization buffers containing 5 mM $MgCl_2$ was applied, the fluorescence intensity decreased. We attribute it to photo bleaching by continuous laser scanning. For hybridization buffer containing 50 mM $MgCl_2$ and 500 mM $MgCl_2$, the fluorescence intensity increment of the perfectly matched probes was more than that of single base mismatched probes. The noncomplementary molecular beacon probe MB1 used as negative control also showed decrease of fluorescence.

Figure 2.7 displayed the corresponding results of the molecular beacon arrays immobilized on the PAA films. The hybridization results in hybridization buffer containing 5 mM $MgCl_2$ show that the fluorescence intensity increment of the perfectly matched probes is about two folds of that of the single base mismatched probes. The perfectly matched probes and single base mismatched probes can be easily distinguished from the image. With the increment of $MgCl_2$ concentration, the ratio of fluorescence intensity increment of the perfectly matched probes to that of the single base mismatched probes also increased, even if the increase rate was low when the $MgCl_2$ concentration was higher than 50 mM.

Figure 2.8 displayed the corresponding results of the molecular beacon arrays immobilized on the agarose films. The results are similar to those of the PAA immobilized ones. But the fluorescence increments are larger than those of the PAA film immobilized ones. The results may be attributed to the relatively lower immobilization capacity of the PAA films as discussed in section 2.2.3 and the lower quench efficiency of the molecular beacon probes immobilized on PAA film as discussed in section 2.3.2.

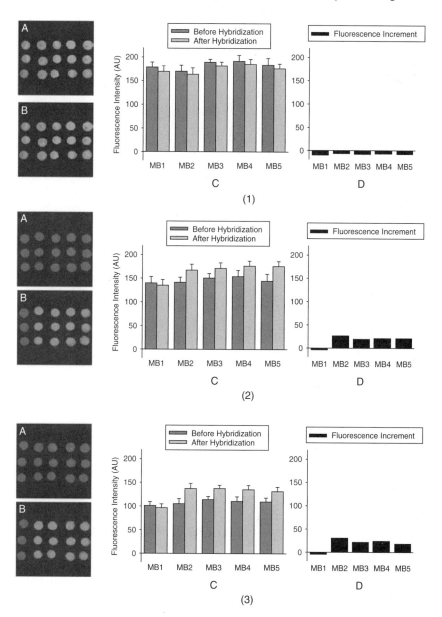

FIGURE 2.6. Images and fluorescence intensities of the molecular beacon array on a glass slide. Hybridization buffer: 10nM T2 in 20 mM Tris-HCl (pH = 8.0) containing (1) 5, (2) 50 and (3) 500 mM MgCl2. Images of (A) and (B) are fluorescence images before and after hybridization, respectively. The plots of (C) and (D) are fluorescence intensity before and after hybridization and fluorescence intensity increment, respectively.

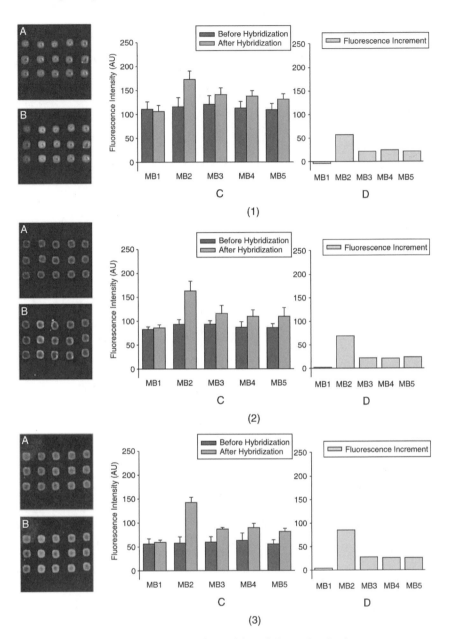

FIGURE 2.7. Images and fluorescence intensities of the molecular beacon array on a PAA film. Hybridization buffer: 10nM T2 in 20 mM Tris-HCl (pH = 8.0) containing (1) 5, (2) 50 and (3) 500 mM MgCl$_2$. Images of (A) and (B) are fluorescence images before and after hybridization, respectively. The plots of (C) and (D) are fluorescence intensity before and after hybridization and fluorescence intensity increment, respectively.

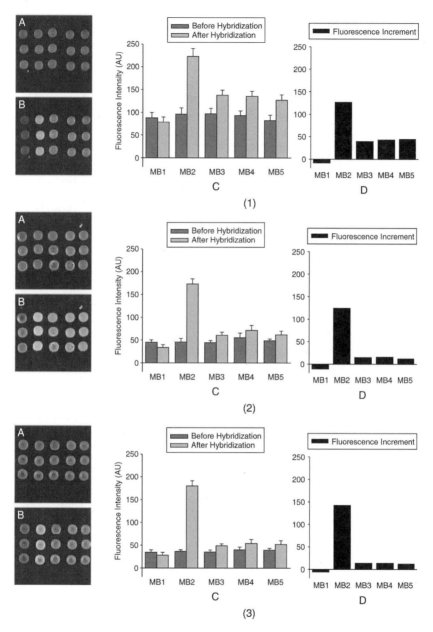

FIGURE 2.8. Images and fluorescence intensities of the molecular beacon array on an agarose film. Hybridization buffer: 10nM T2 in 20 mM Tris-HCl (pH = 8.0) containing (1) 5, (2) 50 and (3) 500 mM $MgCl_2$. Images of (A) and (B) are fluorescence images before and after hybridization, respectively. The plots of (C) and (D) are fluorescence intensity before and after hybridization and fluorescence intensity increment, respectively.

2.3.3.2   Single Nucleotide Mismatch Discrimination Ratio in Different Ion Strength Solutions

The single nucleotide mismatch discrimination ratio (SMR), which can be used to evaluate the ability to identify the single nucleotide mismatch, is defined as (PM-MM)/PM. PM is the average fluorescence increment of perfectly matched probes, and the MM is the average fluorescence increment of one base mismatched probes. The mismatch discrimination ratio versus $MgCl_2$ concentration was plotted in Figure 2.9.

For molecular beacon arrays immobilized on the glass slides, the SMR ratio increases from 0.09 (10 mM $MgCl_2$) to 0.3 (500 mM $MgCl_2$). The fluorescence increment is too low to get reliable ratios when the hybridization is performed in buffers containing 1mM and 5mM $MgCl_2$.

For molecular beacon arrays immobilized on the PAA films, the SMR ratio is 0.37 in 1mM $MgCl_2$, increasing to 0.63 in 10 mM $MgCl_2$ and 0.68 in 50mM $MgCl_2$. Further increasing the concentration of $MgCl_2$ will not change the SMR.

For molecular beacon arrays immobilized on the agarose films, the SMR is 0.4 in 1mM $MgCl_2$, increasing to 0.67 in 5 mM $MgCl_2$ and 0.81 in 10 mM $MgCl_2$. This ratio will increase to 0.9 in 50 mM $MgCl_2$ and there is little more increment with higher concentration of $MgCl_2$.

For solution data, the discrimination ratio is 0.98 in 1 mM $MgCl_2$ and increased to 0.99 in higher concentration of $MgCl_2$.

From this plot, we could see the SMRs of the molecular beacon arrays immobilized on the PAA films and the agarose films are similar to those in homogeneous solution. The results indicated that the molecular beacon in

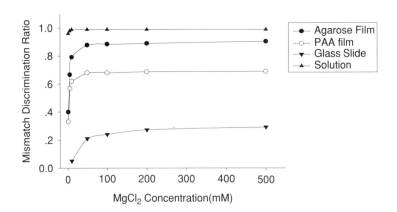

FIGURE 2.9. Mismatch discrimination ratios (SMR) of the different molecular beacons in bulk solution, immobilized on the glass slides, the PAA films and the agarose films, respectively. Discrimination ratio is defined as (PM-MM)/PM. PM is the fluorescence increment of perfectly matched probes, and the MM is the average fluorescence increment of three one-base mismatched probes. The hybridization solution is 10 nM target T2 in 20 mM Tris-HCl (pH = 8.0) with different concentration of $MgCl_2$.

solution and immobilized on the PAA films and the agarose films holds the high specificity while the glass slide immobilized one is not satisfied.

The above results can be explained by the following discussion:

Firstly, it is due to incomplete quenching of molecular beacon. From the results of annealing experiments of the molecular beacon arrays in section 2.3.2, we can find that a large part of the glass immobilized molecular beacons are not quenched. These unquenched molecular beacon probes contribute to high background and greatly reduce the number of probes available for conformation change and fluorescence restoration. High quench efficiency of the PAA films and the agarose films immobilized molecular beacons contributes to the high mismatch discrimination ratio. Moreover, the high immobilization capacity of the PAA films and the agarose films greatly improves the number of immobilized molecular beacons, which will provide more molecular beacons for conformation change and large fluorescence increment.

Secondly, the immobilization of the molecular beacon changes the electrostatic properties and the local ion strength. The environment of the glass immobilized molecular beacon probes is quite different from that they experienced in bulk solution. Even high concentration hybridization buffer is used; it is difficult to counteract the interfacial effect. The PAA films and the agarose films can provide a favorable solution-like environment and hybridization in low ion strength hybridization buffer can produce reliable results. High ion strength hybridization buffers will facilitate the formation of secondary structure of PCR products and hinder the hybridization with immobilized probes in practical use.

Thirdly, considering the SMR is about 0.3 to 0.7 for immobilized linear probes employed in most of the microarray technology (31), the SMR of the molecular beacon arrays immobilized in the hydrogel microporous films are satisfying. The thermodynamic balance between the hairpin structure and the duplex formed by molecular beacon with target ensures the high specificity of the molecular beacon in target recognition and discrimination.

2.4 Hybridization Kinetics of Molecular Beacon Arrays

The real-time hybridization process at room temperature was investigated. Hybridization solutions of different ion strength buffers, target concentrations were applied to the molecular beacon arrays immobilized on the different substrates.

2.4.1 Hybridization Kinetics in Different Ion Strength Buffers

Time serial images after applying 10 nM target T2 in 10 mM $MgCl_2$ hybridization buffers to the molecular beacon arrays immobilized on the glass slides, the PAA films and the agarose films were collected. The fluorescence intensity increment were calculated by subtracting the fluorescence background (fluorescence intensity before hybridization) from the fluorescence intensity at different time and plotted in Figure 2.10 (A), (C) and (E),

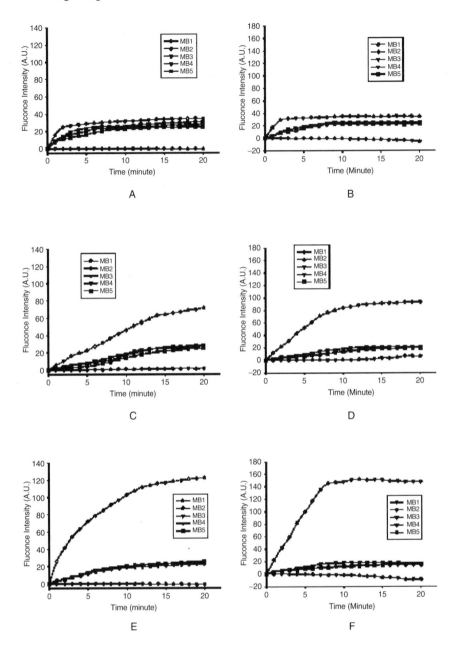

FIGURE 2.10. Fluorescence increment versus time after applying 10 nM target T2 in hybridization buffers containing 10 mM (left panel) and 500 mM (left panel) MgCl₂ to the molecular beacon arrays immobilized on the glass slides (A,B), the PAA films (C,D) and the agarose films (E,F).

respectively. These figures indicate that the hybridization process of the molecular beacon arrays immobilized on glass slides is so fast that 90% of the reaction completed in 4 minutes, while the reactions are much slower for the PAA films and the agarose films immobilized ones and the reaction completed in 18 minutes and 14 minutes, respectively.

The corresponding fluorescence increment versus time after applying 10 nM target T2 in hybridization buffers containing 500mM $MgCl_2$ to the molecular beacon arrays immobilized on the glass slides, the PAA films and the agarose films were shown in Figure 2.10 (B), (D) and (F), respectively. The results indicate the hybridization reactions completed in a shorter time and the fluorescence increment is larger with the increase of $MgCl_2$ concentration.

We define the hybridization time as the time needed for the fluorescence intensity increment of perfect marched molecular beacon probe reaches 90% of saturated value. The relationship between the hybridization time and $MgCl_2$ concentration was illustrated in Figure 2.11.

The results in Figure 2.11 indicate that the hybridization reaction time decreases with the increase of the $MgCl_2$ concentration. The hybridization time decreased from 25 minutes, 25 minutes and 13 minutes at 5 mM $MgCl_2$ to 3 minutes, 9 minutes and 7 minutes at 500 mM $MgCl_2$ for the glass slide, the PAA films and the agarose films immobilized arrays, respectively. The effect of $MgCl_2$ is apparent when the concentration is lower than 50mM.

The results will be discussed as follows. Firstly, the hybridization reaction time of the glass slide immobilized arrays is much shorter than that of the PAA films and the agarose films immobilized ones. It is because the

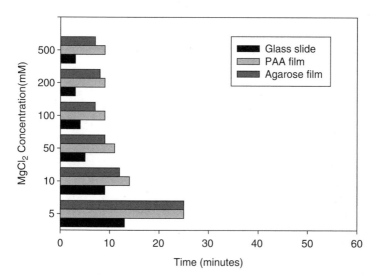

FIGURE 2.11. Hybridization time of the molecular beacon arrays immobilized on the different substrates. The hybridization solution containing 10 nM target T2 in 20 mM Tris-HCl (pH = 8.0) and different concentration of $MgCl_2$.

hybridization process in the hydrogel microporous films is mass transport limited reaction and the mass transport process is slower than the hybridization process. One of the main advantages of the glass slides based microarray technology over the traditional filter hybridization is the high diffusion rate and the short reaction time. Secondly, the hybridization reaction time decreases with the increase of the ion strength. Since the increase of the ion strength will facilitate the duplex formation and stabilization and the depletion of the target at the surface, the diffusion of target to the surface will be accelerated and the hybridization reaction will complete in a shorter time.

2.4.2   Hybridization Kinetics in Different Target Concentration Solutions

The above results indicated the agarose film immobilized molecular beacon arrays performed well in signal intensity and single nucleotide mismatch discrimination ratio.

To investigate the relationship between the fluorescence intensity and target concentration, different concentration targets in hybridization buffer containing 20 mM Tris-HCl (pH = 8.0) and 10 mM $MgCl_2$ were applied to the agarose film immobilized molecular beacon arrays.

The fluorescence intensity increment of the perfect matched probes versus time is shown in Figure 2.12. The hybridization reactions are faster when higher concentration targets are applied.

2.5   Application of Molecular Beacon Arrays

A polymorphism in codon 158 of the human ApoE gene (34), which plays a key role in the transport and metabolism of plasma cholesterol and triglyc-

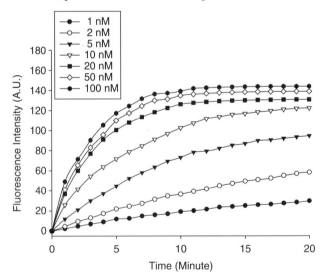

FIGURE 2.12. Fluorescence intensity increment versus time of the molecular beacon array immobilized on the agarose films after adding the hybridization solution containing different concentration target T2 in 20 mM Tris-HCl (pH = 8.0) and 10 mM $MgCl_2$.

erides, was employed to investigate the performance of molecular beacon arrays in complex environment.

2.5.1 Human Genome DNA Extraction and PCR Amplification

Human genomic DNA was extracted from whole blood cells by phenol-chloroform method.

A 218-bp DNA fragment containing the single nucleotide polymorphism of codon 158 of human apoE gene was amplified with the primer sequences: 5'-TCCAAGGAGCTGCAGGCGGCGCA (Forward) and 5'-GCCCCG-GCCTGGTACACTGCCA (Reverse) as previously described (34).

The symmetric PCR reactions were performed in a total volume of 25 μL containing 75 mM Tris-HCl (pH = 9.0), 20 mM ammonium sulfate, 0.1 mL/L Tween, 1.5 mM $MgCl_2$, 500 nM of each primer, 200 μM dNTPs, 100 mL/L DMSO, and 1 unit of Taq polymerase. Amplification condition consisted of an initial 10-min denaturation at 94 °C followed by 40 cycles of 30 s of denaturing at 94 °C, 30 s of annealing at 65 °C, and 30 s of extension at 70 °C.

The asymmetric PCR mixture contained all the reaction components in identical amount as that in symmetric PCR except that forward primer concentration was 50nM. Amplification condition was the same as the symmetric PCR.

Fragmented PCR products were prepared as symmetric PCR reactions described above with the exception of using 160 μM dTTP and 40 μM dUTP instead of 200 μM dTTP. Amplification condition was the same as symmetric PCR except that the annealing temperature was 55°C. The PCR products were fragmented by adding 2U of UNG and incubating at 37°C for 60 minutes, followed by heating the solution to 95°C for 5 minutes to inactivate the enzyme.

Aliquots of 2 μL PCR products were electrophoresed on 2% agarose gel and visualized by ethidium bromide staining. The Electrophoretic image is shown in Figure 2.13 (A). Lane 1 and lane 2 are 218-bp symmetric and asymmetric PCR products, respectively. Lane 3 and lane 4 are symmetric PCR products amplified from the PCR mixture containing dUTP before and after fragmentation with UDG, respectively. In lane 4 the fragmented PCR product appears as a smear.

2.5.2 Genotyping by PCR-RFLP

The apoE genotype was indentified by a polymerase chain reaction restriction fragment length polymorphism (PCR-RFLP) method. PCR-RFLP was carried out by adding 2 units of BstH2 I ((prototype *Hae* II, Sibenzyme Ltd.,Russia) to 20 μL symmetric PCR products for 2 h at 65 °C and analyzed by 4% agarose gel. The Restriction site of BstH2 I is RGCGC∧Y (R: purine, adenine and/or guanine; Y: pyrimidine, thymine and/or cytosine).

The Electrophoresis image of the PCR products was shown in Figure 2.13 (B). Lane 1 and lane 2 depict the 218-bp PCR product before and after BstH2 I digestion, respectively. Lane 2 depicts a C/T heterozygote.

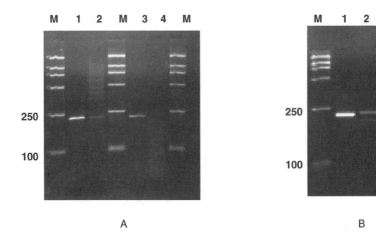

FIGURE 2.13. (A) Electrophoretic image: Lane 1 and lane 2 are symmetric and asymmetric PCR products, respectively. Lane 3 and lane 4 are PCR products before and after fragmentation with uracil-N-glycosylase, respectively. (B) Electrophoretic image: Lane 1 and lane 2 depict the 218-bp PCR product before and after HaeII digestion, respectively. Mutation from C to T causes the loss of restriction site. Lane 2 depicts a C/T heterozygote.

2.5.3   Hybridization of PCR Product to Molecular Beacon Arrays

20μL asymmetric and fragmented PCR products were added to 80μL hybridization buffers (20 mM Tris-HCl (pH = 8.0) and 10 mM MgCl$_2$) and pumped to hybridization cell for 30 minutes.

The hybridization images of asymmetric and fragmented PCR products were displayed in Figure 2.14 (A) and (B), respectively. The fluorescence intensity increments are shown in Figure 2.14 (C) and the discrimination ratio were calculated to be 0.61 and 0.64. Though the discrimination ratio is lower than that of the oligonucleotide targets at the same concentration of MgCl$_2$, it is good enough for practical applications.

2.6   Conclusions

We investigated and compared the immobilization, annealing, hybridization and application of molecular beacon arrays on different substrates, and demonstrated the excellent performance of hydrogel films as the immobilization substrate for molecular beacon microarray.

Molecular beacon array has shown many advantages over conventional microarray. Firstly, no target labeling is needed. Labeling is an important step in most of the microarray-based target preparing protocols. It is not only time consuming, and rather expensive, but also can change the levels of targets originally present in the sample. The use of molecular beacon allows one to eliminate target labeling. Secondly, no washing step is required. Since the

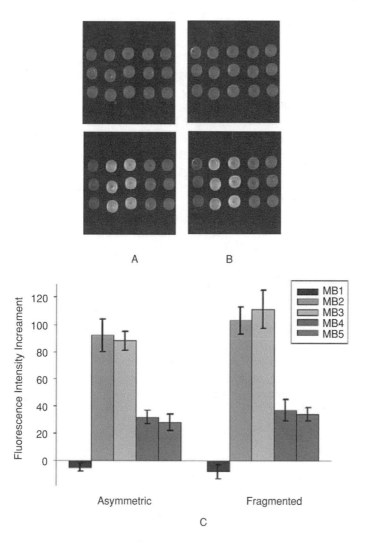

FIGURE 2.14. Fluorescence images before (up) and after (down) hybridization with asymmetric (A) and fragmented (B) PCR products, respectively. C) The fluorescence intensity increments after hybridization with asymmetric and fragmented PCR products.

unlabeled targets will not contribute to the fluorescence background and specificity of the molecular beacon probes are guaranteed by thermodynamic property of hairpin structure, washing step is not necessary. The high background caused by washing problem in conventional porous film microarray is resolved. It will also simplify the miniature device, such as lab-on-chip, design without considering washing problems in small volume. Moreover, the hybridization process can be easily monitored in real time and more reliable results can be obtained from the hybridization dynamic curves. Thirdly, high specificity of molecular beacon ensures the single oligonucleotide mismatch can be easily detected. The presence of hairpin structure maximizes the specificity of molecular beacon probes. Furthermore, unlike linear probes, molecular beacon probes are insensitive to mismatch type and position, which greatly simplify probe design.

We investigated the annealing and hybridization process of molecular beacon array on different substrates in different hybridization buffers. For glass slide, the interfacial effects destabilize the hairpin structure of molecular beacon and lead to high background. Hydrogel films, combining the fast reaction speed of glass slide and solution-like environment provides an ideal support for molecular beacon arrays. Low fluorescence background after annealing and high mismatch discrimination ratio promise great capacities for practical applications of hydrogel film immobilized molecular beacon array. Further experiment with PNA molecular beacons may take the advantages of uncharged backbone of PNA and further improve the single nucleotide mismatch discrimination ratio.

The instrumentation for fabricating molecular beacon microarrays and signal detection are compatible with the state of art microarray technique. Other high density microarray fabrication technique, such as light-directed synthesis, liquid dispersing method and molecular stamping, allows fabricate the high density molecular beacon microarray economically.

It is expected that molecular beacon microarrays can perform high-throughput mutation analysis and disease diagnosis in a parallel, cost saving and label-free way.

3.   Microarray Based Melting Curve Analysis Method for Single Nucleotide Mismatch Detection

3.1   Melting Curve Analysis Methods

DNA fragments can be distinguished from each other by their melting properties. The melting properties of a short piece of DNA, such as a PCR product, are very strongly influenced by the size, sequence composition, and mismatch bases. Various melting curve based analysis methods were developed in the last few years (12). In these methods, melting can be performed by an increasing gradient of some denaturing agent, such as temperature, electronic field, ion strength, etc. In this section, several melting curve analysis methods were briefly reviewed.

3.1.1   Homogeneous Melting Curve Analysis Methods

A number of melting curve analysis methods has been developed that allow fully homogeneous assays to be performed.

Melting curve analysis of single nucleotide polymorphism (McSNP) is a method combining a classic approach, restriction enzyme digestion, with a melting curve analysis method (35). After PCR, products are digested with the appropriate restriction enzyme for the SNP of interest. This creates different length and melting temperature fragments between the two SNP alleles. Then melting curve measured by slowly heating DNA fragments in the presence of the dsDNA-specific fluorescent dye SYBR Green I. As the sample is heated, fluorescence rapidly decreases when the melting temperature of a particular fragment is reached. By analyzing the melting curve, the genotype of the sample can be determined.

Another approach of melting curve analysis method is based on fluorescence resonance energy transfer (FRET) (36). It involves two adjacent, fluorescently labeled probes, one overlapping the SNP position and acting as a donor while the other acting as an acceptor (quencher). The quench efficiency is strongly dependent on the distance of the donor and the acceptor. The single nucleotide mismatch at the SNP position will significantly reduce the melting temperature and the genotype can be determined.

Though many efforts have been devoted to the automatization of melting curve analysis in homogeneous solution, it is difficult to meet the high-throughput requirement in post-genome era.

3.1.2   Heterogeneous Melting Curve Analysis Methods

Dynamic allele specific hybridization (DASH) is a fast, cheap, robust and accurate genotyping method, which is suitable for medium scale genetic association studies (37). DASH involves a PCR with a biotinylated forward primer, immobilizing the resulting PCR products to a 96-well streptavidin-coated plate, denaturing away the reverse strand and probing the region of interest. After that, the dsDNA-specific fluorescent dye SYBR Green and the probe are added. Then the probe is gradually melted away, and since a mismatching probe melts at a lower temperature compared with the matched probe, the samples can be genotyped. DASH is a low cost and flexible method for SNP detection. However, like dot-blot, only one to two assays can be performed with each sample.

Microarray based melting curve analysis method was firstly reported by Mirzabekov et al. (38-39). The PAA gel pad immobilized oligonucleotide probes were used to identification of β-thalassemia mutations. With the melting curve analysis methods, the base changes and the homozygous and heterozygous β-thalassemia mutations can be reliably identified. Mirzabekov et al also investigated the thermodynamic properties for perfect and mismatched short oligonucleotide (8 bp) in the gel pad with equilibrium melting curves and compared the results with the solution data.

Another approach is to use electronic field instead of the temperature to denature the duplex. The NanoChip® Workstation provided by Nanogen Inc. can perform this kind of electronic stringency hybridization of the target DNA to the array, a precisely controlled negative electric field is used to dehybridization of the target from the mismatched capture probes and facilitate the discrimination of the matched and mismatched probes (40).

The disadvantage of the gel pad based system is the complex fabrication procedure and the high cost of the gel pad substrate. The Nanochip system also needs a high coat workstation and biochips.

3.2   Melting Curves of Duplexes in Solution and on Microarrays

3.2.1   Microarray Experiment Setups

Melting experiments on microarray were performed in real time on an experimental setup illustrated in Figure 2.15. The setup consists of detection optics, a hybridization chamber, a water bath temperature controlling system and an XY stage.

In our system, Leica confocal microscope TCS SP is used as detecting system. Compared with non-confocal system, low focus depth of confocal optics can effectively reject unwanted fluorescence from the solution and improve the signal-background ratio. The fluorescence images were collected in a scanning mode and the time for image collection was about 4 second (4 accumulation each image). The fluorescence change during the image collection can be ignored.

The hybridization chamber is composed of a 100 µL hybridization cell, an XY stage and a peristaltic pump and a valve to control the hybridization solution and buffer flowing through the pipe. The microarray can be pressed against the hybridization cell during the experiment.

Around the hybridization cell, the stage is hollowed to form a water jacket which connects a water bath reservoir.

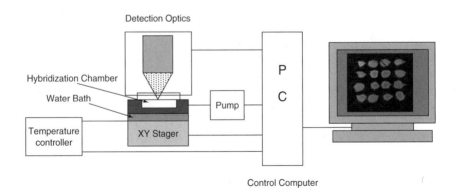

FIGURE 2.15. Experiment setup for melting curve experiments on microarrays.

3.2.2   Softwares

Fluorescence images were analyzed by ImageJ version 1.27. ImageJ is a public domain Java image processing program inspired by NIH Image. It can be freely downloaded from http://rsb.info.nih.gov/nih-image/index.html.

The nonequilibrium melting curve can be obtained by plotting the fluorescence intensities extracted by ImageJ versus temperature. The dissociation temperature (T_d) of the matched and mismatched duplexes can be determined from the nonequilibrium melting curve. We use recently issued web-based software to calculate the dissociation temperature. The software is available at http://stahl.ce.washington.edu. The software is explained in a detailed in an article published in (41).

3.2.3   The Probe and Target Sequences

The 20mer probe and 20mer, 17mer and 14mer unlabeled and fluorescein labeled probes and targets used in this section were listed in Table 2.2.

3.2.4   Melting Curves of Duplexes in Solution

Melting experiments in solution were performed by measuring absorption hypochromism on a Shimadzu 2100 UV-VIS spectrophotometer equipped with a circulated water bath. Melting curves were measured when the temperature of the solution increased from 26 °C to 90 °C at the rate of 0.5 °C/min. Annealing curves were measured when the temperature of the solution decreased from 90 °C to 26 °C at the rate of 0.5 °C/min. Absorbance date at 260nm were collected at 1 °C intervals.

The perfect matched and single central base mismatched probes and targets were diluted to the final concentration of 1μM each strand in a melting buffer containing 10mM PBS (pH=7.0) and 1M sodium chloride.

TABLE 2.2.   The probe and target sequences

Symbol	Sequence
P20G	5′-NH$_2$-AG GAG GCT A**GT** TCT CTC AGG
P20C	5′-NH$_2$-AG GAG GCT A**CT** TCT CTC AGG
P20A	5′-NH$_2$-AG GAG GCT A**AT** TCT CTC AGG
P20T	5′-NH$_2$-AG GAG GCT A**TT** TCT CTC AGG
T20C	3′-TC CTC CGA T**CA** AGA GAG TCC
T17C	3′-CTC CGA T**CA** AGA GAG TC
T14C	3′-TC CGA T**CA** AGA GAG
F20C	3′-FAM-TC CTC CGA T**CA** AGA GAG TCC
F20A	3′-FAM-TC CTC CGA T**AA** AGA GAG TCC
F17C	3′-FAM-CTC CGA T**CA** AGA GAG TC
F14C	3′-FAM-TC CGA T**CA** AGA GAG

Bold underline indicates the single nucleotide mismatch position.

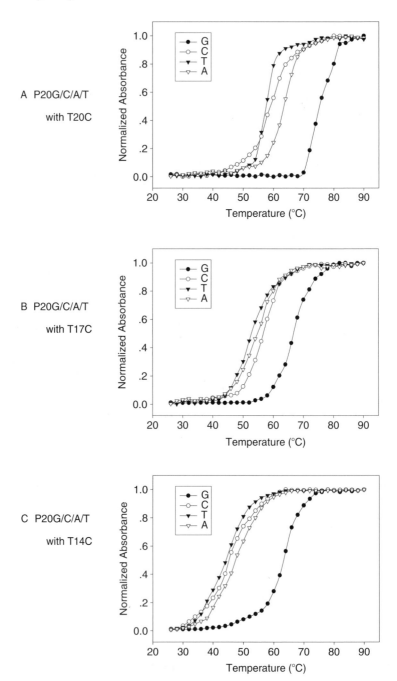

FIGURE 2.16. Normalized melting curves of perfect matched and one central base mismatched 20mer probe / 20mer target, 20mer probe /17mer target and 20mer probe/ 14mer target in solution.

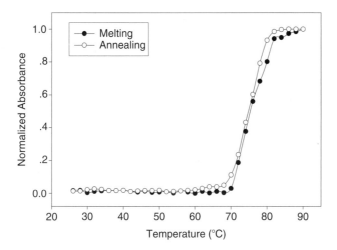

FIGURE 2.17. Normalized melting and annealing curves of the 20mer perfectly matched probe/target (P20G/T20C) in solution.

To compare the melting curves of different duplexes, we normalized the absorbance data by

$$f = (A_{260} (T_{min}) - A_{260}(T))/ (A_{260}(T_{min}) - A_{260}(T_{max}))$$

and plotted f versus temperature in Figure 2.16.

The normalized melting and annealing curves of the 20mer perfectly matched duplex, P20G and T20C, in solution were shown in Figure 2.17. The absence of hysteresis suggests that these curves were measured under equilibrium conditions and the melting process in the experiment was slow enough to reach the equilibrium at each temperature. The melting temperature, defined as the middle point of the melting transition, of different duplexes in solution was calculated and listed in Table 2.3.

The average melting temperatures of 20mer, 17mer and 14mer single central base mismatched duplexes are 60 °C, 54 °C and 45 °C respectively. The melting temperature differences of the 20mer, 17mer and 14mer perfect matched duplexes and one central base mismatched duplexes are 14 °C, 13 °C and 18 °C, respectively.

3.2.5   Melting Curves of Duplexes on Microarrays

The microarrays used for melting curve investigation were prepared as follows. The aminosilane derived glass slides (Cat. No S3003, Dako) were

TABLE 2.3. Melting temperature of different duplexes

	G/C	C/C	T/C	A/C
P20/T20	74	59	58	63
P20/T17	67	56	53	55
P20/T14	63	45	44	48

cleaned with deionized distilled water and incubated in 5% glutaraldehyde in 0.1 M PBS buffer (pH = 7.4) for 2 hours. Then the slides were thoroughly washed twice with methanol, acetone and deionized distilled water, and dried. Spotting solutions were obtained by dissolving oligonucleotide probes in sodium carbonate buffer (0.1 M, pH = 9.0) at the concentration of 100 μM. Pin-based spotting robot PixSys5500 with CMP3 pin was used to perform microarray spotting. After spotting, the glutaraldehyde derived glass slides were incubated at room temperature for 2 hours and at 37 °C for 2 hours and thoroughly washed in 0.1% Tween.

Melting experiments on microarrays were performed with the instrument introduced in section 3.2.1. The hybridization solutions were prepared by dissolving the targets F20C, F17C and F14C in a hybridization buffer (10mM PBS (pH = 7.0) and 1M sodium chloride) to the final concentration of 400nM. Before the melting experiments, the hybridization solutions were pumped into the hybridization cell for 30 minutes.

Fluorescence images during the melting process were collected when the temperature in the hybridization cell increased from 25 °C to 70 °C at the rate of 0.5 °C/min. Fluorescence images during the annealing process were collected when the temperature in the hybridization cell decreased from 70 °C to 25 °C at the rate of 0.5 °C/min. The fluorescence images were collected every 2 °C and they were accumulation of four scanning to reduce the electronic noise. The fluorescence intensity versus temperature is plotted in Figure 2.18.

The melting and annealing curves of the 20mer perfectly matched duplex P20G/F20C, and one base mismatched duplex P20T/F20C on microarray were shown in Figure 2.19. Compared with results in solution shown in Figure 17, the apparent hysteresis in annealing curves suggests that the melting curves were not measured at the equilibrium state. It can be seen from the Figure 2.19 that the mismatched targets anneal at a slower rate than the matched on do.

The dissociation temperature T_d, distinguished from the melting temperature determined from the equilibrium melting curves, is defined as the temperature at which half of the duplex were dehybridized. With the web based software provided by Washington University, microarray melting data can be normalized and the dissociation temperature can be calculated. The results provided by the software were listed in Table 2.4.

The average melting temperatures of 20mer, 17mer and 14mer one central base mismatched duplexes are 40.4 °C, 34.2 °C and 32.4 °C respectively. The melting temperature differences of the 20mer, 17mer and 14mer perfect matched duplexes and one central base mismatched duplexes are 4.4 °C, 5.2 °C and 5.8 °C, respectively.

3.2.6 Discussion

The results in solution and on microarrays will be compared and explained by the following discussion.

Firstly, comparing Figure 2.16 and Figure 2.18, the melting curves of the duplexes on microarrays are broadened and depressed relative to the

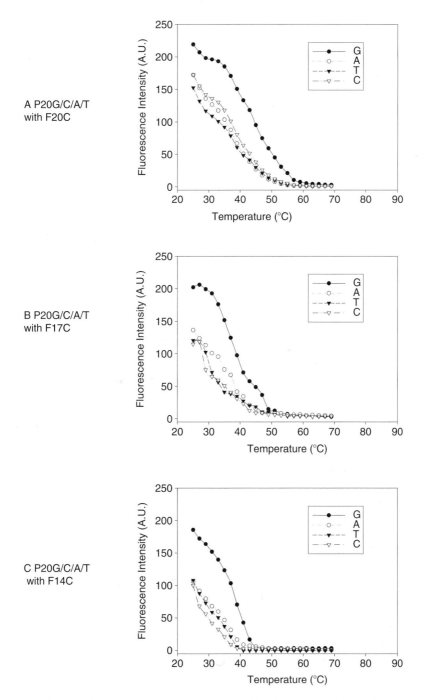

FIGURE 2.18. Melting curves of perfect matched and one central bas mismatched 20mer/20mer, 20mer/17mer and 20mer/14mer probe/target on microarrays.

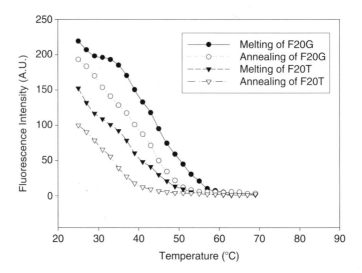

FIGURE 2.19. Melting and annealing curves of the 20mer perfectly matched probe/target (P20G/T20C) and one base mismatched probe/target (P20G/T20T) on a microarray.

corresponding duplexes in solution. The melting properties are greatly affected by the local dielectric environment and ion strength. The low local ionic strength at the liquid-solid interface, which is induced by the low dielectric constant siloxane layer and the high negative charged phosphate backbones, alters the melting behavior of duplexes on the microarrays.

Secondly, compared with the solution data, the melting curves on microarrays move to lower temperature, which indicates the lower thermal stability of duplexes on microarrays. It can also be attributed to the low ion strength and the high negative charges at the interface.

Thirdly, the duplexes in solution reach equilibrium state at each temperature, while the duplexes on microarrays do not reach at the same heating rate. The results indicate it may take several hours to get equilibrium curves on microarray, which makes the diagnostic applications of equilibrium curves analysis impractical. In next section, we will report the non-equilibrium melting curves analysis, which can give reliable results in about one hour.

Finally, the melting temperature difference between the perfect matched duplexes and single central base mismatched duplexes is decreased with increase of duplex length. In solution, the difference is more than 10 °C.

TABLE 2.4. Dissociation temperature of different duplexes

	G/C	C/C	T/C	A/C
P20/F20	45.0	38.8	41.7	40.8
P20/F17	39.4	34.0	32.8	35.9
P20/F14	38.2	29.8	32.7	34.6

Considering the sharpness of melting curves, it is easy to distinguish the matched duplexes from the single central base mismatched duplexes. In contrast, on microarrays, the dissociation temperature difference is about 5 °C. The broadened melting curves make it more difficult to distinguish the matched duplexes from the mismatched ones. In state-of-the-art microarray technology, hundreds to thousands of oligonucleotide probes are immobilized on the glass slides. It is really difficult to normalized thermodynamic properties and hybridization conditions of these probes. Melting curves analysis may provide a solution to the contradiction of high throughput and high reliability.

3.3   Application in Detecting HBV Mutations

Hepatitis B is a global health problem with a considerable morbidity and mortality. Some mutations of HBV appear to be closely related with hepatocellular carcinoma (HCC), one of the most common human cancers (42-43).

In this section, we constructed a microarray containing probes targeting five sets of these mutations. The GC contents and the melting temperature of the probes are over a wide range. The results demonstrate the excellent single nucleotide mismatch discrimination ability of nonequilibrium melting curve analysis method.

3.3.1   The Design of Oligonucleotide Probes

We chose five sets of probes targeting hepatitis B virus mutations related to hepatocellular carcinoma (HCC) at nt positions of 531(T to G), 546(C to T) and 587(G to A) in SHBsAg protein and 1762 (A to T) and 1764 (G to A) in core promoter region.

The 17 base oligonucleotide probes were modified with amino group at 5′ end for immobilization. The mutation position is at the center position to maximize the mismatch discrimination. The GC contents of probes ranged from 29.4% (5 GC in 17 base) to 64.7% (11 GC in 17 base). The probes symbol, sequence, GC contents, mutation position and type were listed in Table 2.5. A 5′ end amino group modified and 3′ end TAMRA modified oligonucleotide probe of 17T was used as control probe in the experiment.

3.3.2   Fabrication of HBV Microarrays

The preparation of the slides and microarray spotting were described in Section 5.3 with the exception that the concentration of control probe was 20μM. The layout of the microarray is illustrated in Figure 2.20 (A). The double labeled control probe was spotted in duplex format.

3.3.3   Amplification of HBV DNA

Sample of HBV DNA was isolated from serum by phenol-chloroform extraction and ethanol precipitation. Segment the viral genome was amplified using

TABLE 2.5. The probe sequences, GC contents and nt postions

Probe	Probe Sequence	GC	NT Position
P1	NH$_2$-GAGCAGGA**A**TCGTGCAG	10	531 wide type
P2	NH$_2$-GAGCAGGA**C**TCGTGCAG	11	531 (T to G)
P3	NH$_2$-GCAGTTTC**C**GTCCGAAG	10	587 wide type
P4	NH$_2$-GCAGTTTC**T**GTCCGAAG	9	587(G to A)
P5	NH$_2$-ACATAGAG**G**TTCCTTGA	7	546 wide type
P6	NH$_2$-ACATAGAG**A**TTCCTTGA	6	546(C to T)
P7	NH$_2$-ACAAAGA**CC**TTTAACCT	6	1762 1764 wide type
P8	NH$_2$-ACAAAGA**CC**A**TTAACCT	6	1762(A to T)
P9	NH$_2$-ACAAAGAT**C**T**TTAACCT	5	1764(G to A)
P10	NH$_2$-ACAAAGAT**C**A**TTAACCT	5	1762(A to T) 1764(G to A)
F	NH$_2$-TTTTTTTTTTTTTTTTTTT-TAMRA		Control Sequence

Bold underline indicates the single nucleotide mutation of HBV.

duplex PCR with primers GTTGCCCGTTTGTCCTCT (forward) and GATGTTGTACAGACTTGGCC (reverse), and GGCATACTTCAAA-GACTGTG (forward) and GAAGGAAAGAAGTCAGAAGG (reverse). The PCR reactions were performed in a total volume of 25 μL containing 1 U Taq polymerase (Takara Shuzo Co. Ltd., Japan), 1×PCR buffer, 0.2 mM

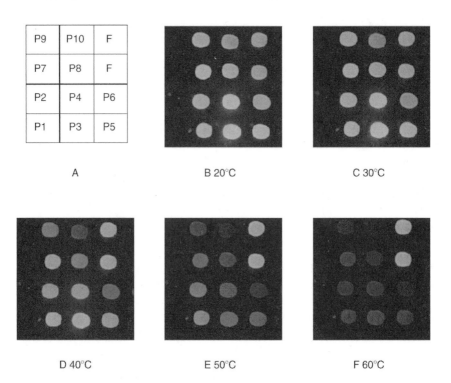

FIGURE 2.20. Probes Layout (A) and Fluorescence Image of the HBV microarray at different temperature (B)-(F).

dATP, dCTP, dGTP and 0.16 mM dTTP and 0.04mM SpectrumRed-dUTP (Vysis Co. USA), 1µl DNA template sample and 500 nM of each primer. Amplification condition consisted of an initial 10-min denaturation at 94 °C followed by 30 cycles of 30s of denaturing at 94 °C, 30 s of annealing at 52 °C, and 30s of extension at 72 °C.

3.3.4   Melting Curve Measurement of the HBV Microarray

The PCR product was applied to the HBV microarray at 20 °C for 10 minutes. Then the microarray was mounted on the hybridization chamber containing hybridization buffer ((10mM PBS (pH=7.0) and 1M sodium chloride). Melting experiment was carried out with the temperature increasing from 20 °C to 60 °C at a rate of 2 °C/min.

The fluorescence images were collected every minute by laser scanning confocal microscope. The fluorescence was excited with 568nm line of Kr-Ar ion laser and the collection was performed with the standard TRITC filter.

The fluorescence images of the microarray at 20-60 °C are shown in Figure 2.20 (B) to (F). The fluorescence intensity of the TAMRA labeled control probes shows no apparent change, which indicates the photobleaching during experiment is neglectable.

Figure 2.20 (B), we can see the fluorescence intensity difference between low GC content matched and mismatched probes. The fluorescence intensity of two base mismatched p10 and one base mismatched probe p8 and p9 is much lower than that of perfect matched probe p7. The perfect matched probe p5 is brighter than one base mismatched p6. There are no difference of the higher GC contents matched and mismatched probe p1, p2, p3 and p4.

When the temperature is increased to 30 °C, as shown in Figure 2.20 (C), the two base mismatched probe p10 is hard to see. And the fluorescence contrast between p5 and p6 is improved.

Further increase the temperature to 40 °C, as shown in Figure 2.20 (D), the low GC contents mismatched probes: p6 and p8, p9, p10 are hardly to see. The fluorescence intensities of higher GC content matched probe p1 and p3 are higher than those of mismatched p2 and p4, respectively.

At 50 °C in Figure 2.20 (E), only the highest GC content perfect matched probe p1 and fluorescence labeled control probes can be seen.

At 60 °C in Figure 2.20 (F), only the fluorescence labeled control probes are discernable.

The fluorescence intensity versus temperature of four location of HBV mutation is plotted in Figure 2.21. The fluorescence intensities are normalized with the average intensities of two control probes to minimize the effects of laser fluctuate and fluorescence bleaching during the continuous scanning.

The melting curve data were submitted to the web-based dissociation temperature calculation software and the returned results are listed in Table 2.6.

The single nucleotide mismatch discrimination ratios at different temperatures were calculated and shown in Figure 2.22. We can see from the Figure 2.22 and Table 2.6, the largest difference of fluorescence intensities reaches

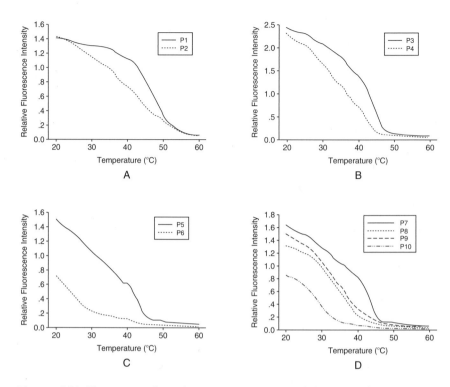

FIGURE 2.21. Fluorescence intensity versus temperature of the HBV microarray.

when the temperature is between the dissociation temperature of the perfect matched and one base mismatched duplex. The results can be further confirmed by the plot of single nucleotide mismatch discrimination ratio versus temperature in Figure 2.22. The highest discrimination ratios of p1 to p2, p3 to p4, p5 to p6, p7 to p8 and p7 to p9 are at 45 °C, 45 °C, 42 °C, 41 °C and 40 °C, respectively.

3.3.5   Discussion

We reported a microarray based non-equilibrium melting curve analysis method and its application in detecting HBV mutations in this section. The results indicate that the single nucleotide mismatch with a wide range of GC contents on the same microarray can be readily and reliably distinguished. With this method, the hardware requirements of microarray systems in clin-

TABLE 2.6. Dissociation temperature of each probe

P1	P2	P3	P4	P5	P6	P7	P8	P9	P10	P1
47.3	42.3	42.9	36.5	39.3	26.9	42.3	35.1	34.9	29.7	47.3

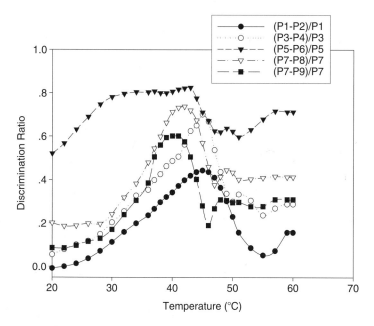

FIGURE 2.22. Single nucleotide mismatch discrimination ratio versus temperature of the HBV microarray.

ical applications were greatly decreased. The results are derived from a serial of images at different temperature instead of a single hybridization image, so even with microscope equipped with low precision temperature controller will give reliable conclusions. Moreover, this method will resolve the problem of normalization of the hybridization conditions of large number of probes in high density microarray. Several hybridization images at different temperatures will easily provide reliable sequence information of large number of different targets.

3.4 Conclusions

We investigated and compared the melting properties of oligonucleotide duplexes in solution and on microarrays. Though many publications reported the thermodynamic properties of duplex in solution with different compositions and mismatch types and positions, few data are available for duplexes on microarrays. Investigating the thermodynamic parameters and understanding the factors influencing the melting properties at the solid-liquid interface are necessary and fundamental for the application of microarray technology. Our results indicate the duplexes on microarrays show low thermal stabilities and the melting curves of them are greatly broadened and depressed by the surface effects.

One important factor effect the clinical application of microarray is the reliability of single nucleotide mismatch discrimination. Surface effects reduce the fluorescence difference between the perfect matched and single base mismatched duplexes, which makes it difficult to get reliable results from a single hybridization image. Our application of the non-equilibrium melting curve analysis method in HBV mutation detection showed great single nucleotide mismatch discrimination ability over a wide range of GC contents on the same microarray, which indicate a wide application of this method in microarray composed of probes with different length, base composition and mismatch positions. The potential of this method in detecting unbalanced alleles and mutations are under investigation.

Acknowledgements. This research work is supported by National Natural Science Foundation (project No. 60121101), the State Key Fundamental Research Scheme (973 project No.G1998051200), and the State Key High-Technology Scheme (863 project No. 2002AA2Z2004).

References

1. Lander ES. Array of hope. NAT GENET. 1999, 21S: 3-4.
2. Michael JH. DNA microarray technology: devices, systems, and applications. Annu Rev Biomed Eng, 2002, 4: 129-153.
3. Blohm DH, Guiseppi-Elie A. New developments in microarray technology. Curr Opin Biotechnol, 2001, 12: 41-47.
4. Fodor SPA, Read L, Pirrung M, Stryer L, Lu AT, and Solas D. Light-directed, spatially addressable parallel chemical synthesis. Science, 1991, 251: 767-773.
5. Fodor SPA, Rava RP, Huang XC, Pease AC, Holmes CP, Adams CL. Multiplexed biochemical assays with biological chips. Nature, 1993, 364: 555-557.
6. Okamoto T, Suzuki T, Yamamoto N. Microarray fabrication with covalent attachment of DNA using Bubble Jet technology. Nat Biotechnol, 2000, 18 (4): 438-441.
7. Xiao PF, He NY, He QG,, Zhang CX, Wang YW, Lu ZH, Xu JQ. DNA microarray synthesis by using PDMS molecular stamp (II) -Oligonucleotide on-chip synthesis using PDMS stamp. Science in China Series B-Chem., 2001, 44: 442-448.
8. Schena M, Shalon D, Davis RW, Brown PO. Quantitative monitoring of gene expression patterns with a complementary DNA microarray. Science, 1995, 270: 467-470.
9. Collins FS, Guyer MS, Chakravarti A. Variations on a theme: cataloging human DNA sequence variation. Science, 1997, 278: 1580-1581.
10. Hurles M Are 100,000 "SNPs" useless? Science, 2002, 298 (5598): U1-U1
11. Syvanen AC. Accessing genetic variation: Genotyping single nucleotide polymorphisms. Nat Rev Genet, 2002, 2 (12): 930-942.
12. Breen G. Novel and alternate SNP and genetic technologies. Psychiatr Genet. 2002,12(2):83-88.
13. Tyagi S, Kramer FR. Molecular beacons: probes that fluoresce upon hybridization. Nat Biotechnol, 1996, 14: 303-308.
14. Kostrikis LG, Tyagi S, Mhlanga MM, Ho DD, and Kramer FR Spectral genotyping of human alleles. Science, 1998, 9:1228-1229.
15. Giesendorf BA, Vet JA, Tyagi S, Mensink EJ, Trijbels FJ, and Blom HJ Molecular beacons: a new approach for semiautomated mutation analysis. Clin Chem, 1998, 44:482-486.

16. Eun AJC and Wong SM Molecular beacons: a new approach to plant virus detection. Phytopathology, 2000, 90:269-275.
17. Fortin NY, Mulchandani A, and Chen W Use of real-time polymerase chain reaction and molecular beacons for the detection of Escherichia coli O157:H7. Anal Biochem, 2001, 289:281-288.
18. Yates S, Penning M, Goudsmit J, Frantzen I, van De Weijer B, van Strijp D, and van Gemen B Quantitative detection of Hepatitis B Virus DNA by real-time nucleic acid sequence-based amplification with molecular beacon detection. J Clin Microbiol, 2001,39:3656-3665.
19. Perlette J and Tan W, Real-time monitoring of intracellular mRNA hybridization inside single living cells. Anal Chem, 2001,73:5544-5550.
20. Li JJ, Fang XH, Schuster SM, and Tan WH. Molecular beacons: a novel approach to detect protein -DNA interactions. Angew Chem Int Ed. 2000, 39: 1049-1052.
21. Tyagi S, Bratu DP, and Kramer FR, Multicolor molecular beacons for allele discrimination. Nat Biotechnol, 1998, 16: 49-53.
22. Tyagi S, Marras SAE, Kramer FR. Wavelength-shifting molecular beacons. Nat Biotechnol, 2000, 18: 1191-1196.
23. Dubertret B, Calame M, Libchaber AJ. Single-mismatch detection using gold-quenched fluorescent oligonucleotides. Nat Biotechnol, 2001,19 (4): 365-370.
24. Liu X and Tan W A fiber-optic evanescent wave DNA biosensor based on novel molecular beacons. Anal Chem. 1999, 71: 5054-5059.
25. Liu X, Farmerie W, Schuster S, Tan W. Molecular beacons for DNA biosensors with micrometer to submicrometer dimensions. Anal Biochem. 2000, 283: 56-63.
26. Brown LJ, Cummins J, Hamilton A, Brown T. Molecular beacons attached to glass beads fluoresce upon hybridisation to target DNA. Chemical Comm. 2000: 621-622.
27. Steemers FJ, Ferguson JA, and Walt DR. Screening unlabeled DNA targets with randomly ordered fiber-optic gene arrays. Nat Biotechnol. 2000, 18: 91-94.
28. Wang H., Li J, Liu HP, Liu QJ, Mei Q, Wang YJ, Zhu JJ, He NY, and Lu ZH. Label-free hybridization detection of a single nucleotide mismatch by immobilization of molecular beacons on an agarose film. Nucleic Acids Research 2002. 30:e61.
29. Bowtell D, Sambrook J, A molecular cloning manual: DNA microarrays. Gold Spring Harbor Laboratory Press. 2002, 61-100.
30. Proudnikov D, Timofeev E, Mirzabekov. Immobilization of DNA in polyacrylamide gel for the manufacture of DNA and DNA-oligonucleotide microchips. Anal Biochem. 1998, 259(1):34-41.
31. Guschin,D., G.Yershov, A.Zaslavsky, Gemmell A, Shick V, Proudnikov D, Arenkov P, Mirzabekov A.. Manual manufacturing of oligonucleotide, DNA, and protein microchips. Analytical Biochemistry. 1997, 250: 203-211.
32. Timofeev E, Kochetkova SV, Mirzabekov AD, Florentiev VL. Regioselective immobilization of short oligonucleotides to acrylic copolymer gels. Nucleic Acids Res. 1996 ,24(16):3142-3148.
33. Afanassiev,V., V.Hanemann, and S.Wolfl. Preparation of DNA and protein micro arrays on glass slides coated with an agarose film. Nucleic Acids Research. 2000, 28: E66.
34. Zivelin,A., Rosenberg,N., Peretz,H., Amit,Y., Kornbrot,N. and Seligsohn,U. Improved method for genotyping apolipoprotein E polymorphisms by a PCR-based assay simultaneously utilizing two distinct restriction enzymes. Clin. Chem., 1997, 43:1657-1659.

35. Akey JM, Shriver MD. Melting curve analysis of SNPs (McSNP): A simple gel-free low-cost approach to SNP genotyping and DNA fragment analysis. CLIN CHEM. 2000, 46 (11): 28

36. Bullock GC, Bruns DE, Haverstick DM. Hepatitis C genotype determination by melting curve analysis with a single set of fluorescence resonance energy transfer probes. CLIN CHEM. 2002, 48 (12): 2147-2154

37. Howell WM, Jobs M, Gyllensten U, et al. Dynamic allele-specific hybridization-A new method for scoring single nucleotide polymorphisms. NAT BIOTECH-NOL. 1999, 17 (1): 87-88

38. Drobyshev A, Mologina N, Shik V, Pobedimskaya D, Yershov G, Mirzabekov A. Sequence analysis by hybridization with oligonucleotide microchip: identification of beta-thalassemia mutations. Gene. 1997, 25, 188(1): 45-52.

39. Fotin AV, Drobyshev AL, Proudnikov DY, Perov AN, Mirzabekov AD. Parallel thermodynamic analysis of duplexes on oligodeoxyribonucleotide microchips. Nucleic Acids Res. 1998, 15; 26(6): 1515-21.

40. Gilles PN, Wu DJ, Foster CB, et al. Single nucleotide polymorphic discrimination by an electronic dot blot assay on semiconductor microchips. Nat Biotechnol. 1999, 17(4): 365-70.

41. Urakawa H, Noble PA, El Fantroussi S, Kelly JJ, Stahl DA. Single-base-pair discrimination of terminal mismatches by using oligonucleotide microarrays and neural network analyses. Appl Environ Microbiol. 2002,68(1):235-44

42. Hou,J.L., Wang,Z.H., Cheng, J.J. et al. Prevalence of naturally occurring surface gene variants of hepatitis B virus in nonimmunized surface antigen-negative Chinese carriers. Hepatology, 2001, 34: 1027-1034.

43. Chen,W.N. and Oon,C.J. Human hepatitis B virus mutants: significance of molecular changes. Febs Letters, 1999, 453: 237-242.

3

Miniaturized Multiplexed Protein Binding Assays

MARKUS F. TEMPLIN, OLIVER POETZ, JOCHEN M. SCHWENK, DIETER STOLL, AND THOMAS O. JOOS
NMI Natural and Medical Sciences Institute at the University of Tuebingen, Markwiesenstrasse 55, 72770 Reutlingen, Germany.

Abstract: Lorem ipsum dolor sit amet, consectetuer adipiscing elit aliquam erat volutpat. Ut wisi enim ad minim veniam, quis nostrud exerci tation ullamcorper suscipit lobortis nisl ut aliquip ex ea commodo consequat

Key words: Greeking.

1. Introduction

The fundamental principles of miniaturised and parallelised microspot ligand-binding assays were described more than a decade ago. In the 'ambient analyte theory', Roger Ekins and co-workers [1] explained why microspot assays are more sensitive than any other ligand-binding assay. At that time, the high sensitivity and enormous potential of microspot technology had already been demonstrated using miniaturised immunological assay systems.

Nevertheless, the enormous interest that microarray-based assays evoked came from work using DNA chips. The possibility of determining thousands of different binding events in one reaction in a massively parallel fashion perfectly suited the needs of genomic approaches in biology. The rapid progress in whole-genome sequencing (e.g. [2, 3]) and the increasing importance of expression studies [expressed sequence tag (EST) sequencing] was matched with efficient *in vitro* techniques for synthesising specific capture molecules for ligand-binding assays. Oligonucleotide synthesis and PCR amplification allow thousands of highly specific capture molecules to be generated efficiently. New trends in technology, mainly in microtechnology and microfluidics, newly established detection systems and improvements in computer technology and bioinformatics were rapidly integrated into the development of microarray-based assay systems. Now, DNA microarrays, some of them built from tens of thousands of different oligonucleotide probes per square centimetre, are well-established high-throughput hybridisation systems that generate huge sets of genomic data within a single experiment. Their use for the analysis of single nucleotide polymorphism's and in expression profiling has already changed pharmaceutical research, and their use as diagnostic tools will have a big impact on medical and biological research.

As known from gene expression studies, however, mRNA level and protein expression do not necessarily correlate [4, 5]. Protein functionality is often dependent on post-translational processing of the precursor protein and regulation of cellular pathways frequently occurs by specific interaction between proteins and/or by reversible covalent modifications such as phosphorylation. To obtain detailed information about a complex biological system, information on the state of many proteins is required. The analysis of the proteome of a cell (i.e. the quantification of all proteins and the determination of their post-translational modifications and how these are dependent on cell-state and environmental influences) is not possible without novel experimental approaches. High-throughput protein analysis methods allowing a fast, direct and quantitative detection are needed. Efforts are underway, therefore, to expand microarray technology beyond DNA chips and establish array-based approaches to characterise proteomes [6-9].

2.   The Microspot – A Concept

The ambient analyte assay theory shows that miniaturised ligand-binding assays are able to achieve a superior sensitivity. A system that uses a small amount of capture molecules and a small amount of sample can be more sensitive than a system that uses a hundred times more material. Ekins and co-workers [10] developed a sensitive microarray-based analytical technology and proved the high sensitivity of the miniaturised assay. With this system, analytes, such as thyroid stimulating hormone (TSH) or Hepatitis B surface antigen (HbsAG), could be quantified down to the femtomolar concentration range (corresponding to 10^6 molecules ml^{-1}). Miniaturisation is the key to understanding the principle of miniaturised binding assays. Capture molecules are immobilised to the solid phase only in a very small area, the *microspot* – although the amount of capture molecules present in the system is low, a high density of molecules in the microspot can be obtained.

During an assay, target molecules, or analytes, are captured by the microspot but the number of capture–target complexes is low owing to the small area of the microspot. As a result, the capture process does not change the concentration of the target molecules in the sample significantly, even for targets present in low concentration and for binding reactions that occur with a high affinity. This is true if <0.1/K of capture molecules get immobilised, where K is the affinity constant of the binding reaction. These conditions, termed ambient analyte assay, allow measurements where the amount of the target or analyte captured from solution directly reflects its concentration in the assay system.

Interestingly, the concentration measurement under ambient analyte conditions makes the system independent of the actual volume of sample used

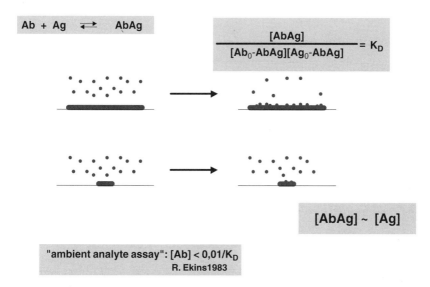

FIGURE 3.1. **The microspot.** The major difference between a micro- and a macrospot is not the size but the fact that the number of capture molecules in the spot is relatively small. Therefore, the capture of analyte by the molecules in the spot leads only to a small decrease of free analyte. Its concentration (Ag0-AgAb) is close to the initial value (Ag0). In an assay were also [AbAg] is low compared to [Ab0] (i.e. the measured signal is less then 1/10 of the maximal signal) the amount of capture –target complexes is directly proportional to concentration of the analyte. Such a system working is under these "ambient analyte assay conditions" and acts as a concentration sensor. Therefore, it doesn't measure the absolute amount analyte present.

and it can give results that combine high sensitivity with low sample consumption. The sensitivity that can be obtained is high for two reasons.

First, the binding reaction occurs at the highest possible target concentration. Second, the capture-molecule–target complex is found only in the small area of the microspot, resulting in a high local signal (Figure 3.1). Capture molecules are immobilised in a constant surface density onto spots that have an increasing spot size. With increasing spot size, the total amount of capture molecules present in an assay increases, as does the sum signal obtained from the spot. The signal density, however, starts to decrease with increasing spot size because the amount of target starts to become a limiting factor. The capture process leads to a significant reduction of target concentration in solution and at the same time the probe–target complexes get distributed over a larger area. As a result, the maximal signal that can be obtained from any point in a spot is decreased. Decreasing the spot size will decrease the overall signal per microspot but the signal density will increase for smaller spots (Figure 3.2). Below a certain spot size, the signal density approaches an optimum (ambient

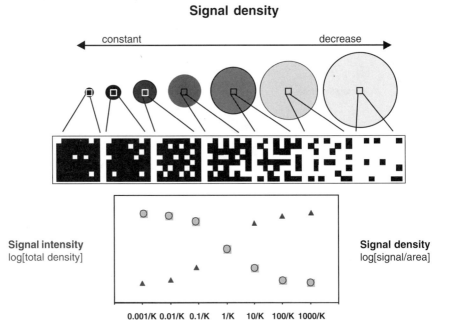

FIGURE 3.2. **Signal and signal density in microspots.** Signal density (signal/area, relative intensities, log scale) and signal (total intensity, log scale) of captured targets in microspots are shown for different concentrations of capture molecules. The capture molecules are immobilised with the same surface density on all spots. The signal (total signal) increases with increasing amount of capture molecules at growing spot size. When most of the targets are captured from the solution the signal reaches its maximum. By contrast, signal density (signal/area) increases with decreasing amount of capture molecules (decreasing spot size), reaching a constant level when the capture molecule concentration is <0.1/K (K is the association constant). Under these ambient analyte conditions, target concentration in solution is minimally altered by the amount of captured targets on the microspot. The figure was adapted from [10].

analyte conditions; amount of target not limiting) and will stay approximately constant with any further decrease in spot size. Therefore, the highest signal intensities and optimal signal-to-noise ratios can be achieved in small spots.

3.   From DNA to Protein

Biochips have become important tools for life science research. At present mainly DNA chips or microarrays are used routinely. Protein arrays are still in an early phase of development, but the increase in the number of publications

TABLE 3.1. Microarray applications of DNA and proteins with respect to their properties

Properties	DNA	Protein
Activity Prediction	Possible. Based on primary nucleotide sequence	Hardly possible yet. Bioinformatics is working on prediction models based on sequence homologies, structure prediction etc.
Amplification	Established (→ PCR)	Not available
Functional State	Denatured, no loss of activity → can be stored dry	3D Structure is important for protein activity → keep hydrated all the time
Interaction Affinity	high	Dependening on Protein: Range from very low to high
Interaction Specificity	high	Dependening on Protein: Range from very low to high
Interaction Sites	1 by 1 interaction	Multiple active interaction sites
Structure	Uniform, stable	Individual types with individual strability

on protein microarrays clearly demonstrates their large potential. The different technologies that had to be developed for DNA chip applications are now well established and microarray surfaces, arraying and detection systems are available. However, when switching from DNA to protein microarrays one has to keep in mind that DNA and proteins are very different types of molecules (Table 3.1).

4. Tools for the Generation of Protein Arrays

4.1 Chip Surface and Protein Attachment

Proteins can be immobilised using non-covalent surface interactions with hydrophobic (nitrocellulose, polystyrene) or positively charged (poly-Lysine, aminosilane) surfaces. These materials are well known from established protein assays such as ELISA or Western blotting. Membranes possess a large surface area and have high protein binding capacity but suffer from background problems due to auto-fluorescence and unspecific protein binding. To counteract this problem, glass slides can be prepared as ultraflat devices with a minimal degree of fluorescence. However, glass shows poor protein binding capacity and therefore surface modifications are required to facilitate protein binding. Established surface treatments for DNA microarrays are aminosilane, poly-lysine, or aldehyde surface activation [11]. While all of these modifications have been shown to work for proteins, sophisticated surface chemistries are developed to meet the specific needs for immobilising and stabilising proteins on microarrays. Good examples are modified structures based on self assembling monolayers (Zeptosens AG, Switzerland; Zyomix Inc, USA etc.) and hydrogel modifications [12] that are used to prevent the immobilised proteins from drying out.

Non-covalent attachment of protein capture molecules might be to weak to prevent the loss of capture molecules during the assay procedure. Therefore, the need for covalent, spatially defined protein immobilisation procedures might increase in future. So far, the covalent attachment of proteins on microarrays is usually performed using a variety of chemically activated surfaces (e.g. aldehyde, epoxy, active esters [13-15]). Specific bimolecular interactions (e.g. streptavidine – biotin [16-18]), his-tag – nickel-chelates [19] or the formation of stable complexes between phenylboronic acid labelled analytes and salicylhydroxamic modified surfaces (Prolinx Inc., USA) have been used to immobilise capture molecules on surfaces.

4.2   Arraying Devices

The size of microspots is usually found in the range from 50 μm to 500 μm diameter with a spot to spot distance of 70 to 1000 μm. The production of protein arrays on chip surfaces is done using dedicated microspot arrayers. Contact printing devices are equipped with needles that place sub nanoliter volumes directly onto the chip-surface. Alternatively, non-contact deposition technologies (capillaries, ink-jet) are employed to deposit nanoliter to picoliter droplets onto the surface. TopSpot™ (IMIT, Germany), a microstructured device where arrays of tiny capillaries (up to 96) are directly linked to distinct macroscopic fluid reservoirs [20] is an interesting example of such an arrayer. A piezo actuator hits the back of the microcapillary device, thus generating a steep air pressure ramp to the open upper side of the liquid reservoirs. This air pressure change results in parallel dispensing of microdroplets featuring volumes down to 1 nl from each capillary tip. New developments are: micropatterned protein arrays produced by photolithographic methods [21], scanning probe based lithography [22] or lithography technologies [23-27]. Microcontact printing (μCP) allows the printing of delicate biomolecules together with a tiny hydrogel spot onto surfaces [28, 29]. Electrospray deposition (ESD) of protein solutions on slightly conductive surfaces [30] might be another future possibility to handle delicate proteins.

4.3   Readout

A number of different detection technologies have been discussed and employed for microarray experiments. So far, the detection of captured targets is mainly performed by fluorescence using CCD-cameras or laser scanners with confocal detection optics. A very sensitive alternative to confocal optics with regard to signal intensity, linearity, signal-to-noise ratio and background is the application of planar waveguide excitation devices combined with CCD cameras or photomultipliers as detectors [31-34]. Capture molecules are immobilised on a thin (100-200 nm) film (planar wave guide) which consists of a high-refractive index material (e.g. Ta_2O_5) deposited on a transparent support. A laser beam is optically coupled via diffractive grading into the planar waveguide. The light is propagated in the thin layer and creates a strong, surface confined evanescent electromagnetic field. The penetration depth of this evanescent field into the adjacent medium is limited to about 200 nm. Thus, only surface confined fluorophores are excited and emit fluorescent light. Fluorophores in the bulk medium are not excited and therefore not detectable. A CCD-camera is used to detect fluorescent light with high spatial resolution. Parallel excitation and parallel detection of binding events on different spots is performed in a highly selective and sensitive way, even in solution (Figure 3.3).

As an alternative to planar microarrays, bead based assays in combination with flow cytometry have been developed to perform multiparametric immunoassays. In bead based assay systems the capture molecules must be immobilised on addressable microspheres. Each capture molecule for each individual immunoassay is coupled to a distinct type of microsphere and the

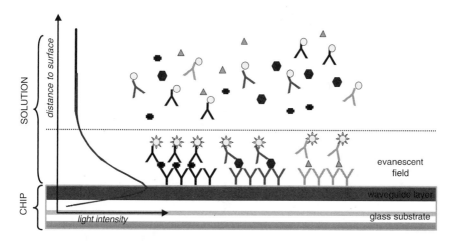

FIGURE 3.3. **Planar wave guide technology.** Capture antibodies are immobilised on the planar wave guide in a microarray format. A pre-incubated cocktail mix of sample and fluorescently labelled detection antibodies is added onto the planar wave guide microarray. The formed sandwich immuno-complex can be detected without washing away unbound fluorescently labelled detection antibodies. Laser light of the desired wavelength is coupled into the planar wave guide and generates an evanescent field, which extends only a few hundred nanometers into the solution. Therefore, only the surface-bound fluorophores are selectively excited. Individual signals correlating to the amount of captured analytes on each microspot are monitored by means of imaging detectors, such as CCD-cameras. Here an example of a sandwich immoassay is shown. Three different capture antibodies are immobilised and are detect their antigen.

immunoassay reaction takes place on the surface of the microspheres. Dyed microspheres with discrete fluorescence intensities are loaded separately with their appropriate capture molecules. The different bead sets carrying different capture probes can be pooled as necessary to generate custom bead arrays. Bead arrays are then incubated with the sample in a single reaction vessel to perform the immunoassay (Figure 3.4).

Product formation of the targets with their immobilised capture molecules is detected with a fluorescence based reporter system. Targets can either be labelled directly by a fluorogen or detected by a second fluorescently labelled

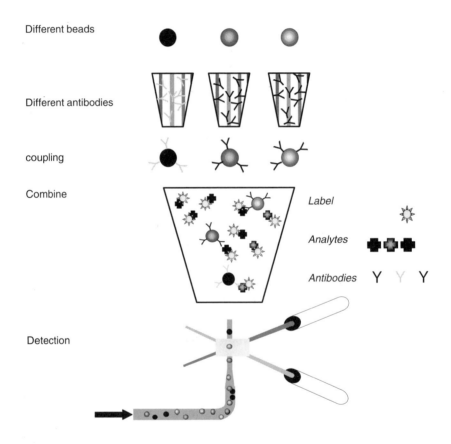

FIGURE 3.4. **Bead-based flow cytometric technology.** Capture antibodies are immobilised on distinct bead populations of microspheres. Different bead sets loaded with specific capture antibodies are pooled together and incubated with the sample together with fluorescently labelled detection antibodies. Formation of the different sandwich immuno-complexes on the different bead populations can be measured and quantified with a flow cytometer. There is no need for washing away unbound fluorophores, as only signals present on the bead surface are measured.

capture molecule. The signal intensities derived from captured targets are measured in a flow cytometer. The flow cytometer first identifies each microsphere by its individual colour code. Second the amount of captured targets on each individual bead is measured by the second colour fluorescence specific for the bound target. This allows multiplexed quantitation of multiple targets from a single sample within the same experiment. Sensitivity, reliability and accuracy are comparable to standard microtiter ELISA procedures. With bead based immunoassay systems cytokines can be simultaneously quantified from biological samples [35, 36]. With colour coded microspheres it is possible to perform simultaneously up to a hundred different types of assays so far (LabMAP system, USA). One big advantage of bead based systems is the individual coupling of the capture molecule to distinct microspheres. Each individual coupling can be perfectly analysed and optimised. Furthermore only quality controlled batches with defined capture loading will be used for multiparameter immunoassays. If additional parameters have to be included into the assay, only the new types of loaded beads have to be added to the bead array used for the assay.

Improved fluorescent labels like luminescent quantum dots have been developed recently which may turn out to become important tools for multi-colour detection on protein arrays in the near future. These nanocrystals are of interest, because their fluorescence emission wavelength can be continuously tuned by changing the particle size. In addition, quantum dots have narrow, symmetrical emission peaks and are highly stable against photobleaching [37, 38]. Radioactivity [39], chemiluminescence [40], mass spectrometry using surface enhanced laser desorption ionisation (SELDI) [41] or label-free surface plasmon resonance (SPR) detection systems [42, 43] were also used for bimolecular interaction analyses on surfaces. Additionally, SPR based detection can be directly combined with mass spectrometric identification of captured molecules [44]. Electrochemical detection or quartz crystal microbalance (QCM) detection systems [45-47] which have been established for biosensor applications, might be useful alternatives for microarray assays. They are, however, so far not very common for microarray application.

5.   The Protein Array

Due to the complementary nature of the DNA, the capture probes which are immobilised for DNA microarrays can easily be predicted on the basis of the sequence of interest. These capture DNA molecules can easily be generated using well established technologies like the PCR amplification of defined cDNAs or standard oligonucleotide synthesis. Alternatively, oligonucleotides can be directly synthesised on the solid support [48, 49]. The analysis of RNA or DNA targets involves the extraction from the cell, frequently an amplification by PCR, labelling and hybridisation to the immobilised complementary capture probes.

In the protein world, it is difficult or even impossible to predict protein capture molecules from the primary amino acid sequence. This is due to the very diverse and individual molecular structures, which define the biological activity of each protein. In addition, proteins cannot be amplified via PCR. Technologies for the generation, identification, production and purification are necessary to generate protein microarrays and large numbers of highly specific capture molecules.

5.1   Capture Molecules

Large numbers of highly specific capture molecules showing a high affinity to their target are prerequisite for the identification and quantification or proteins using array based approaches. Different types of possible capture agents are shown in Figure 3.5. Not only protein capture molecules could be useful for the development of protein arrays or the screening for protein function. DNA-protein interactions can be analysed on DNA or oligonucleotide arrays. Peptide arrays, built from large sets of individual peptides or peptide libraries can be useful tools for screening for unknown enzymatic activities or for selection of antibodies [50-52].

Recently it was shown, that arrays of small organic molecules synthesised by established solid phase combinatorial chemistry can be used for high-throughput screening for interactions between organic compounds and proteins [53]. This could be an interesting future technology for high-throughput drug compound screening in pharmaceutical industry.

5.2   Antibodies

Antibodies are highly specific targeting agents and very valuable tools for *in vitro* and *in vivo* diagnostic applications. With the advent of monoclonal antibody technology, the utilisation of antibodies as essential tools dramatically in almost every field of biological sciences has increased enormously [54]. Monoclonal antibodies have become key components in a vast array of clinical laboratory diagnostic tests. But antibodies whether polyclonal or monoclonal have some disadvantages in terms of generation, cost, and overall applications. The continuous culture of hybridoma cells that produce monoclonal antibodies offers the potential of an unlimited supply of reagent, when compared with the rather limited supply of polyclonal antibodies. The feature of a continuous supply enables the standardisation of both reagent and assay technique, but the production of monoclonal antibodies is often time consuming, laborious and therefore expensive. To overcome these limitations, some molecular biological approaches such as the phage display or ribosomal display technique and the development of alternative types of binders, i.e. aptamers, have been started. Phage antibody-display has shown to be a very effective tool for the fast generation of antibodies. The development of large primary antibody libraries and, even more, the development of

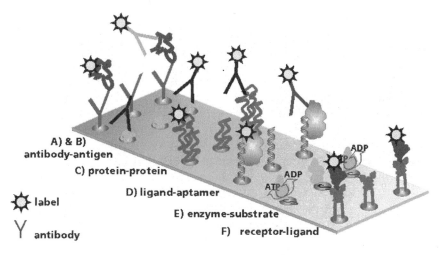

FIGURE 3.5. **Different classes of molecules can act as capture molecules in microarray assays.** Each specific interaction of a distinct pair of biomolecules can be used to design an interaction assay. The illustration shows examples of different capture molecules and some of their typical applications for protein microarray assays. Antigen-antibody interactions, where either the antigen (A) or a specific antibody (B) is immobilized are well known in the field of diagnostics. Detection is performed on the basis of the labeled analyte (antigen or antibody) itself or by labeled secondary antibodies. Specific protein-protein interactions (C) on protein microarrays can be measured directly using labeled analyte protein or by detection of captured protein using a secondary antibody. Synthetic oligonucleotides or peptides, termed aptamers (D), selected from random combinatorial synthetic libraries for high affinity and selectivity to their target molecules, are useful molecules for the design of capture microarrays. Arrays built from DNA fragments or oligonucleotides can be used for the determination of specific DNA-protein interactions as well. Enzymatic processing of immobilized substrates on microarrays can also be monitored in a microarray format (E). Immobilization of receptor ligands (e.g. oligosaccharides, hormones, etc.) can be used for receptor binding studies (F).

entirely synthetic combinatorial antibody libraries [55,56] -enables the selection of antibodies against nearly any target in a relatively short time (i.e. weeks). Many of these antibodies exhibit binding affinities with a K_D in the nM range. Maturation strategies allow the improvement of this affinity to a K_D in the pM range.

Phage-antibody display involves the fusion of an antibody gene to the gene for a phage coat protein. The antibodies encoded by this gene are displayed on the phage surfaces. From large repertoires of antibody variable region (V) genes, displayed on phage particles, a specific antibody population can be selected by multiple rounds of affinity purification on the desired target antigens. Different phage- and phagemid vectors systems have been used to display the antibodies. E.g., antibody fragments have been displayed on the surface of filamentous phage fused to the product of gene VIII [57]. However, by far the most successful fusion partner to date is g3p, the product of the gene gIl1. Three to five copies of g3p exist on each phage particle. Antibody fragments can be displayed, apparently equally well, either fused to the amino-terminus of the mature protein or at the amino-terminus of a truncated g3p, lacking the first two amino-terminal domains (reviewed in [58]). It has to be taken into account that most antibody fragments which are displayed on phagemid particles are monovalent [59]. The valency of displayed antibodies can be increased either by using a phage vector or a helper phage with the gII gene deleted. This results in a gIIIp-protein population that consists almost completely of gIIIp-antibody fusion proteins [60], which can later be propagated with high titres [61]. Antibody fragments can be displayed as single-chain variable region fragments (scFv) and as Fab fragments. Displaying scFv, the heavy chain and light chain variable regions are fused by a linker and the carboxyl-terminus of the scFv polypeptide is fused to gIIIp [62]. scFv fragments are encoded by relatively short genes, which leads to libraries that are genetically stable and easy to handle. scFvs form predominantly monomers, but sometimes they can also form dimers and trimers with higher molecular weight. This can complicate the selection and characterisation. But such diabodies may be bivalent molecules with enhanced avidity [63]. Fab fragments are displayed by fusing either the light or the heavy chain carboxyl-terminus to gIIIp. The partner chain is then expressed unfused and forms an intact Fab after secretion into the periplasmatic space [64]. Fabs usually show a lower degree of dimerisation than scFvs. This makes them easier to characterise. Their larger genes, however, may make Fab libraries less stable.

Primary phage libraries are divided into two groups based on the different sources of the antibody genes. First, 'natural' or rearranged antibody V-gene repertoires derived from human or animal donors – here the immune system (B-cells) is used to create the diversity of the V-genes – and second, synthetic antibody V-gene repertoires which are constructed *in vitro*. For natural phage libraries either immunised [65,66] or "naïve" [67] donors can be utilised. Repertoires from immune donors phage-antibody libraries made from donors naturally mounting an immune response can generate antibodies with high affinity and specificity. Immune libraries are highly biased towards V-genes that encode antibodies against the immunogen, so that relatively small (10^5 clones) libraries can be successfully selected. In addition, many of the genes will encode affinity matured antibodies, thus increasing the number of high affinity antibodies represented in the library. The drawbacks concern the little control over the immune response and the existing tolerance mechanisms against self antigens. The major advantage of repertoires from non-immunised donors over libraries from immunised donors is that a single library of sufficient largeness and diversity can be used for all antigens with no limitation concerning self-immunogenic, non-immunogenic and toxic antigens. Selection from large primary libraries generally ends up in many different antibody populations with subnanomolar to submicromolar affinities [68,69]. For most purposes the affinities of the isolated antibodies will be well suited, however, for specific applications there may exist the need to improve these primary lineages. The improvement of this antibody lineages can be achieved by the construction, selection and screening of secondary phage display libraries, a process analogous to 'affinity maturation'.

Secondary repertoires are essentially synthetic libraries based on a lead or candidate antibody that was identified selecting a primary library. Criteria such as potency, affinity, cross-reactivity, expression level, germline homology etc. have to be considered. Potential target sequences have been selected from libraries of mutants diversified at heavy and/or light chain CDR3 and higher affinity binders have been generated [70]. Saturation mutagenesis in combination with affinity selection of CDR3 of heavy and light chain [70] – they structurally comprise the antigen binding site – can lead to antibodies with dissociation constants below 100 pM. Further sequence analysis of human antibodies emerging in the primary and secondary immune responses also suggests other key residues for potential affinity maturation [71]. Using randomised codons a complete library with only six codons would theoretically have to contain 2.5×10^9 clones. Variant libraries would have to be even lager to ensure complete representation. Therefore, several strategies have been developed to identify residues with functional rather than structural roles as promising targets for codon randomisation.

Since large and highly diverse repertoires have to be engaged to perform successful antibody phage display there is a need for appropriate selection strategies and for efficient screening technology. For most applications, the selection can be performed effectively by panning phage on antigen-coated plastic ware. Limitations for such an approach are the need of purified antigen, the detection of native protein antigen, the identification of high affinity clones despite avidity effects and discrimination between clones of similar affinity. Affinity chromatographic methods (i.e. biotin-streptavidine system) in combination with sequential elutions have been used to separate antibody populations with different affinities [72]. Depletion and/or subtraction methods have been used to deplete antibodies against non-target antigens and to select antibodies against complex antigens [72]. It is very often possible and desirable to select multiple different antibodies to the same target which bind via different epitopes. The demands on antibody screening increase in terms of quality (e.g. functional screening) and quantity (high throughput). The appropriate system to identify antibodies with the optimal characteristics has not only to be fast, robust and sensitive but also amenable to automation and miniaturisation. With respect to these demands, micro array technology on solid or fluid phase is very promising. Protein or peptide microarrays combining the inherent possibility to perform multiplexed assays with minimal sample consumption at enhanced sensitivity meet these requests for antibody screening and validation perfectly. Furthermore, antibody engineering allows the manipulation of genes encoding antibodies so that the antigen binding domain can even be expressed intracellulary. These intracellular antibodies are termed 'intrabodies' and demonstrate new roads to therapeutic abilities and screening strategies taking place directly in the living cell [73, 74].

5.3   Aptamers – An Alternative Class of Binders

The development of a specific methodology made it possible to isolate oligonucleotide sequences with the capacity to recognise virtually any class of target molecules with high affinity and specificity. Such oligonucleotides derived from an *in vitro* evolution process called SELEX (systematic evolution of ligands by exponential enrichment) are referred to as "aptamers". They have the potential to build up a class of molecules that reveal antibodies in both therapeutic and diagnostic applications. Although they are molecularly different from antibodies, they mimic the molecular recognition properties of antibodies in a variety of diagnostic formats. The demand for diagnostic assays is increasing and aptamers could potentially fulfil molecular recognition needs in those assays. Compared with the skilled antibody technology, aptamer research is still at the beginning but it is progressing at a fast pace. Aptamers may play a key role either in conjunction with, or in place of antibodies. Originally, aptamers have been evolved to bind proteins associated with distinct disease states. This led to the development of many powerful antagonists of such proteins. If aptamers contain modified nucleoside triphosphates (e.g. PNA [75] or LNA [76]) their resistance to nucleases is normally enhanced and can be kept for longer time periods in the circulation of animal disease models, especially when conjugated to vehicles of higher molecular weight. In such approaches, the aptamers can inhibit physiological functions of their target proteins.

Aptamers immobilised in a microarray format on solid surfaces can be used as diagnostic tools. They will become rapidly available because the SELEX protocol has been successfully automated [77, 78]. The use of photocross-linkable aptamers will allow the covalent attachment of aptamers to their cognate proteins. This allows rigid wash procedures prior to detection and therefore ratios lowers the unspecific assay background and enhances signal-to-noise ratios. Finally, protein staining with any reagent, which distinguishes functional groups of amino acids from those of nucleic acids will give a direct readout of proteins on the solid support.

In the meantime, not only oligonucleotides, PNAs or LNAs but also short peptides which are generated by mRNA display technology [79], are called aptamers. The mRNA display *in vitro* selection technique allows the identification of peptide aptamers to specific target proteins. Polypeptide libraries with a complexity of 10^{13} random peptides can be generated. Aptamers with dissociation constants as low as 5 nM have been isolated from such peptide libraries. Peptide aptamer affinities are therefore comparable to those found for monoclonal antibody-antigen complexes without sophisticated engineered scaffolds. Given a sufficient length and diversity, high-affinity peptide aptamers can be obtained directly from random non-constrained peptide libraries [80]. In principle every protein or peptide showing a selective, specific high affinity binding to a substrate can be used as a template for recombinant production of capture molecules. The only prerequisite is, that mutations in the binding site do not affect protein stability to much. Based on these prerequisites other protein templates were used for the recombinant production of capture molecules. Examples for such molecules are the so called receptins [81]. Receptins represent binding molecules of microbial origin including a number of key microbial proteins involved in host-parasite interactions and in virulence. Common features of these proteins are their specific binding properties for mammalian proteins. Well known binder molecules engineered based on this protein group are the so called affibodies. Affibodies are selected from naive or constructed combinatorial libraries derived from the *alpha*-helical receptor domain of protein A of *Staphylococcus aureus* [82-85]. Such affibody libraries can also be subjected to affinity maturation procedures and multimerisation to generate binders with high affinity.

6. Applications of Protein Microarrays

6.1 *Immunoassays for Diagnostics*

Miniaturised and parallelized immunoassays are of general interest for all diagnostic applications, where several parameters in an individual sample have to be determined simultaneously from a limited amount of material. Besides microarray based systems, bead-based assays can be used for such diagnostic applications, especially when the number of parameters of interest is comparably low.

Mendoza et al. [86] described a microarray based approach capable to perform high-throughput, enzyme-linked immunosorbent assays. This system consists of an optically flat glass plate with 96 wells separated by a Teflon mask. Within each well more than a hundred capture molecules are immobilised in a microarray format. Sample incubation and washing steps can be performed with an automated liquid pipettor. The microarrays are quantitatively imaged with a scanning charge-coupled device detector. With marker antigens the feasibility of multiplex detection of arrayed antigens in a high-throughput fashion was demonstrated. Other microarray based approaches have been published by Silzel et al. [87] who demonstrated that multiple IgG subclasses detection can be performed using microarray technology. Arenkov et al. [12] performed microarray sandwich immunoassays and direct antigen of antibody detection experiments using a modified polyacrylamide gel as a substrate for the immobilisation of capture molecules. In our laboratory, we used microarray technology to screen for autoantibodies present in patient sera. Eighteen different autoantigens, commonly used as diagnostic markers for autoimmune diseases like systemic rheumatic diseases, were immobilised together with control proteins in a microarray format. Arrays were incubated with patient sera and bound autoantibodies were detected by a peroxidase labelled secondary antibody and a chemiluminescent reaction monitored by a CCD camera. The microarray based assay allowed the parallel identification of different types of autoantibodies present in the sample. From less than one fl of a patient serum the autoantibody titers were determined with high accuracy [40]. Only minimal unspecific binding and no cross-reactivity to non-specific proteins could be observed. Sandwich immunoassays were also miniaturised and parallelised and performed in a microarray format. This was recently demonstrated by the parallel determination of 24 different cytokines from conditioned media and patient sera with high specificity and sensitivity [88].

Planar microarrays or bead based multiplexed arrays are both very well suited for multiplexed immunoassays. Accurate quantification and control of assay performance and reproducibility can be achieved by including several positive and negative controls and/or internal calibration standards. Thus microarray or bead based assay have enormous potential to become robust and reliable diagnostic assays.

6.2   Protein Microarrays for Proteomics

A proteome represents the physiological state of a cell on the protein level. The analysis of such protein patterns and their correlation with intracellular or extracellular parameters (Figure 3.6) is one of the major challenges for protein chemists today. 2D-PAGE combined with mass spectrometry is the most common technology used for proteome analyses to date. Due to the inadequacies of these technologies, alternative analytical methods are needed to allow for an accurate analysis of a proteome.

Microarrays, which are well established for RNA expression or single nucleotide polymorphism analysis, may also be suitable for protein analysis. Different molecules binding or interacting with proteins can be immobilised in a microarray format, and complex interaction analysis studies can subsequently be performed (Figure 3.6). Microarrays of highly specific capture molecules might be used for direct protein identification and quantification out of complex protein mixtures isolated from cells or tissue samples. The principle of differential display analysis of proteins from different sources on such microarrays (Figure 3.6A) was recently shown [89]. The results suggest that protein microarrays are a convenient tool to characterise patterns of variation in hundreds of thousands of different proteins. The generation and characterisation of highly specific antibodies for high density antibody arrays, the main prerequisite for direct differential protein display analysis, is a big challenge, so far. Modern technologies for the generation of large capture molecule libraries combined with methods for the generation of highly

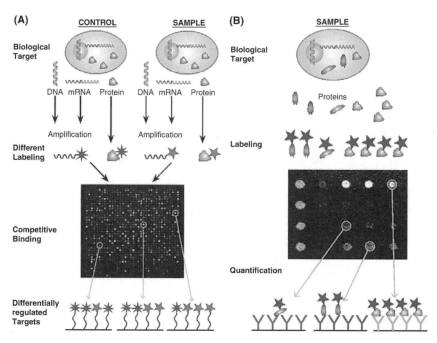

FIGURE 3.6. **Comparative and quantitative proteome analysis.** (A) Comparative genome and proteome analysis. Subtractive or comparative strategies are most efficient for discovering gene activity in a physiological or pathological context. Equivalents of the proteome (e.g. mRNA, proteins or DNA) of phenotypically different sources are isolated and subjected to cross-matched analysis. In general, the isolated molecules are labelled differentially, either with radioisotopes or fluorescent dyes. The labelled targets represent different phenotypes in two distinct colours. Differences in sequence or expression pattern can be examined with a comparative analysis using an equal mixture of the differentially labelled targets. For that purpose the scans of both fluorescent dyes are matched and the resulting colour of each spot is visualised. For example, for an up-regulated protein expression in 'treated situation' one would expect to see the respective spot coloured as the label of treated sample, for down-regulation the label of the control situation is expected (red or green spots in the array picture). For an unchanged expression level, an intermediate colour of the spots (yellow spots in the array picture) is obtained. (B) Parallel quantification of proteins. Different proteins labelled with fluorochromes can be detected in parallel with a microarray-based assay. Specific capture antibodies immobilised in an array interact with their respective target proteins present in the solution. The resulting signal intensity correlates with the amount of captured target. Within each microarray, different kinds of control spots can be included, such as positive and negative control spots and/or internal calibration spots. This will allow accurate signal quantification.

diverse non-redundant expression libraries, as the so-called Unigene-Uniprotein sets [90], for antibody screening, will be valuable tools to pass this challenge.

An alternative approach for chip-based proteome analysis is the SELDI (surface enhanced laser desorption and ionisation) technology, which utilises mass spectrometry as a read out system to analyse differential protein expression on spot arrays [91,92]. Cell extracts derived from different sources are incubated on different spots of the same adsorptive surface chemistry (e.g. cation/anion exchange material, hydrophobic surfaces). After washing away unbound proteins, the whole variety of non-specifically captured target proteins can be analysed by SELDI mass spectroscopy. The mass spectrum shows the different molecular weights of the captured proteins. The comparison of two MS data sets generated from two different samples immediately identifies the differentially expressed proteins. In some cases the differentially displayed proteins can be identified immediately by their molecular weights, but usually these proteins have to be enriched by affinity chromatography and identified by methods known from protein analysis (e.g. Edman sequencing, Western blot, digest mass fingerprinting) [93]. This sophisticated expensive technology is an easy to handle tool for fast screening for differences in total protein content. As the detector sensitivity of time of flight mass analysers is decreasing with increasing molecular weights, SELDI can be a valuable supply to 2D-PAGE technology, which usually has difficulties in the detection of small proteins and peptides. But the identification of the individual protein differences might be a laborious process. Furthermore, sensitivity is lower than with fluorescence-labelled captured targets on antibody microarrays.

The most direct approach towards an understanding of gene function and regulation is to analyse the biochemical activities of the individual proteins [94,95]. Systematically probing for biochemical activities of every protein is another possible strategy for proteome analysis. An alternative way is to produce protein products in a high throughput manner and to analyse their functions in parallel using protein microarrays. Such microarrays could provide data on direct nucleic acid-protein, protein –protein, ligand –receptor or enzyme-substrate interactions, which are required for an interpretation of protein networks within biological systems. The variety of different microarray-based assays useful for proteome analyses is impressively illustrated by recent publications where the analysis of such interactions is discussed (most important Zhu et al. [96]).

Studies on DNA-protein interactions in a microarray format were performed by Bulyk et al [97] who created microarrays of double-stranded oligonucleotides. Initially, high-density microarrays of single-stranded oligonucleotides primers were produced using Affymetrix technology. Subsequently, these single-stranded oligonucleotide microarrays were converted into double-stranded oligonucleotide (dsDNA) microarrays by an enzymatic primer extension reaction. Double-stranded oligonucleotide arrays were then incubated with restriction enzymes cleaving specifically distinct dsDNA sequences. No DNA-cleavage occurred, when the dsDNA was enzymatically methylated prior to the incubation with the specific restriction enzymes. This example showed that dsDNA arrays can be further modified biochemically in order to study specific DNA-protein interactions. In general DNA-protein interaction assays could be useful for the characterisation and identification of DNA-binding proteins, such as e.g. transcription factors. Heng Zhu and co-workers [96] overexpressed and purified the corresponding proteins of 5800 cloned open reading frames of yeast as GST/HisX6-fusion products. All these proteins were printed onto nickel-coated slides at high spatial density. These proteome chips were used to analyse protein-nucleic acid interactions and several other types of activities.

Enzyme-substrate arrays have been described for different kinds of enzymes such as e.g. restriction enzymes, peroxidase, phosphatase and protein kinases [98,99]. In a proof of concept experiment MacBeath and Schreiber [11] immobilised three different kinase substrates in a microarray onto a planar glass surface. Identical microarrays were incubated individually with one specific kinase together with radioactively-labelled ATP. Each substrate was phosphorylated only by its specific kinase. In a more advanced approach, Zhu et al. [39] analyzed the activities of 119 of the 122 known or predicted protein kinases from *saccheromyces cerevisiae* for 17 different substrates. They used microwell plates with substrates covalently linked to individual microwells. The overexpressed, purified kinases were subsequently incubated on these microwell arrays along with radioactively labelled ATP. After finishing the reaction, kinases and the non-incorporated radioactive ATP were washed away and the arrays were analyzed for phosphorylated substrates by a phosphoimager. Using this approach, novel activities of individual kinases were identified. Sequence comparison of enzymes, which phosphorylate tyrosine residues, revealed that they often share common amino acid residues around their catalytic region.

In the field of protein-protein interaction assays dot-blot filter arrays were used to screen for specific interactions of immobilised proteins with other proteins. Filters were prepared from highly purified fully-active recombinant proteins. Used filters were recycled after the assays and were re-used many times with different targets. This so-called universal protein array (UPA) system is a very effective method of screening protein interactions at low cost. Specific protein-protein interactions were detected between a radioactively-labelled human p52 GST fusion protein and immobilised capture proteins like nucleoline or a serine-argenine protein fraction isolated from HeLa cells [100]. In addition interactions of DNA, RNA, or low molecular weight ligands with the immobilised molecules were also detected. In principle, the UPA arrays could easily be miniaturised. Proteome chips generated from recombinant protein probes of all 5800 open reading frames of yeast were tested by probing for protein-protein interactions, protein-lipid interactions and protein-nucleic acid interactions [96].

To test for protein-protein interactions, the yeast proteome was probed with biotinylated calmodulin. Many known CamKinases [101] and calcineurins [102] were identified. Additionally, 33 new potential binding partners of calmodulin were found which have a potential binding motif. Furthermore, Zhu et al. presented for the first time a genome-wide analysis of proteins interacting with phospholipids. Six types of liposomes of different composition were used to identify a total of 150 different protein targets including integral membrane proteins, peripherally-associated proteins and many others. Many of the uncharacterised proteins are predicted to be membrane associated, indicating that they preferentially bind specific phospholipids *in vivo*. This study clearly demonstrated the advantage of a proteome chip approach. An entire proteome can be prepared and directly screened *in vitro* for a wide variety of activities including protein-drug interactions and protein-lipid interactions, which might not be accessible by other approaches. Preparation of protein arrays of 10-100,000 proteins for global high throughput proteome analysis in humans and other eucaryotes is feasible using similar procedures.

6.3   Protein Microarrays – Promising Tools for Drug Screening

Data from genomic or proteomic analyses can also be used for screening for new drugs in the pharmaceutical industry. Microarrays of immobilised proteins and of small organic compounds might be powerful tools for future high-throughput drug screening technologies. For receptor-ligand assays small organic molecules produced by combinatorial solid phase chemistry were immobilised in a microarray format. Single resin beads from combinatorial synthesis were placed in 96-well plates and the organic molecules were chemically released from the beads. The organic molecules were diluted, spotted and covalently attached on derivatised glass slides. These microarrays, produced by small molecule printing technology, were incubated with fluorescence labelled target proteins in order to identify new ligands [53]. This technology enables parallel high-throughput screening for ligand-receptor interactions but requires only very small quantities of the sample, which could improve screening for active substances in the pharmaceutical industry.

7.   Conclusions and Perspectives

Though the principles of protein microarray technology were described and established years ago, it is only beginning to show its full potential. The number of publications dealing with protein microarray technology is increasing very fast. Several useful applications for protein microarrays were shown with different kinds of experiments. So far, real life applications for protein microarrays were developed mainly for diagnostics. Improvements in the generation of large sets of recombinant proteins and in high throughput generation of capture molecules will further increase the interest in this field. The growing demand for alternative tools for proteome analysis supplementing 2D-PAGE and mass spectrometry will promote the development of high-density capture molecule arrays. Proteomic research, high-throughput drug compound screening and diagnostic applications will be the major fields addressed by protein microarray technologies.

In medical research, protein microarrays will accelerate immune diagnostics significantly by analysing in parallel all relevant diagnostic parameters of interest. The reduction of sample volume is of great importance also, in particular for those applications, where only minimal amounts of samples are available. One example might be the analysis of multiple tumour markers from a minute amount of biopsy material. Furthermore, new possibilities for patient monitoring during disease treatment and therapy will be developed based on this emerging technology. Microarray-based technology beyond DNA chips will accelerate basic research in the area of protein-protein interactions and will allow protein profiling from limited numbers of proteins up to high density array-based proteomic approaches. Protein and peptide arrays will be useful tools for the analysis of enzyme-substrate specificities and for the measurement of enzyme activities on different kinds of substrates in a highly parallel fashion.

The whole field of protein microarray technology shows dynamic growth driven by the increasing genomic information and the growing interest in proteome analyses. The multidisciplinary collaboration of scientists from different fields such as e.g. biology, biochemistry, material sciences or bioinformatics is a prerequisite for the development of robust, reliable protein microarrays for future applications. Protein microarray technology is just leaving its infancy.

Acknowledgments. We thank M. Hartmann, A. Döttinger, C. Vöhringer, R. Knapp, P. Traub, and M. Schrenk for their help and many fruitful discussions.

References

1. Ekins, R.P.: Multi-analyte immunoassay. J Pharm Biomed Anal 7, 155-168 (1989)
2. Lander, E.S et al. Initial sequencing and analysis of the human genome. Nature 409, 860-921 (2001)
3. Venter, J.C. et al.: The sequence of the human genome. Science 291, 1304-1351 (2001)
4. Linck, B., P. Boknik, T. Eschenhagen, F.U. Muller, J. Neumann, M. Nose, L.R. Jones, W. Schmitz & H. Scholz: Messenger RNA expression and immunological quantification of phospholamban and SR-Ca(2+)-ATPase in failing and non-failing human hearts. Cardiovasc Res 31, 625-632 (1996)

5. Gygi, S.P., Y. Rochon, B.R. Franza & R. Aebersold: Correlation between protein and mRNA abundance in yeast. Mol Cell Biol 19, 1720-1730 (1999)

6. Anderson, N.L. & N.G. Anderson: Proteome and proteomics: new technologies, new concepts, and new words. Electrophoresis 19, 1853-1861 (1998)

7. Albala, J.S. & I. Humphery-Smith: Array-based proteomics:High-throughput expression and purification of IMAGE consortium cDNA clones. Curr Opin Pharm Proteomics 680-684 (1999)

8. Emili, A.Q. & G. Cagney: Large-scale functional analysis using peptide or protein arrays. Nat Biotechnol 18, 393-397 (2000)

9. Cahill, D.J.: Protein arrays: a high-throughput solution for proteomics research? Proteomics: A Trends Guide 47-51 (2000)

10. Ekins, R., F. Chu & E. Biggart: Multispot, multianalyte, immunoassay. Ann Biol Clin (Paris) 48, 655-666 (1990)

11. MacBeath, G. & S.L. Schreiber: Printing proteins as microarrays for high-throughput function determination. Science 289, 1760-1763 (2000)

12. Arenkov, P., A. Kukhtin, A. Gemmell, S. Voloshchuk, V. Chupeeva & A. Mirzabekov: Protein microchips: use for immunoassay and enzymatic reactions. Anal Biochem 278, 123-131 (2000)

13. Disley, D.M., P.R. Morrill, K. Sproule & C.R. Lowe: An optical biosensor for monitoring recombinant proteins in process media. Biosens Bioelectron 14, 481-493 (1999)

14. Sanders, G.H.W. & A. Manz: Chip-based microsystems for genomic and proteomic analysis. Trends in Analytical Chemistry 19, 364-378 (2000)

15. Blawas, A.S. & W.M. Reichert: Protein patterning. Biomaterials 19, 595-609 (1998)

16. Nakanishi, K., H. Muguruma & I. Karube: A novel method of immobilizing antibodies on a quartz crystal microbalance using plasma-polymerized films for immunosensors. Anal Chem 68, 1695-1700 (1996)

17. Shriver-Lake, L.C., B. Donner, R. Edelstein, K. Breslin, S.K. Bhatia & F.S. Ligler: Antibody immobilization using heterobifunctional crosslinkers. Biosens Bioelectron 12, 1101-1106 (1997)

18. Dontha, N., W.B. Nowall & W.G. Kuhr: Generation of biotin/avidin/enzyme nanostructures with maskless photolithography. Anal Chem 69, 2619-2625 (1997)

19. Pritchard, D.J., H. Morgan & J.M. Cooper: Patterning and Regeneration of Surfaces with Antibodies. Anal.Chem. 67, 3605-3607 (1995)

20. Ducree, J., H. Gruhler, N. Hey, M. Muller, S. Bekesi, M. Freygang, H. Sandmaiser & R. Zengerle: TOPSPOT -a new method for the fabrication of microarrays. In: Proceedings IEEE Thirteenth Annual International Conference on Micro Electro Mechanical Systems, IEEE, Piscataway, NJ, USA 317-322 (2000)

21. Schwarz, A., J.S. Rossier, E. Roulet, N. Mermod, M.A. Roberts & H.H. Girault: Micropatterning of biomolecules on polymer substrates. Langmuir 14, 5526-5531 (1998)

22. Wadu-Mesthrige, K., X. Song, N.A. Amro & L. Gang-yu: Fabrication and imaging of nanometer-sized protein patterns. Langmuir 15, 8580-8583 (1999)

23. Zhao, X.M., Y.N. Xia & G.M. Whitesides: Soft lithographic methods for nanofabrication. J. Mat. Chem. 7, 1069-1074 (1997)

24. Bernard, A., E. Delamarche, H. Schmid, B. Michel, H.R. Bosshard & H. Biebuyck: Printing patterns of proteins. Langmuir 14, 2225-2229 (1998)

25. McDonald, J.C., D.C. Duffy, J.R. Anderson, D.T. Chiu, H. Wu, O.J. Schueller & G.M. Whitesides: Fabrication of microfluidic systems in poly(dimethylsiloxane). Electrophoresis 21, 27-40 (2000)

26. Unger, M.A., H.P. Chou, T. Thorsen, A. Scherer & S.R. Quake: Monolithic microfabricated valves and pumps by multilayer soft lithography. Science 288, 113-116 (2000)

27. James, C.D., R.C. Davis, L. Kam, H.G. Craighead, M. Isaacson, J.N. Turner & W. Shain: Patterned protein layers on solid substrates by thin stamp microcontact printing. Langmuir 14, 741-744 (1998)

28. Gaber, B.P., B.D. Martin & D.C. Turner: Create a protein microarray using a hydrogel "stamper". Chemtech 29, 20-24 (1999)

29. Kane, R.S., S. Takayama, E. Ostuni, D.E. Ingber & G.M. Whitesides: Patterning proteins and cells using soft lithography. Biomaterials 20, 2363-2376 (1999)

30. Morozov, V.N. & T.Y. Morozova: Electrospray deposition as a method for mass fabrication of mono- and multicomponent microarrays of biological and biologically active substances. Anal Chem 71, 3110-3117 (1999)

31. Lundgren, J.S., A.N. Watkins, D. Racz & F.S. Ligler: A liquid crystal pixel array for signal discrimination in array biosensors. Biosens 15, 417-421 (2000)

32. Pawlak, M., E. Grell, E. Schick, D. Anselmetti & M. Ehrat: Functional immobilization of biomembrane fragments on planar waveguides for the investigation of side-directed ligand binding by surface-confined fluorescence. Faraday Discuss 111, 273-288; discussion 331-243 (1998)

33. Rowe, C.A., L.M. Tender, M.J. Feldstein, J.P. Golden, S.B. Scruggs, B.D. MacCraith, J.J. Cras & F.S. Ligler: Array biosensor for simultaneous identification of bacterial, viral, and protein analytes. Anal Chem 71, 3846-3852 (1999)

34. Joos T.O., Stoll D. and Templin M.F. Miniaturised multiplexed immunoassays. Curr Opin Chem Biol, 6:76-80 (2002)

35. Carson, R.T. & D.A. Vignali: Simultaneous quantitation of 15 cytokines using a multiplexed flow cytometric assay. J Immunol Methods 227, 41-52 (1999)

36. Chen, R., L. Lowe, J.D. Wilson, E. Crowther, K. Tzeggai, J.E. Bishop & R. Varro: Simultaneous Quantification of Six Human Cytokines in a Single Sample Using Microparticle-based Flow Cytometric Technology. Clin Chem 45, 1693-1694 (1999)

37. Chan, W.C.W. & S.M. Nie: Quantum dot bioconjugates for ultrasensitive non-isotopic detection. Science 281, 2016-2018 (1998)

38. Bruchez, M., M. Moronne, P. Gin, S. Weiss & A.P. Alivisatos: Semiconductor nanocrystals as fluorescent biological labels. Science 281, 2013-2016 (1998)

39. Zhu, H., J.F. Klemic, S. Chang, P. Bertone, A. Casamayor, K.G. Klemic, D. Smith, M. Gerstein, M.A. Reed & M. Snyder: Analysis of yeast protein kinases using protein chips. Nat Genet 26, 283-289 (2000)

40. Joos, T.O., M. Schrenk, P. Hopfl, K. Kroger, U. Chowdhury, D. Stoll, D. Schorner, M. Durr, K. Herick, S. Rupp, K. Sohn & H. Hammerle: A microarray enzyme-linked immunosorbent assay for autoimmune diagnostics. Electrophoresis 21, 2641-2650 (2000)

41. Merchant, M. & S.R. Weinberger: Recent advancements in surface-enhanced laser desorption/ionization-time of flight-mass spectrometry. Electrophoresis 21, 1164-1177 (2000)

42. Nelson, B.P., T.E. Grimsrud, M.R. Liles, R.M. Goodman & R.M. Corn: Surface plasmon resonance imaging measurements of DNA and RNA. Analytical Chemistry 73, 1-7 (1999)

43. Brecht, A., R. Burckardt, J. Rickert, I. Stemmler, A. Schuetz, S. Fischer, T. Friedrich, G. Gauglitz & W. Goepel: Transducer-based approaches for parallel binding assays in high throughput screening. J. Biomol. Screening 1, 191-201 (1996)

44. Nelson, R.W., D. Nedelkov & K.A. Tubbs: Biosensor chip mass spectrometry: a chip-based proteomics approach. Electrophoresis 21, 1155-1163 (2000)

45. Pei, R.J., Z.L. Cheng, E.K. Wang & X.R. Yang: Amplification of antigen-antibody interactions based on biotin labeled protein-streptavidin network complex using impedance spectroscopy. Biosensors & Bioelectronics, 16, 355-361 (2001)

46. Renken, J., R. Dahint, M. Grunze & F. Josse: Multifrequency Evaluation of Different Immunosorbents on Acoustic Plate Mode Sensors. Anal.Chem. 68, 176-182 (1996)

47. Rickert, J., T. Weiss, W. Kraas, G. Jung & W. Goepel: A new affinity biosensor: self-assembled thiols as selective monolayer coatings of quartz crystal microbalances. Biosens. Bioelectron. 11, 591-598 (1996)

48. Fodor, S.P., R.P. Rava, X.C. Huang, A.C. Pease, C.P. Holmes & C.L. Adams: Multiplexed biochemical assays with biological chips. Nature 364, 555-556 (1993)

49. Collins, F., S. & B. Phimister: The chipping forecast. Nat. Genet. 21 suppl., 61 p (1999)

50. Reineke, U., R. Volkmer-Engert & J. Schneider-Mergener: Applications of peptide arrays prepared by the SPOT-technology. Curr Opin Biotechnol 12, 59-64 (2001)

51. Toepert, F., J.R. Pires, C. Landgraf, H. Oschkinat & J. Schneider-Mergener: Synthesis of an Array Comprising 837 Variants of the hYAP WW Protein Domain. Angew. Chem. Int. Ed. Engl. 40, 897-900 (2001)

52. Reineke, U., A. Kramer & J. Schneider-Mergener: Antigen sequence- and library-based mapping of linear and discontinuous protein-protein-interaction sites by spot synthesis. Curr Top Microbiol Immunol 243, 23-36 (1999)

53. MacBeath, G., A. Koehler & S. Schreiber: Printing small molecules as microarrays and detecting protein-ligand interactions en masse. J.Am.Chem.Soc. 121, 7967-7968 (1999)

54. Borrebaeck, C.A.: Antibodies in diagnostics-from immunoassays to protein chips. Immunol Today 21, 379-382 (2000)

55. Gao, C., S. Mao, C.H. Lo, P. Wirsching, R.A. Lerner & K.D. Janda: Making artificial antibodies: A format for phage display of combinatorial heterodimeric arrays. Proc Natl Acad Sci U S A 96, 6025-6030 (1999)

56. Knappik, A., L. Ge, A. Honegger, P. Pack, M. Fischer, G. Wellnhofer, A. Hoess, J. Wolle, A. Pluckthun & B. Virnekas: Fully synthetic human combinatorial antibody libraries (HuCAL) based on modular consensus frameworks and CDRs randomized with trinucleotides. J Mol Biol 296, 57-86 (2000)

57. Clackson, T. & J.A. Wells: In vitro selection from protein and peptide libraries. Trends Biotechnol 12, 173-184 (1994)

58. Hoogenboom, H.R., A.P. de Bruine, S.E. Hufton, R.M. Hoet, J.W. Arends & R.C. Roovers: Antibody phage display technology and its applications. Immunotechnology 4, 1-20 (1998)

59. Skerra, A. & A. Pluckthun: Assembly of a functional immunoglobulin Fv fragment in Escherichia coli. Science 240, 1038-1041 (1988)

60. Bass, S., R. Greene & J.A. Wells: Hormone phage: an enrichment method for variant proteins with altered binding properties. Proteins 8, 309-314 (1990)

61. Barbas, C.F., A.S. Kang, R.A. Lerner & S.J. Benkovic: Assembly of combinatorial antibody libraries on phage surfaces: the gene III site. Proc Natl Acad Sci U S A 88, 7978-7982 (1991)

62. McCafferty, J., A.D. Griffiths, G. Winter & D.J. Chiswell: Phage antibodies: filamentous phage displaying antibody variable domains. Nature 348, 552-554 (1990)

63. Griffiths, A.D., M. Malmqvist, J.D. Marks, J.M. Bye, M.J. Embleton, J. McCafferty, M. Baier, K.P. Holliger, B.D. Gorick & N.C. Hughes-Jones: Human anti-self antibodies with high specificity from phage display libraries. Embo J 12, 725-734 (1993)

64. Hoogenboom, H.R., A.D. Griffiths, K.S. Johnson, D.J. Chiswell, P. Hudson & G. Winter: Multi-subunit proteins on the surface of filamentous phage: methodologies for displaying antibody (Fab) heavy and light chains. Nucleic Acids Res 19, 4133-4137 (1991)

65. Clackson, T., H.R. Hoogenboom, A.D. Griffiths & G. Winter: Making antibody fragments using phage display libraries. Nature 352, 624-628 (1991)

66. Burton, D.R., C.F. Barbas, 3rd, M.A. Persson, S. Koenig, R.M. Chanock & R.A. Lerner: A large array of human monoclonal antibodies to type 1 human immunodeficiency virus from combinatorial libraries of asymptomatic seropositive individuals. Proc Natl Acad Sci U S A 88, 10134-10137 (1991)

67. Marks, J.D., H.R. Hoogenboom, T.P. Bonnert, J. McCafferty, A.D. Griffiths & G. Winter: By-passing immunization. Human antibodies from V-gene libraries displayed on phage. J Mol Biol 222, 581-597 (1991)

68. Xie, M.H., J. Yuan, C. Adams & A. Gurney: Direct demonstration of MuSK involvement in acetylcholine receptor clustering through identification of agonist ScFv. Nat Biotechnol 15, 768-771 (1997)

69. Vaughan, T.J., A.J. Williams, K. Pritchard, J.K. Osbourn, A.R. Pope, J.C. Earnshaw, J. McCafferty, R.A. Hodits, J. Wilton & K.S. Johnson: Human antibodies with sub-nanomolar affinities isolated from a large non-immunized phage display library. Nat Biotechnol 14, 309-314 (1996)

70. Schier, R., A. McCall, G.P. Adams, K.W. Marshall, H. Merritt, M. Yim, R.S. Crawford, L.M. Weiner, C. Marks & J.D. Marks: Isolation of picomolar affinity anti-c-erbB-2 single-chain Fv by molecular evolution of the complementarity determining regions in the center of the antibody binding site. J Mol Biol 263, 551-567 (1996)

71. Barbas, C.F., 3rd, D. Hu, N. Dunlop, L. Sawyer, D. Cababa, R.M. Hendry, P.L. Nara & D.R. Burton: In vitro evolution of a neutralizing human antibody to human immunodeficiency virus type 1 to enhance affinity and broaden strain cross-reactivity. Proc Natl Acad Sci U S A 91, 3809-3813 (1994)

72. Hernaiz, M., J. Liu, R.D. Rosenberg & R.J. Linhardt: Enzymatic modification of heparan sulfate on a biochip promotes its interaction with antithrombin III. Biochem Biophys Res Commun 276, 292-297 (2000)

73. Duff, R.J., S.F. Deamond, C. Roby, Y. Zhou & P.O. Ts'o: Intrabody tissue-specific delivery of antisense conjugates in animals: ligand-linker-antisense oligomer conjugates. Methods Enzymol 313, 297-321 (2000)

74. Lecerf, J.M., T.L. Shirley, Q. Zhu, A. Kazantsev, P. Amersdorfer, D.E. Housman, A. Messer & J.S. Huston: Human single-chain Fv intrabodies counteract in situ huntingtin aggregation in cellular models of Huntington's disease. Proc Natl Acad Sci U S A 98, 4764-4769 (2001)

75. Soomets, U., M. Hallbrink & U. Langel: Antisense properties of peptide nucleic acids. Front Biosci 4, D782-786 (1999)
76. Braasch, D.A. & D.R. Corey: Locked nucleic acid (LNA): Fine-tuning the recognition of DNA and RNA. Chem Biol 8, 1-7 (2001)
77. Brody, E.N., M.C. Willis, J.D. Smith, S. Jayasena, D. Zichi & L. Gold: The use of aptamers in large arrays for molecular diagnostics. Mol Diagn 4, 381-388 (1999)
78. Brody, E.N. & L. Gold: Aptamers as therapeutic and diagnostic agents. J Biotechnol 74, 5-13 (2000)
79. Wilson, D.S., A.D. Keefe & J.W. Szostak: The use of mRNA display to select high-affinity protein-binding peptides. Proc Natl Acad Sci U S A 98, 3750-3755 (2001)
80. Blum, J.H., S.L. Dove, A. Hochschild & J.J. Mekalanos: Isolation of peptide aptamers that inhibit intracellular processes. Proc Natl Acad Sci U S A 97, 2241-2246 (2000)
81. Kronvall, G. & K. Jonsson: Receptins: a novel term for an expanding spectrum of natural and engineered microbial proteins with binding properties for mammalian proteins. J Mol Recognit 12, 38-44 (1999)
82. Nord, K., O. Nord, M. Uhlen, B. Kelley, C. Ljungqvist & P.A. Nygren: Recombinant human factor VIII-specific affinity ligands selected from phage-displayed combinatorial libraries of protein A. Eur J Biochem 268, 4269-4277 (2001)
83. Nord, K., E. Gunneriusson, J. Ringdahl, S. Stahl, M. Uhlen & P.A. Nygren: Binding proteins selected from combinatorial libraries of an alpha-helical bacterial receptor domain. Nat Biotechnol 15, 772-777 (1997)
84. Gunneriusson, E., P. Samuelson, J. Ringdahl, H. Gronlund, P.A. Nygren & S. Stahl: Staphylococcal surface display of immunoglobulin A (IgA)- and IgE-specific in vitro-selected binding proteins (affibodies) based on Staphylococcus aureus protein A. Appl Environ Microbiol 65, 4134-4140 (1999)
85. Gunneriusson, E., K. Nord, M. Uhlen & P. Nygren: Affinity maturation of a Taq DNA polymerase specific affibody by helix shuffling. Protein Eng 12, 873-878 (1999)
86. Mendoza, L.G., P. McQuary, A. Mongan, R. Gangadharan, S. Brignac & M. Eggers: High-throughput microarray-based enzyme-linked immunosorbent assay (ELISA). Biotechniques 27, 778-780, 782-776, 788 (1999)
87. Silzel, J.W., B. Cercek, C. Dodson, T. Tsay & R.J. Obremski: Mass-sensing, multianalyte microarray immunoassay with imaging detection. Clin Chem 44, 2036-2043. (1998)
88. Huang, R.P.: Simultaneous detection of multiple proteins with an array-based enzyme-linked immunosorbent assay (ELISA) and enhanced chemiluminescence (ECL). Clin Chem Lab Med 39, 209-214 (2001)
89. Haab, B.B., M.J. Dunham & P.O. Brown: Protein microarrays for highly parallel detection and quantitation of specific proteins and antibodies in complex solutions. Genome Biol 2, 4.1-4.13 (2001)
90. Cahill, D.J.: Protein and antibody arrays and their medical applications. J Immunol Methods 250, 81-91 (2001)
91. Weinberger, S.R., T.S. Morris & M. Pawlak: Recent trends in protein biochip technology. Pharmacogenomics 1, 395-416 (2000)
92. Davies, H.A.: The ProteinChip System from Ciphergen: a new technique for rapid, micro-scale protein biology. J Mol Med 78, B29 (2000) MacBeath, G. & S.L. Schreiber: Printing proteins as microarrays for high-throughput function determination. Science 289, 1760-1763 (2000)

93. von Eggeling, F., H. Davies, L. Lomas, W. Fiedler, K. Junker, U. Claussen & G. Ernst: Tissue-specific microdissection coupled with ProteinChip array technologies: applications in cancer research. Biotechniques 29, 1066-1070 (2000)

94. Zhu, H. & M. Snyder: Protein arrays and microarrays. Curr Opin Chem Biol 5, 40-45 (2001)

95. Caveman: "I'll have a genome with chips, please". By Caveman. J Cell Sci 113, 3543-3544. (2000)

96. Zhu, H., M. Bilgin, R. Bangham, D. Hall, A. Casamayor, P. Bertone, N. Lan, R. Jansen, S. Bidlingmaier, T. Houfek, T. Mitchell, P. Miller, R.A. Dean, M. Gerstein & M. Snyder: Global analysis of protein activities using proteome chips. Science 293, 2101-2105. (2001)

97. Bulyk, M.L., E. Gentalen, D.J. Lockhart & G.M. Church: Quantifying DNA-protein interactions by double-stranded DNA arrays. Nat Biotechnol 17, 573-577 (1999)

98. Disley, D.M., P.R. Morrill, K. Sproule & C.R. Lowe: An optical biosensor for monitoring recombinant proteins in process media. Biosens Bioelectron 14, 481-493 (1999)

99. Sanders, G.H.W. & A. Manz: Chip-based microsystems for genomic and proteomic analysis. Trends in Analytical Chemistry 19, 364-378 (2000)

100. Rowe, C.A., S.B. Scruggs, M.J. Feldstein, J.P. Golden & F.S. Ligler: An array immunosensor for simultaneous detection of clinical analytes. Anal Chem 71, 433-439 (1999)

101. Hook, S.S. & A.R. Means: Ca(2+) / CaM-dependent kinases: from activation to function. Annu Rev Pharmacol Toxicol 41, 471-505 (2001)

4

MEA-Based Spike Recording in Cultured Neuronal Networks

YASUHIKO JIMBO, NAHOKO KASAI, KEIICHI TORIMITSU, AND TAKASHI TATENO*
*NTT Basic Research Laboratories, NTT Corporation; *Osaka University*

Abstract: A Micro-Electrode Array (MEA) based neuronal spike recording system is described. MEAs are cell-culture dishes with multiple embedded micro-electrodes on their surface. Multi-site recording of electrical signals enable us of visualizing spatially propagating activity patterns in cultured neuronal networks. Here in this chapter, we first introduce basic properties of MEAs, from the viewpoints of their technically advantageous features. Then some examples of actual recording are shown. Monitoring of spontaneous activity during development of cortical networks demonstrates their capability of long-term non-invasive recording. Characteristics of evoked responses can be modified by repetitive electrical stimulation through the embedded electrodes. Based on the high temporal resolution of this system, spatially propagating action potentials in a single neuron are visualized. Finally, combination of MEA recording with fluorescence imaging is illustrated.

Key words: MEA, multi-site recording, neuronal spike, plasticity.

1. Introduction

Sparse-coding is one of the keys in biological information processing, particularly in the central nervous system. To see spatio-temporal signals in neural networks, various technical tools have been extensively explored. Optical recording using CCDs or photodiode arrays is one of the most promising methods and some interesting results have been already reported [1, 2]. The Micro-Electrode Array (MEA), which we introduce in this chapter, is another possibility. MEAs were first introduced by Gross [3] and Pine [4], and since then considerable technical improvements have been achieved.

A MEA is a dish for cell culture, on the surface of which multiple micro-electrodes are embedded. The unique advantages of MEAs are as follows: (1) non-invasiveness, (2) high time-resolution, (3) capability of multi-site stimulation. Signals generated by electrically active cells, such as neurons and heart muscles, can be detected extracellularly. Extracellular recording through these embedded electrodes causes no effects on cultured cells, thus long-term monitoring of the activity can be achieved [5-7]. "Non-invasiveness" means no-interaction with other measurement tools. Well-established electrophysiological tools like patch-clamp recording or Ca sensitive dye recording can be carried out simultaneously with the MEA recording. The time resolution is basically determined by the sampling rate of the total system. Recent progress in electronic devices provides us with quite high-speed A/D converting systems, which are sufficiently fast for visualizing most of the electrophysiological signals. The final advantage, capability of multi-site stimulation is probably the most useful property of MEAs. Cells cultured on the MEA can be stimulated by electrical voltage or current injected through the embedded electrodes. Stimulation intensity and timing can be precisely controlled and the stimulation sites can be selected among any combination of the sites [8].

In the following sections, first, basic aspects of MEA recording are summarized. Then, some examples of actual recordings are described mainly from the technical viewpoints. For demonstrating long-term noninvasive recording, developmental changes in spontaneous activity of rat cortical neurons are shown. Their activity is continuously monitored for two months. Then the effects of repetitive electrical stimulation on their evoked responses are demonstrated. Reliability as well as reproducibility of the evoked responses increased by the repetitive stimuli. We believe that this kind of modification observed in *in vitro* systems reflects some basic aspects of the mechanisms underlying learning and memory in the brain [9-11]. Finally, two more recordings are illustrated. One is the visualization of action-potential propagation patterns in a single neuron. Neurons are cultured on densely packed electrode patterns. Using such kind of patterns, action potentials in a single neuron can be detected at several sites. Due to its high temporal resolution, time delay of spikes in a single neuron could be measured. Another is simultaneous recording of electrical activity and intracellular calcium transients. Imaging technology has been extensively improved recently, both in acquisition hardware and dye materials. Thus the combination of MEA recording with imaging will be useful tools in the field of experimental neuroscience. The results shown here demonstrate that periodic Ca transients, which are the characteristic properties of the cortical networks, are synchronized with the electrical bursts.

2.   Fabrication and Recording Setup

2.1   Fabrication of MEAs

MEAs are fabricated by basic photolithography. The typical size of neuronal cell body is 10-30 μm, which is much larger than the basic pitch of recent fabrication technology (less than 1 μm). Thus the fabrication process is quite simple. Optically transparent materials are most widely used: glass substrates, ITO (indium-tin-oxide) electrodes, and insulation films. The typical fabrication process is as follows. First, ITO layer is deposited by sputtering on the glass substrates. Then this conductive layer is etched to form electrode patterns. Finally, passivation film is cast except the recording sites. The number of recording sites, their size and distance, and the spatial alignment are determined by the photo-mask design, and we can freely design the mask-patterns depending on the purpose of individual experiments. Typically, substrates with 64 recording sites (8×8 matrix) of 10-30 μm squares and 50-1000 μm distance are used. To reduce interface impedance, local platinization of the recording sites is often carried out. Some MEAs have been already commercially available [12]. Application of the standard semiconductor materials has also been reported for integrating electronic circuits [13, 14]. Particularly to reduce fabrication costs, plastic substrates are also tried.

2.2   Cell Culture

Though it seems to be a hard task to contact biological materials with hard substrates like MEAs, practically cell culture is quite easy. After sterilization of the dishes, materials of high affinity with the target cells, such as poly-lysine and laminin, are coated on the surface. This kind of coating provides quite good surface for cell cultures. Figure 4.1 shows an example of cultured neurons on a MEA. In this example, cortical neurons were taken from E18 Wistar rat embryos. After digestion by papain and mechanical trituration, dissociated cells were plated on the dish and cultured for a week. Half of the culture medium, the basic composition of which is DMEM (Dulbecco's modified Eagle's medium) with 5% FBS and 5% horse serum, added with insulin and antibiotics, was changed twice a week. Use of the conditioned medium was quite effective for long-term cultures. We could keep them in a good condition for more than six months.

2.3   Recording Electrical Signals

Signals of electrically active cells derive from their membrane potentials. To record membrane potentials directly, a pair of electrodes is necessary, and one of them must be set inside the cell. MEA recording is extracellular recording in principle. Local electrical fields produced in the volume conductor by transient current flow due to biological activity are detected (see inset in Figure 4.2). Thus the intensity of the signal is not large, three orders smaller than the original membrane potential changes. Typical extracellular spike intensity is around 100 μV, while the actual action potential amplitude is about 100 mV. The correspondence between whole cell current clamp signals and MEA recording is shown in Figure 4.2. The clamped cell was electrically stimulated by current injection through the whole-cell

FIGURE 4.1. A MEA substrate and cultured cortical neurons on it.

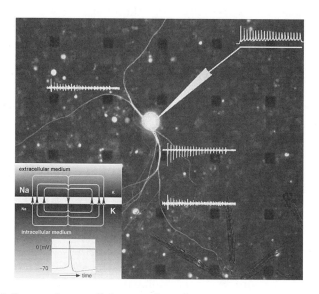

FIGURE 4.2. Intra and extracellular recording of neuronal spikes.

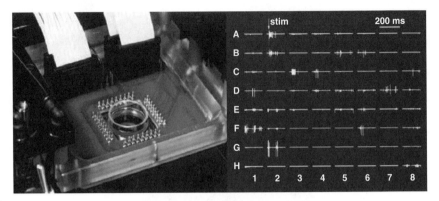

FIGURE 4.3. A connection board and an example of 64-channel MEA recording.

pipette and spike trains were elicited. In this example, extracellular spike signals, which show good correspondence with the whole cell recording, were recorded at three sites.

The electrical signals detected at the substrate electrodes are transferred to recording setup through the connection board. Sixty-four gold pins on the board, each of which contains a small spring in its inside, push the contact pads of the MEA substrate by air pressure (see Figure 4.3). The signals are then amplified, filtered, A/D converted and stored on hard disks. It is reported that the highest frequency component in the extracellular signals is around 6 kHz [15]. Thus the standard setting of band-pass filtering is 100 Hz to 10 kHz, and the sampling rate is 25-50 kHz. An example of the raw recordings at the 64 recording sites is shown in Figure 4.3. In this case, a stimulation pulse was applied from the site A2 and the evoked responses were recorded. Electrical stimulation, particularly stimulation from multiple sites is the unique and useful property of MEA recordings. To achieve precise control of stimulation sites, intensity and timings, careful consideration on the electrode/electrolyte interface properties is necessary. This issue was discussed in detail elsewhere [8]. The result shown here is an example obtained based on those considerations. Clear evoked activity is recorded at the stimulation site.

3.   MEA-Based Recording

3.1   Developmental Changes in Spontaneous Activity

As far as extracellular recording of spontaneous activity is concerned, MEA recording has very little interaction with cultured cells on the surface. This means that MEA recording has the potential of long-term noninvasive monitoring of activity. We applied this to continuous monitoring of developmental changes in spontaneous activity. It is well-known that cultured cortical networks generate synchronized spontaneous bursts, particularly in a low Mg^{2+} condition [16]. To achieve continuous recording under standard culture conditions, the recording setup and a medium perfusion system were directly connected to the samples kept in the CO_2 incubator. Culture medium was constantly perfused at a quite slow rate (0.1 ml/min). We could avoid contamination by this closed-chamber setting and could maintain the culture for two months at a stable recording condition.

Recording was started at five days *in vitro* (DIV). The results are summarized in Figure 4.4. First, weakly coupled signals of low signal to noise ratio (S/N) were detected. Then the number of active sites, intensity of the signals, as well as their synchronicity among recording sites increased with culture days. This probably reflects the increases in the synaptic density, maturation of ion channels and receptors. However, these gradual changes in the properties were not monotonous. After about two weeks *in vitro*, their spontaneous firing started to diminish and almost disappeared once. And then, they restarted. The properties of individual bursts and their spatial propagation patterns were not identical between before and after this silent period. The mechanisms producing this silent period are still under investigation. After this silent period, the restarted spontaneous activity increased monotonously and showed almost stable properties after about one month. The synchronized bursts were produced at about 1 Hz, the initiation sites were not

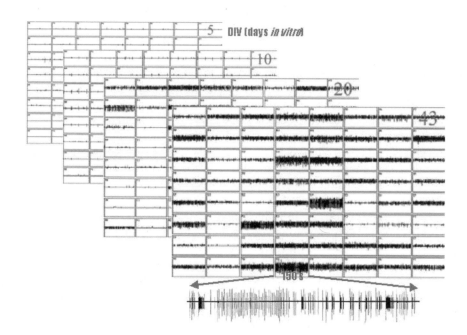

FIGURE 4.4. Developmental changes of spontaneous activity in cultured cortical networks.

constant, and the spatial propagation patterns could be categorized into a few groups. Some asynchronous components were mixed with the dominant synchronized bursts.

One of the technical problems in this long-term continuous monitoring was the amount of data. As described before, the sampling rate of at least 25 kHz is necessary for spike analysis. In the case of 25 kHz sampling of 16 bit resolution for 64 channels, about 3 MB raw data are generated every second (10 GB per hour and 240 GB per day). Thus data reduction is mandatory. For this purpose, we used on-line spike detection. Neuronal spikes were extracted by the standard threshold method. The parameters describing extracted spikes are stored. Raw data of a certain time window that contained spikes were extracted. These extracted data and spike describing parameters were stored on a hard disk and then transferred to DVD (digital video disk) media.

3.2   Effects of Focal Electrical Stimulation

It is widely accepted that fundamental structures of neuronal connections are determined based on the genetically-coded information. However, fine tuning of their connections, as well as modification of synaptic weights are probably carried out through the inputs from their surrounding environments. So-called developmental plasticity, and also learning and memory could be the results of this kind of "experiences". In the brain or neuronal networks, the inputs from the environments are dominantly expressed as spatio-temporal patterns of spikes. Thus artificial induction of evoked activity by electrical stimulation could produce modification in the neuronal network characteristics.

Learning and memory has been one of the most widely investigated themes in the field of recent neuroscience. Long-term potentiation and depression (LTP/LTD) are the most promising candidates for synaptic plasticity, and lots of LTP/LTD-related research works have been already done [17, 18]. However, little has been reported about how these synaptic modifications are integrated and reflected in the network activity. The reason for this was dominantly due to the lack of experimental tools for measuring network activity. The MEA-based stimulation/recording system can be one of the useful and promising methods.

Here in this section, we introduce modification of evoked response characteristics induced by repetitive focal stimulation. The samples used for this experiment were cultured rat cortical networks, the same as described before. Cortical neurons were cultured for two months before starting stimulation experiments. As mentioned above, cortical cultures show quite stable properties both in spontaneous and evoked activity after one month *in vitro*. The samples were kept in the CO_2 incubator and evoked responses were recorded. The stimuli were applied from a single site of the 64 substrate electrodes. A single biphasic pulse of 250 mV was applied every ten second for one hour. Total 360 evoked responses were recorded and analyzed. Figure 4.5 shows the

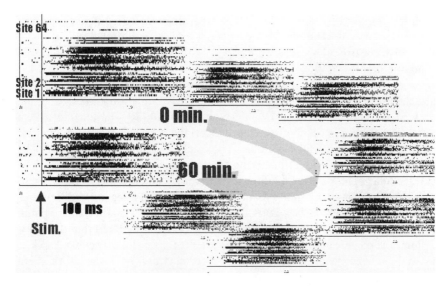

FIGURE 4.5. Gradual changes in the evoked responses induced by repetitive application of electrical stimulation.

transition of evoked response properties. In this figure, raster plots of the spikes detected at all the 64 sites are shown. We can see two major modifications in these plots. One is the increase in propagation velocity of the evoked responses. The velocity can be roughly estimated in the raster plots from the slope of the initial spikes detected at each of the recording sites. The slope was steeper in the final state than that in the initial one. This suggested the gradual increase in the propagation velocity. The other was the change in the activity of the upper area of the plots. Spikes in the upper area clearly recorded in the initial state, almost disappeared in the final state. These results suggested that the network gradually adapted to the input stimulus.

The responses evoked from other sites were also modified by this repetitive focal stimulation. We could find a common rule dominated these modifications. The details were described elsewhere [9, 10].

3.3   Other Applications

In this final section, we introduce two other applications of MEA-based recording. The first one is visualization of spike propagation in a single neuron. A neuronal action potential is a voltage pulse of about 100 mV in amplitude and 1 ms in duration. Thus, to visualize its intracellular propagation, less than 0.1 ms time resolution is mandatory. As mentioned before, time resolution of the MEA recording is basically determined by the sampling rate of the system. In our case, the maximum rate is 50 kHz, by which we can obtain required temporal resolution. For this experiment, we used densely packed electrode patterns. The electrodes were 10 μm squares and their distances were 50 μm. Cortical neurons were plated on the dishes and cultured for more than one month.

A whole-cell pipette containing a fluorescence dye (Lucifer Yellow) was attached to one of the neurons on the dish. The stimulation pulses were then applied through the pipette to elicit action potentials. The elicited action potentials were simultaneously recorded by the attached whole-cell pipette and through the substrate electrodes. The results are shown in Figure 4.6. The shape of the recorded cell including fine neurites is clearly visualized. The extracellular spikes corresponding to the whole-cell recorded action potentials were observed at more than 20 sites. We could see time differences among the extracellularly recorded signals and could visualize their spatial propagation in the cell. The superimposed traces demonstrated that the intracellular propagation patterns were quite stable. Extracellularly evoked spikes, as well as synaptically triggered spikes were also analyzed and all of them showed the same kind of stability.

The other is the combination with image acquisition. Recent progress in imaging techniques is quite attractive. Both image-acquisition hardware and staining dyes have been much improved. Not only intracellular ion concentration but also fine morphology of the cells, even synaptic events could be visualized. The MEA recording has no interaction with the optical recording. Thus any of these dye recordings can be applied simultaneously with the MEA recording.

As an example, we illustrate recording intracellular Ca transients. Cortical neurons cultured on the MEA were stained with Fluo-4, a well-known Ca indicator. Then the medium was changed to low-Mg^{2+} containing solution. It was previously reported that cultured cortical neurons showed Ca oscillation

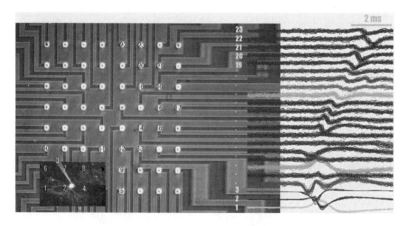

FIGURE 4.6. Visualization of spike propagation in a single neuron.

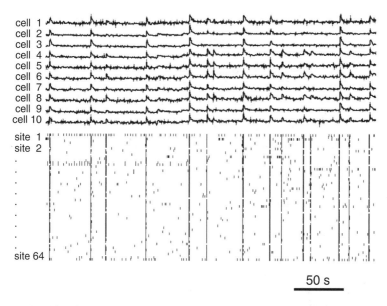

FIGURE 4.7. Simultaneous recording of electrical activity and intracellular Ca transients.

under reduced Mg^{2+} conditions [16]. In our experiments shown in Figure 4.7, the neurons produced highly synchronized Ca transients. The upper ten traces are the Ca transients in the selected ten neurons. The rests are the raster plots of the spikes recorded at the 64 substrate electrodes. Corresponding to most of the synchronized Ca transients, we can see synchronized bursts in the electrical activity. This suggested that the previously reported Ca oscillation was probably the results of electrical burst activity. The combination of MEA recording and imaging techniques will be more widely used in near future.

4.   Summary

In this chapter, MEA-based recording of neuronal activity was introduced mainly from the technical aspects. The predominant advantage of the MEA is in its noninvasiveness. Due to this property, MEAs are suitable for long-term recording and are applicable to combination with other methods such as optical recording. In the field of neuroscience, wide range of applications will be done in the future. The interface between neuronal tissue and integrated electronic devices is the issue not only in the basic neuroscience but also in the clinical medicine. Artificial cochlea [19] was already introduced in practical use and retinal implants [20] have also been extensively investigated. This kind of research works will be one of the important themes in the tissue engineering.

References

1. Shoham D., Glaser D. E., Arieli A., Kenet T., Wijnbergen C., Toledo Y., Hildesheim R., Grinvald A., Imaging cortical dynamics at high spatial and temporal resolution with novel blue voltage-sensitive dyes, Neuron 24, pp.791-802, 1999
2. Gogan P., Schmiedel-Jakob I., Chitti Y., Tyc-Dumont S., Fluorescence imaging of local membrane electric fields during the excitation of single neurons in culture, Biophys J. 69, pp. 299-310, 1995
3. Gross G. W., Rieske E., Kreutzberg G. W., Meyer A., A new fixed-array multi-microelectrode system designed for long-term monitoring of extracellular single unit neuronal activity in vitro, Neurosci. Lett., 6, pp. 101-105, 1977
4. Pine J., Recording action potentials from cultured neurons with extracellular microcircuit electrodes, J. Neurosci. Meth. 2, pp. 19-31, 1980
5. Gross G. W., Schwalm F. U., A closed flow chamber for long-term multichannel recording and optical monitoring, J. Neurosci. Meth. 52, pp. 73-78, 1994
6. Potter S. and DeMarse T., A new approach to neural cell culture for long-term studies, J. Neurosci. Meth. 110, pp. 17-24, 2001
7. Mukai Y., Shiina T., Jimbo Y., Continuous monitoring of developmental activity-changes in cultured cortical networks, Trans. IEEJ C122, pp. 1481-1489, 2002
8. Jimbo Y., Kasai N., Torimitsu K., Tateno T., Robinson H. P. C., A system for MEA-based multi-site stimulation, IEEE Trans. BME, in press
9. Jimbo Y., Tateno T., Robinson H. P. C., Simultaneous induction of pathway-specific potentiation and depression in networks of cortical neurons, Biophys. J. 76, pp. 670-678, 1999
10. Tateno T., Jimbo Y., Activity-dependent enhancement in the reliability of correlated spike timings in cultured cortical neurons, Biol. Cybern. 80, pp. 45-55, 1999
11. Shahaf G., Marom S., Learning in networks of cortical neurons, J. Neurosci. 21, pp. 8782-8788, 2001
12. http://www.multichannelsystems.com/,
    http://www.med64.com,
    http://www.plexoninc.com/
13. Offenhausser A., Knoll W., Cell-transistor hybrid systems and their potential applications, Trend. Biotechnol. 19, pp. 62-66, 2001
14. Pancrazio, J. J., Bey, P. P., Jr., Cuttino, D. S., Kusel, J. K., Borkholder, D. A., Shaffer, K. M., Kovacs, G. T. A., Stenger, D. A., Portable cell-based biosensor system for toxin detection, Sens. Act. B Chem. 53, pp. 179-185, 1998
15. Najafi K., Solid-state microsensors for cortical nerve recordings, IEEE Eng. in Med. Biol. Magaz., pp. 375-387, 1994
16. Robinson H. P. C., Kawahara M., Jimbo Y., Torimitsu K., Kuroda Y., Kawana A., Periodic synchronized bursting and intracellular calcium transietnts elicited by low magnesium in cultured cortical neurons, J. Neurophysiol. 70, pp. 1606-1616, 1993
17. Bennett M. R., The concept of long term potentiation of transmission at synapses, Prog. Neurobiol. 60, pp. 109-137, 2000
18. Bi G., Poo M., Synaptic modification by correlated activity: Hebb's postulate revisited, Annu Rev Neurosci. 24, pp. 139-166, 2001
19. Watts L., Kerns D. A., Lyon R. F., Mead C. A., Improved implementation of the silicon cochlea, IEEE J. Solid-State Circuits SC-27, pp. 692-700, 1992
20. Stett A., Barth W., Weiss S., Haemmerle H., Zrenner E., Electrical multisite stimulation of the isolated chicken retina, Vision Res. 40, pp. 1785-1795, 2000

5

Cell-Transistor Hybrid Systems
Electrogenic Cells as Signal Transducing Elements Coupled to Microelectronic Devices

Sven Ingebrandt and Andreas Offenhäusser
Forschungszentrum Jülich GmbH, D-52425 Jülich, Germany

Abstract: Due to a number of advances in cell and tissue culture in combination with more sensitive methods to transduce biological signals, it has become increasingly feasible to detect unknown toxicity or pharmacological effects by using cells or tissues which are electrically coupled to microfabricated microelectrode arrays (MEAs) or field-effect transistors (FETs). In order to identify the contributions of the various cell signals we have investigated the coupling of cardiac myocytes with FETs. On the other side such systems can also be used to study the very basics of distributed information processing by interfacing cultured neuronal networks with microelectronic devices. In contrast to the totally random network structure, which is grown on a homogeneous substrate we are more interested in the formation and characterization of small well defined network architectures. By using the so-called micro-contact-printing (µCP) technique we are able to manipulate the adhesion sites of individual neurons and control the direction and outgrowth of dendrites and axons. Rat neurons grown on these pattern formed simple neuronal circuits containing electrical and chemical synapses, which was shown by patch-clamp measurements.

Key words: Field-effect transistor, microelectrode, cell-electronic interface, cellular patterning.

1. Introduction

The combination of biological signal processing elements such as membrane proteins, whole cells or even tissue slices with electronic transducers for the detection of physical signals creates functional hybrid systems that bring together the living and the technical worlds. Functional coupling of physiological processes with micro- and nanoelectronic devices will have great impact for a wide range of applications. The high sensitivity and selectivity of biological recognition systems with a man-made signal-detection and process-ing system will open up exciting possibilities for the development of new biosensors as well as for new approaches in neuroscience and computer science.

One of the most important reasons for the use of living cells is to obtain functional information, such as the effect of a stimulus on a living system. Information at the cellular level can yield insight into mechanisms of bio-chemical compound action, enabling not only detection but also classifica-tion. This has led to the development and validation of a number of cell-based biosensor concepts. In the field of biosensors one has realized that cellular proteins exhibit enhanced stability when expressed in cell systems, as compared to coating this proteins on electrochemical or optical transducers. In addition cultured excitable cells have been proposed as a cellular trans-ducing system, where changes in spontaneous or evoked action potentials reflect functional changes associated with the exposure to pharmacological compounds. These cells are usually cultured on multisite recording arrays to allow long-term observation from many cells at the same time.

In addition, the coupling of a 2D cellular network with an extracellular recording system will allow the structure-function relationship of such a net-work to be studied in detail. For extracellular signal recording from electro-genic cells *in vitro*, the use of microfabricated microelectrode arrays (MEAs) or field-effect transistors (FETs) is of increasing interest. Sufficient electrical coupling between the cell and the electrode for extracellular signal recording is achieved only when a cell or a part of a cell is located directly on top of the electrode. Electrical signals recorded by these devices show lower signals and a higher noise level (owing to a weaker coupling to the (gate) electrode com-pared to the patch pipette), but can be monitored for weeks. In addition recordings can be done simultaneously with the current array design on 16 or 64 sites! As an example for the usefulness of this cell-transistor hybrid system the influence of certain drugs could be demonstrated.

The need for further manipulations of the cell adhesion and organization at the micron level becomes evident if one looks at a randomly seeded net-work of nerve cells. Usually neurons adhere to the substrate developing den-drites and axons and forming a totally random network structure. However, we are more interested in the formation and characterization of well defined network architectures and, hence, have developed strategies that allow us to manipulate the adhesion sites of individual neurons and control the direction and outgrowth of dendrites and axons. This can be achieved at the level of the adhesion layer, the interface between the substrate (microelectronic device) and the adhering cell. Using the so-called micro-contact-printing (μCP) technique gives us the required control over the network formation. In the μCP approach a stamp made from poly(dimethylsiloxane) (PDMS) is "inked" with the specific peptide or with the whole extracellular matrix pro-tein like laminin and the pattern imprinted onto the substrate, which was pre-coated with the linker molecules used for the covalent attachment of the adhesion promoter.

2.   Field Effect Transistor and Microelectrode Arrays

In order to fulfil the need for selective long-term cell-transducer interfaces *in vitro*, microtechnology is used for the development of planar arrays with large numbers of field-effect transistors or metalelectrodes in the size of an individual cell. These arrays usually consist of a culture chamber in which the chips are embedded. In case of metal-electrode arrays (MEAs) conductive leads (usually gold, some investigators use the transparent conductor indium-tin-oxide (ITO)) are covered with an insulating layer and the tips of the electrodes are obtained by deinsulating of the ends using photolithographic techniques and etching or laser techniques. The first MEA was developed by Thomas et al. and used for recording from cardiac cells [1]. Later, the groups of Gross, Pine and others developed MEAs for recording and stimulation of neuronal networks [2, 3]. On the other hand field-effect transistor arrays have been developed to record the electrical signals from cells. Modifications in standard field effect transistor fabrication processes lead to devices with a metal-free gate electrode. A variation of these devices is the so-called ion-sensitive field-effect transistor (ISFET) where the properties of the gate dielectric is changed to yield higher sensitivity for certain ions as reviewed by Sibbald [4]. The first field-effect transistor for the recording of electrical cell signals was developed by the group of Bergveld at the University of Twente [5]. Later, Fromherz group and our group developed FETs for the recording from individual neurons and cardiac cells [6, 7]. In Figures 5.1 and 5.2, an example of a FET array, as used in our group, is depicted. This chip consists of a rectangular pattern of 16 gate electrodes with a gate size of 18×2 μm^2. The distance between the centres of the gates is 200 μm.

MEA electrodes must have a reasonable low impedance to allow the detection of the small extracellular signals (10 to 1000 μV). Traditionally, this has been achieved by electroplating of porous platinum ("platinum black"). This is not very durable, and thus the impedance can rise when MEAs are reused and during long-term culturing. Recently, very tough, low-impedance coatings have been created by several groups sputtering titanium nitride or iridium oxide [8, 9]. To circumvent these problems related to impedance we have connected metal electrodes directly to the gate of low-noise FETs [10] where the metal electrode acts as an extended gate electrode [EGE]. In Figures 5.3 and 5.4 an example of a EGE which is very similar to a MEA is shown. It consists of a rectangular electrode pattern of 64 gold electrodes with a diameter of 6 μm. The distance between the centres of the electrodes is 200 μm.

3.   Cell-Electronic Interface

A classical method for interfacing cells with (peripheric) electronics is intracellular recording and stimulation using impaled or attached glass pipettes. These techniques not only permit a detailed study of the cellular behaviour, but also

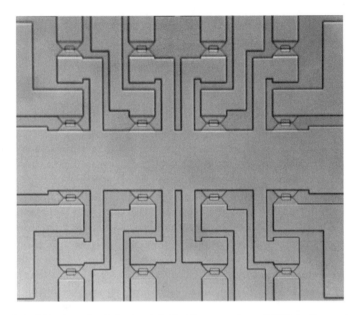

FIGURE 5.1. Micrograph of the 4 × 4 field-effect transistor (FET) in the centre of the chip. In the middle there is the common source connection of every individual FET. The drain connections are leading from the centre to the metal contacts at the outer part of the chip.

FIGURE 5.2. Assembled and encapsulated FET. The chip with the 4×4 transistor array is in the middle of a mini cell culture dish.

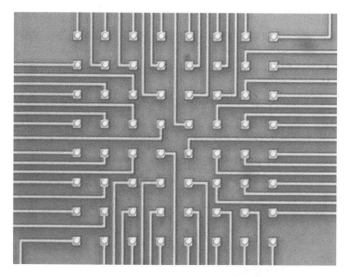

FIGURE 5.3. Micrograph of the 8 × 8 electrode array in the centre of the chip. The metal connections are leading from the electrode tips to the contact pads at the outer part of the chip.

FIGURE 5.4. Assembled and encapsulated EGE. The chip with the 8×8 electrode array is in the middle of a mini cell culture dish. The chip connectors and jacks for the Ag/AgCl bath electrode and the stimulation inputs are included.

allow the observation of a small cellular network using several pipettes. Less conventional is the application of voltage sensitive dyes to the cell membrane.

However, for both practical and scientific purpose the ideal cell-electronic interface would provide a large number of single cell-electrode contacts for long-term use. The electrical function of the cell is primarily determined by the electrical properties of the cellular membrane, which consists of a phospholipid bilayer, in which integral and peripheral proteins perform a variety of functions such as electrical and chemical signalling, and transport of ions. Due to differences in ionic concentrations, a resting potential exists. The electrical properties were identified by Hodgkin and Huxley as being caused by ion specific voltage-gated conductances, each in series with an equilibrium potential (Figure 5.5A).

The first model of the neuron-electrode contact was proposed by the group of Pine at Caltech in 1989 [11]. The cell was presented by a soma of which the membrane area was divided into two parts: a lower patch which is sealed to the electrode and an upper patch which is in free contact with the culture medium (Figure 5.5B). Later this model was improved and it was demonstrated that non-uniform channel distributions were necessary to explain extracellular recording [13] with the point contact model, due to the failure in local compensation of capacitive and ionic current densities.

The metal electrode acts as an electrochemical transducer of current carried by ions and current carried by electrons and vice versa. This charge flow results in a potential difference over the electrode-electrolyte interface, which facilitates the recombination of ions back into the metal. Besides these charge transfer mechanisms, which depends nonlinearily on the electrode-electrolyte

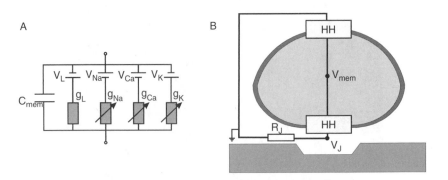

FIGURE 5.5. (A) Hodgkin-Huxley model of the electrical properties of an electrogenic membrane using conductances of the voltage gated channels, the ionic equilibrium potentials and the membrane capacitance C_{mem}. (B) Point-contact model: the attached and free membrane of the cell are described by the Hodgkin-Huxley elements (cf A). The junction is described by the seal resistance R_J. The extracellular voltage V_J is recorded by the transistor respectively gate electrode and the intracellular voltage V_{mem} can be monitored by a patch-clamp electrode.

potential, the linear properties are ascribed to the existence of a Helmholtz layer (capacitive behaviour) and to diffusion (capacitive and resistive behaviour) of ions into the electrolyte.

The transistor gate can be treated as an electrolyte/oxide/semiconductor (EOS) structure and has been applied as capacitive transducer for the high sensitive detection of surface charge at the oxide surface. The electrical response of the electrolyte-oxide-semiconductor (EOS) system can be interpreted analogous to the metal-oxide-semiconductor (MOS) structure. An external bias potential or changes in surface charge density result in changes in the space charge region of the semiconductor. Assuming that there is no charge transfer or ion diffusion due to the isolating properties of the gate oxide the gate electrode shows a pure capacitive behaviour.

4.   Coupling Cells to Microelectronic Devices

Prior to cell seeding, the chip surface needs to be functionalised by coating that promotes cell adhesion and outgrowth and – whenever necessary – allows for cell patterning (see below). Many strategies have been developed in order to achieve this goal: physisorption of extracellular matrix proteins like fibronectin or laminin, physisorption of positively charged polymers like polylysine or nitrocellulose, and construction of supramolecular architectures ranging from simple self-assembled monolayers to complex supramolecular interfacial multi-layer architectures including polymer layers [14], whole adhesion protein layers, or just peptides of it that expose a specific amino acid sequence known to interact with the integrin receptors of the target cells [15].

The chip surface thus prepared can then be used as a substrate for in vitro cell culture. We work with cardiac myocytes both prepared from neonatal (age 1–3 days) or from embryonic (embryonic day 18, E 18) rats as well as neurons prepared from various regions (hippocampus, cortex, brain stem) of the brain of embryonic rats (E 16 – E 18).

Cardiac myocytes offer the particular advantage that after random seeding they grow on the chip surface to a confluent monolayer, establishing both mechanical and electrical contacts between neighboring cells (syncytium). Moreover, after a few days in culture they start spontaneously contracting and firing action potentials, at first totally randomly. After some time one cell becomes the "pacemaker" unit and triggers all others of that population to generate a synchronized mechanical and electrical excitation pattern with a more or less regular repetition rate of about 1 Hz. As a result an excitation wave travels across the cell layer with a velocity in the range of m/sec. The electrical part, the action potential, can be monitored from an individual cell by a classical micropipette recording unit and shows the well-understood voltage-time profile depicted in Figure 5.6A, resulting from the time-dependent contributions of Na^+, K^+ and Ca^{2+}-currents across the cell membrane.

For recordings with our FET arrays, the chip is connected to a 16-channel preamplifier, which can be mounted under the microscope [16]. The offset currents arising from the driving conditions of all 16 channels are compensated and the recorded signals are amplified by a gain of 100. Band pass filtering is applied by using a 3 kHz low-pass (3 dB) and a 1.8 Hz high-pass unit (6 dB), which is included in the amplification system. From the transfer characteristics of each transistor, the corresponding gate-source voltage V_{GS} can be calculated, which corresponds to the voltage V_J in the cell-transistor junction area (Figure 5 B). It has been found that extracellular action potentials with amplitudes ranging from several 100 μV up to several mV could be recorded and that these voltages are at least ~ 5–10 times greater than the background noise.

Figure 5.6B shows a detailed view of such an extracellular AP-recording which was recorded simultaneously with the intracellular signal shown in Figure 5.6A. In order to explain the shape of the extracellular AP it is necessary to take into account the capacitive coupling of the intracellular signal to the gate, and the influence of ionic currents through the ion channels in the membrane in contact with the gate (Figure 5.5B). The capacitive part of the signal is mainly determined by the time constant ($\tau = C_M R_J$) of the resistor-capacitor (RC)-circuit, which is formed by the membrane capacitance C_M in combination with the seal resistance $R_J = 1/G_J$ acting as a highpass filter. As a consequence

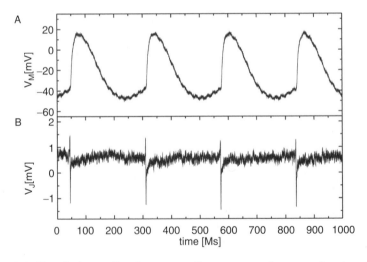

FIGURE 5.6. Electrical recording from rat cardiac myocytes after seven days in culture. (A) Intracellular recordings were made from cells grown several μm away from the recording site of the FET. The fast rise of the intracellular voltage causes first a sharp increase followed by a long lasting decrease in the drain-source current (I_{DS}) of the transistor. The extracellularly recorded trace is scaled as drain-source current (I_{DS}) and as effective voltage in the junction area (V_J) determined by the transfer characteristics. (B) The corresponding extracellular recording shows a sharp peak in the rising phase of the AP and a long-lasting negative signal part.

only the high frequency parts of the signal can pass and the extracellular signal is proportional to the first derivative of the intracellular signal. However, the negative part of the signal cannot be explained by this simple RC-circuitry because the decrease in the intracellular signal is slower than the fast rise at the beginning of the AP. Therefore, the negative part should be much smaller. One possible explanation is the influence of ionic currents flowing across the seal resistance ($R_J = 1/G_J$) and causing the appropriate voltage drop (V_J). Based on the model circuitry discussed previously (Figure 5.5 B), one possibility to explain the shape of the signal is based on the assumption that an increased ionic current density through the membrane in contact with the gate, compared with that in the overall membrane exists, because an equally distributed ionic current density would result in a diminishing signal. The positive spike at the beginning of the signal can then be explained by the stimulation signal received by the cell from its neighbouring cell [17].

As an example for the use of this cell-transistor hybrid system as a whole-cell sensor we have demonstrated the influence of certain drugs known to stimulate changes in the various membrane currents of the cardiac myocytes, like the well-established cardio-stimulants (isoproterenol, ISO and arterenol bitartrate (norepinephrine), NA) and relaxants (verapamil, VP and carbamylcholine, CARB). For the drug application a simple protocol was employed: first a basal recording of cultured myocytes on a particular FET was performed for several seconds, followed by completely replacing the standard recording solution with one that contains either ISO, NA, VP, or CARB. After recordings in the presence of drugs for 60s have been obtained, the cell layer was washed gently 5 times at 1 min intervals [18].

In Figure 5.7, the effects of norepinephine(NA) on the extracellular signal shape is shown. Once the drug is administered, the activity of the ion-channels in the membrane is enhanced. We were able to detect the signal shape change due to the drug response by comparing the signal shapes before and after drug administration. NA binds to a G-protein coupled receptor, which increases Ca^{2+}-current activity. Figure 5.7A and 5.7B shows the extracellular recordings before and after administration of NA and clearly indicates the enhanced Ca^{2+}-current activity.

5. Micro-Manipulating Cell Populations

The need for further manipulations of the cell adhesion and organization at the micron level becomes evident if one looks at a randomly seeded network of nerve cells. An example is given in Figure 5.8. Rat brain stem neurons after 4 days in culture randomly adhered to the substrate developed "healthy" dendrites and axons, the "receivers" and "transmitters" for cell-cell communication, but formed a totally random network structure. In a separate experiment with two microelectrodes attached to two neighboring cells it was possible to demonstrate that these cell populations, indeed, were able to

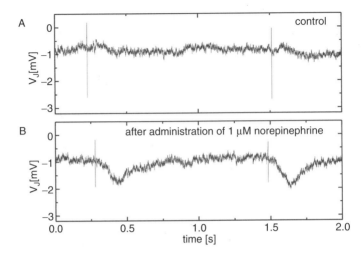

FIGURE 5.7. (A) Electrical recording from rat cardiac myocytes measured by field-effect transistors using standard recording solution. (B) Change in extracellular signal shape after addition of norepinephrine.

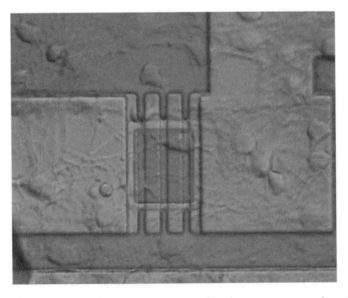

FIGURE 5.8. Micrograph of a random network of brain stem neurons cultured on the surface of a FET chip. One neuronal cell resides completely on the (triple) gate electrode of the transistor.

establish synaptic contacts to each other and, hence, were able to communicate to each other. The obtained hybrid neuronal network/ FET array structure thus could be used to monitor certain electrical network activities and the influence of external stimuli that modulate its activity pattern.

However, our focus of interest is more in the formation and characterization of well defined network architectures. Therefore, we have to develop strategies that allow us to manipulate the adhesion sites of individual neurons, control the direction and outgrowth of dendrites and axons. In addition we have to guarantee for the formation of synaptic contacts between the axon of one cell to the dendrites of other cells in a fully controlled way.

We achieve this control at the level of the adhesion layer, the interface between the microelectronic device and the adhering cell. We could demonstrate that photolithographic patterning of the cell-repulsive or adhesion layer [19] or, more recently, using the so-called micro-contact-printing (µCP) technique, indeed, give us the required control over the network formation. In the µCP approach a stamp made from poly(dimethylsiloxane) (PDMS) is "inked" with the specific biomolecule and the pattern is transferred onto the substrate which could be pre-coated with a linker molecules used for the covalent attachment of the biomolecule [15].

The result of the patterned functionalization of the neuron cell behavior is given in Figure 5.9: By the appropriate geometric design of the stripe pattern

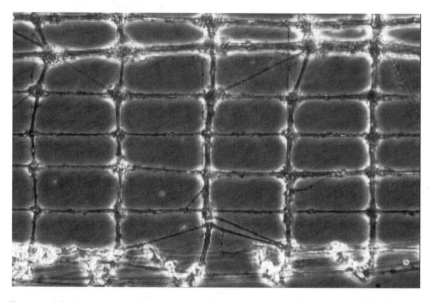

FIGURE 5.9. Controlled adhesion of neuronal cells and guided outgrowth of their dendrites and axons on a patterned substrate. The micrograph shows the response of brain stem neurones after 7 days in culture to a laminin grid pattern with 2 µm wide lines and 10 µm wide nodes crossing in 50/100 µm.

of the stamp at the micron level (which includes, among other factors, the right width of the functionalized stripes and the size and shape of the nodes at the crossing point of two stripes) we direct the individual cells to predetermined positions on the substrate. One can see from the side-by-side presentation of the arrangement of the individual gate electrodes of the FET-arrays (Figure 5.1) and the cell pattern on the printed substrate that the stamp was designed to generate cell networks that are at registry with the microelectronic recording device.

Equally important as the control over the cell position is the control over the outgrowth of the dendrites: their growth cones also show a high affinity to the surface functionalization and hence follow the stripes. This guides them directly to the neighbour cell as it is needed for a controlled network formation.

In addition to dissociated neurons the µCP technique can also be used for brain slice cultures as a cell source for cell patterning which has been evaluated using a variety of stamp parameters (track width, node size) [20]. We could demonstrate that the use of whole tissue slices yielded results similar or superior than standard dissociated cell culturing methods. When culturing ultrathin (~250 µm) brain stem slices of rats (E15-E18) on laminin patterned substrates, the proliferating neuronal cells from the slices consistently formed grid-shaped neuronal networks. The interconnections between neighbouring pairs of neurons within these artificial networks could be assessed electrically by double patch-clamp recordings and optically by fluorescence microscopy using microinjected polar tracers (Figure 5.10).

Both functional and ohmic synapses were detected. Strong evidence was found for a correlation between the pathwidth of the applied pattern and the diameter of neurites growing along these paths [21].

6.   Outlook

In order to combine the extracellular recording devices with the fascinating simplicity of microcontact printing (µCP) extracellular matrix proteins, the stamp needs to be aligned to fit to the recording sites of the microelectronic devices. We have therefore developed a setup to perform aligned µCP of extracellular matrix proteins on microelectronic devices in order to guide the growth of electrogenic cells specifically to these sensitive spots. Our system is based on the combination of a fine-placer with redesigned microstamps having a rigid glass cylinder as backbone for attachment in the alignment tool. Alignment is performed moving the device with an optical table under microscopic control of the superimposed images from stamp and device surface. After successful alignment, the stamp is brought into contact with the device surface by means of a high-precision lever. With our setup we are able to pattern with a lateral alignment accuracy of < 2 µm. Using this method aligned

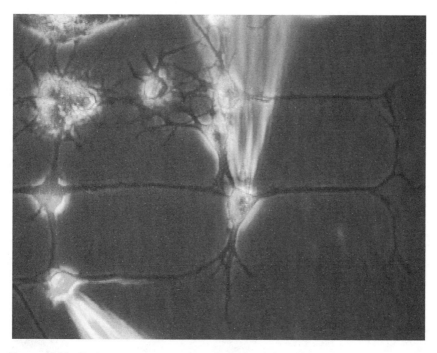

FIGURE 5.10. Brain stem neurons cultured on a laminin grid array. The three neurons are interconnected. The cells were microinjected with fluorescence dyes of different colors.

neuronal growth on patterned devices could be demonstrated using dissociated hippocampal neurons [22].

Among the many possible visions for future directions in research and product development on the basis of the presented achievements the integration with elements of microfluidics are very obvious and actually already "down the road": Any practical use of a cell-based sensor will need a liquid-handling system that has to implement microfluidic design principles. In as much as we will see "lab-on-a-chip" based devices there will be concepts developed that also give us the tools and modules to support our cell-based bio-electronic hybrid system by the required buffer supply, temperature control units, etc.

Our own efforts along these lines currently focus on a very simple feasibility study, combining the FET chip design with the various cells sitting on the recording gate electrodes (with or without lateral patterning) with an 8 micro-channel flow system (Figure 5.11) that will allow us to use multi-pipette dispensers for the simultaneous application of a 8 drugs at a time, e.g., from a combinatorial library.

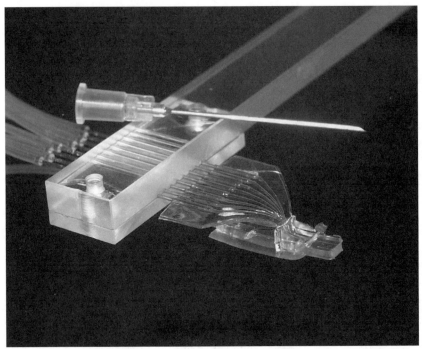

FIGURE 5.11. Polydimethylsiloxane (PDMS) device for patterned application of microfluids to patterned cells. The device consists of 8 lines of a microinjection array integrated in a base flow channel. The device is assembled from multiple PDMS parts [23].

Acknowledgements. Most of this work was done in a long-term collaboration between the Laboratory for Exotic Nanomaterials of the Frontier Research Program at RIKEN in Wako, Japan, and the Max-Planck-Institut für Polymerforschung in Mainz, Germany, which was guided by W. Knoll. It is our great pleasure to acknowledge the very competent and dedicated help of our colleagues within this collaboration: S. Britland, M. Denyer, M. Hara, J. Hayashi, M. Krause, L. Lauer, M. Matsuzawa, J. Rühe, M. Scholl, T. Siepchen, Ch. Sprössler, P. Thiebaud, A. Vogt, and C.-K. Yeung.

References

1. Thomas et al. (1972) Exp. Cell Res. 74, 61-66
2. G. W. Gross et al. (1982) J. Neurosci. Meth. 5, 13-22
3. J. Pine (1980) J. Neurosci. Meth. 2, 19-31
4. A. Sibbald (1983) Proc. Inst. Elec. Eng. I, 130, 233-244
5. P. Bergveld et al. (1976) IEEE Trans Biomed. Eng. 23, 136-144
6. P. Fromherz et al. (1991) Science 252, 1290-1293
7. A. Offenhäusser et al. (1997) Biosens Bioelectron 12, 819-826
8. U. Egert et al. (1998) Brain Res. Protoc. 2, 229-242

9. A. Blau et al. (1997) Biosens. Bioelectron 12, 883-892
10. M. Krause et al. (1999) Sens. Actuators B 70, 101
11. W. G. Regehr et al. (1989) J. Nuerosci. Meth. 30, 91-106
12. R. Weis et al. (1997) Phys. Rev. E 55, 877-889
13. S. Vassanelli et al. (1998) Appl. phys. A 66, 549-563
14. J. Rühe et al. (1999) J. Biomat. Sci. – Polymer Ed. 10, 859-874
15. M. Scholl et al. (2000) J. Neurosci. Meth. 104, 65
16. M. Denyer et al. (1999) Cell Dev. Biol. 35, 352-356
17. C. Sprössler et al. (1999) Phys. Rev. E 60, 2171
18. S. Ingebrandt et al. (2001) Biosens. Bioelectron. 16, 565 -570
19. T. Bohannon et al. (1996) J. Biomat. Sci. – Polymer Ed. 8, 19
20. Yeung et al. (2001) Neurosci. Lett. 305, 147-150
21. Lauer et al. submitted for publication
22. Lauer et al. (2001) IEEE Trans Biomed Eng 48, 838-842
23. Thiebaud et al. (2002) Biosensors&Bioelectronics 17, 87-93

Part II

Microfluidics and Lab-on-a-Chip

6

Microfabricated Chip Electrophoresis Technology for DNA Analysis

FENG XU[1,4], LIHUA ZHANG[1,5], MOHAMMAD JABASINI[1], AND YOSHINOBU BABA[1,2,3]

[1]*Department of Medicinal Chemistry, Faculty of Pharmaceutical Sciences, The University of Tokushima;*[2]*CREST, Japan Science and Technology Corporation (JST), Shomachi, Tokushima 770-8505, Japan;*[3]*Single-Molecule Bioanalysis Laboratory, National Institute of Advanced Industrial Science and Technology (AIST), Hayashi-cho 2217-14, Takamatsu 761-0395, Japan;*[4]*Analytical Instruments Division, Shimadzu Corp., Kyoto, Japan,*[5]*Furuno Electric Co. Ltd., Nishinomiya, Japan*

Abstract: In this report, two microfabricated chip electrophoresis techniques and several application studies were tested for rapid and high-resolution separation of double-stranded (ds)DNA. In one technique, low-viscosity hydroxypropylmethylcellulose-50 (HPMC-50) matrix accompanied by polyhydroxy compounds, such as mannitol, glucose, and glycerol, showed higher resolving power than conventionally and singly used HPMC-50 matrix. The new matrix is easy to be hyphenated in the future µ-TAS platforms. Another technique is through the stepwise (multi-step) filling of different concentrations of one polymer or different types of polymers to achieve high-resolution separation of both short and long DNA fragments simultaneously. The technique has good migration-time reproducibility for the analysis of restriction digest fragments and ladders. The separation application of some PCR products was performed on an Agilent 2100 Bioanalyzer. The reproducibility and accuracy of fragment sizing of a DNA ladder were satisfactory. Fast analysis of DNA polymorphisms on the human Y-chromosome was realized with the analytical time of three genomic polymorphisms on the Y-chromosome (Y *Alu* polymorphism, 47z/*Stu*I and 12 f2) to be within 110 s, respectively. A mixture of nine DNA markers on the human Y-chromosome related to examine the cause of spermatogenic failure was successfully separated with the smallest fragment size difference of 7 bp.

Key words: Microchip electrophoresis, DNA fragments, low viscosity, polyhydroxy additives, linear polymer matrix, stepwise gradient, hydroxypropylmethylcellulose, methylcellulose, polymethylmethacrylate.

1.   Introduction

The successful application of microchip electrophoresis in DNA separation derives from its strong resolving power, quick speed, and easy automation[1,2,3], especially for polymerase chain reaction (PCR) products analysis, restriction digest fragments analysis, and DNA sequencing[4,5,6,7,8,9,10].

Linear polymer-sieving matrices offer many advantages over conventional cross-linked matrices, including ease of operation and low susceptibility of bubble formation in capillaries. Generally, such matrices have high molecular-weight and viscosity[11,12,13], are difficult to be flushed and refilled in microchannels after each run, and are cumbersome to be integrated in μ–TAS platforms[14]. As an alternative, low molecular-weight polymers can form a sieving solution with low viscosity. They are easily replaced after each run and support the repetitive use of microchips. However, low molecular-weight polymers have higher entanglement threshold concentration (the concentration at which the polymer molecules begin to entangle)[15], and relatively limited resolving power[16]. This limitation can be partly overcome by introducing mannitol as an additive in HPMC-5 polymer (M_w 10 000) solution for capillary electrophoresis (CE) separation of pBR322 HaeIII restriction fragments up to 587 bp[15], though the size range needs further extending. The effect of glycerol on the performance enhancement of DNA separation was also reported previously in the separation matrices containing boric acid[17]. However, in microchip electrophoresis, there were still few reports regarding the additive-enhanced separation using low molecular-weight and thus low viscosity polymer solutions. HPMC-50 has a molecular-weight (M_w 11 500) somewhat higher than HPMC-5, and may have a wider separation capability than HPMC-5 while still keeping low viscosity. Herein, HPMC-50 solution with several polyhydroxy compounds (mannitol, glucose, and glycerol) as additives was addressed[10]. The effects of operational variables, such as polymer concentration and additive concentration, on separation in a polymethylmethacrylate (PMMA) electrophoresis microchip were investigated. This is the first part work in this report.

The conventional DNA separation is carried out in a fixed polymer concentration. However, if the distribution of DNA fragments is broad, it is difficult to obtain high resolution for both short and large fragments simultaneously. Several solutions have been proposed in CE, such as mixing different kinds of polymer[18,19], and introducing a field-strength gradient[20,21]. The stepwise variation of a polymer solution in a capillary was also realized by introducing different sieving matrices with electroosmotic flow (EOF) as the propelling force[22]. In the second part of the present report, a stepwise gradient of linear polymer matrices in microchip electrophoresis was described in the absence of EOF[9].

The detection of DNA mutations and polymorphisms is important for the characterization and diagnosis of human genetic diseases. Many techniques of mutation and polymorphism detection have been developed, such as allele-specific amplification, PCR-restriction fragment length polymorphism analysis, single strand conformation polymorphism analysis, heteroduplex assay, and chemical or enzymatic cleavage of mismatches, *etc.* Based on slab-gel electrophoresis, most of these methods are time-consuming[23,24]. CE has also been successfully applied to mutation and polymorphism analysis[12,25]. In this report, fast analyses of three DNA polymorphisms on the human Y-chromosome, Y *Alu* Polymorphism (YAP), 47z/*Stu*I, and 12f2, were carried out by using high-throughput microchip electrophoresis. The reproducibility and accuracy of the size of each allele were validated to prove the reliability of microchip electrophoresis in polymorphism analysis. In addition, the separation resolution is examined by analyzing a mixture of 9 DNA markers on the human Y-chromosome[3].

2.   Materials and Methods

2.1   Instrumentation

Microchip electrophoresis was performed on an SV1100 Microchip CE system (Hitachi Electronics Engineering Co., Ltd., Tokyo, Japan) equipped with an LED detector (exciting at 470 nm and collecting fluorescence at 580 nm), an SV1100-02G high voltage device, and an SV1100a software for data acquisition. The Hitachi PMMA plastic chip had the channel cross-section of 100 μm (width) × 30 μm (depth) × 30 mm (effective separation length). The distances from the crossing point to the reservoir 1 (buffer), reservoir 3 (buffer or analysis well), reservoir 2 (sample outlet) and reservoir 4 (sample inlet) were 5.7, 37.5, 5.2 and 5.2 mm, respectively. Reservoir positions are depicted in Figure 6.1.

Some application studies were carried out on an Agilent 2100 Bioanalyzer (Agilent Technologies, Waldbronn, Germany), using epifluorescent detection with a semiconductor laser emitting at 630 nm. Agilent chips with 12 sample wells, 3 gel-dye mix wells and one external marker well, were made from soda lime glass with the microchannel dimension of 50 μm (width) × 10 μm (depth) × 15 mm (effective separation length).

2.2   Reagents

Reservoir	Injection voltage (V)	Separation voltage (V)
1(buffer)	0	0
2(sample outlet)	600	180
3(buffer or analysis)	0	1300
4(sample inlet)	0	180

Polyhydroxy compounds (mannitol, glucose, and glycerol) were purchased from Sigma (St. Louis, MO, USA). DNA ladders and restriction digest fragments were purchased from different routes, for example, 20 bp ladder and

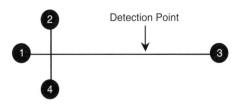

FIGURE 6.1. Schematic diagram of Hitachi PMMA chip with applied injection and separation voltages in each reservoir. The applied field strength for separation was 300 V/cm.

100 bp ladder from GenSura Laboratories (Del Mar, CA, USA), 10 bp ladder and low DNA mass ladder from Life Technologies (Gaithersburg, MD, USA), ΦX174 *Hae*III from Takara Shuzo Co. (Shiga, Japan), and pBR322 *Hae*III from Sigma. They were diluted in deionized water at a concentration of about 1 µg/ml unless otherwise stated and stored at −20 °C until use.

HPMC-50, HPMC-4000 (viscosity of 2% solution at 20 °C, 4000 cP) and methylcellulose (MC) from Sigma were dissolved in 1 × TBE buffer (89 mM Tris-borate, 2 mM EDTA, pH 8.3, Sigma), 1 × TE buffer (89 mM Tris, 8.9 mM EDTA, pH 8.0, Sigma), or TB buffer (50 mM Tris-boric acid), and stored overnight at −4 °C until the solution appeared homogenous and transparent. The polymer solution containing a polyhydroxy additive was prepared by weighing the additive of desired amount and dissolving it in HPMC-50 solutions. Prior to use, the final concentration of 0.5 µg/ml ethidium bromide (EtBr, from Nippon Gene, Tokyo, Japan) or 0.5 µM YOPro-1 (Molecular Probes, Eugene, OR, USA) was mixed into polymer solutions. DNA 500 and 7500 Assay Kits, including dye, gel matrix, DNA marker and ladder solutions, were bought from Agilent Technologies.

2.3   Procedure

The polymer matrix was infused from the buffer well (reservoir 3 in Figure 6.1) into the microchannels of the Hitachi chip using a syringe. Sample injection was accomplished by loading DNA sample in the sample inlet reservoir, applying 600 V at the sample outlet reservoir and grounding other three reservoirs for 60 s. Sample separation at an applied field strength of 300 V/cm was performed by applying 1300 V at the analysis reservoir, 180 V at the two sample reservoirs, and grounding the buffer reservoir for 300 s. The applied field strength other than 300 V/cm was easily adjusted by applying certain voltages at the analysis reservoir and the sample reservoirs. Each measurement was carried out in duplicate at ambient temperature.

A stepwise gradient of polymer matrices was formed by introducing higher viscosity matrix first and lower viscosity matrix later through the analysis reservoir, to prevent the easy movement of lower-viscosity matrix. In addition, the resistance of the higher viscosity matrix can also prevent from mixing with subsequent lower viscosity matrix and form a stable gradient.

Genomic DNAs were prepared according to[26,27,28,29]. The preparation of Agilent chips was carried out according to the instruction of Agilent kits.

3.   Results and Discussion

3.1   Separation in Low Viscosity Matrix and Enhancement of Resolution by Polyhydroxy Additives

The EOF in Hitachi chips was little and could be effectively minimized by HPMC-50 solutions, without using other surface modification agents. Fast separation could be reached at an applied field strength of 300 V/cm, above which the separation efficiency drops drastically, due to the obvious increase of Joule heating in the microchannels.

The 2% HPMC-50 solution can separate 10 of all 11 fragments in the ΦX174 HaeIII sample except for two adjacent fragments 271/281 bp. Further raising HPMC-50 concentration up to 3% cannot resolve 271 bp and 281 bp, either; however the viscosity has increased to a higher value (> 60 cP). Hence, HPMC-50 solution itself owns no enough resolving power for ΦX174 HaeIII.

Subsequently, mannitol, glucose, and glycerol were added into HPMC-50 solutions as additives. An approximately linear relationship between the migration time of fragments and the additive concentration is observed (data not shown). The resolution of all 11 fragments in ΦX174 HaeIII was improved with the increasing of additive concentration, especially for the resolution of a pair of 271/281 bp fragments ($R_{s,271/281}$). Figure 6.2 illustrates the different degree of $R_{s,271/281}$ increasing as a function of the concentrations of additives and HPMC-50. For the three additives, the $R_{s,271/281}$ increased fast at 1.0-2.0% polymer concentration, and slowly at 2.0-3.0% concentration. Nevertheless, the additive concentration to reach optimal separation was fixed, i.e., 8% for mannitol and glucose, and 10% for glycerol, irrespective of the difference in the HPMC-50 concentration. When 2% HPMC-50 and polyhydroxy additives at optimum concentrations were used as the sieving matrices, all the $R_{s,271/281}$ values surpassed two-time the $R_{s,271/281}$ values at 3% HPMC-50 solution (in the absence of additives), which means that polyhydroxy additives can make it possible to separate DNA well even in a lower polymer concentration.

Figure 6.3 illustrates the comparison of microchip separation of ΦX174 HaeIII DNA in 2% HPMC-50 solution, with or without additives. The 271/281 bp fractions were well separated by utilizing the additive-enhanced solutions, and the theoretical plate efficiency (N) reached 3×10^6 plates/m. The whole separation was completed within 170 s at 300 V/cm. Such a high

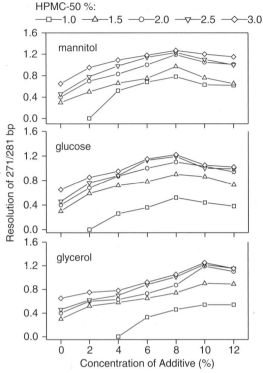

FIGURE 6.2. Resolution of the 271/281 bp fragments *versus* the concentration of polyhydroxy additives (mannitol, glucose, and glycerol) at various HPMC-50 concentrations. Numbers in the legend correspond to the HPMC-50 concentration. Conditions: 0.5 µM YOPro-1 in 1 × TBE buffer; 20 °C; Sample, 1 µg/ml ΦX174 *Hae*III DNA restriction fragments; injection, 600 V at reservoir 2 and 0 V at other three reservoirs for 60 s; applied field strength, 300 V/cm. Reproduced from[10], with permission.

efficiency can well satisfy the separation of other DNA fragments. Figure 6.4 is the separations of another digest fragments, pBR322 *Hae*III, with baseline resolution of fragments differed by only 6 bp (*e.g.*, a couple 51/57 bp) to 8 bp (*e.g.*, a couple 184/192 bp). In these data, the smaller bands, 8, 11, 18 and 21 bp, were not seen owing to very low detectability.

Of three additives, both mannitol and glucose have six hydroxyl groups, while glycerol has only three hydroxyl groups, resulting in the optimal concentration of glycerol higher than those of mannitol and glucose. At their optimal conditions, there is no obvious difference in the resolution and theoretical plate number, meaning that polyhydroxy compounds may own the similar effect on resolution enhancement.

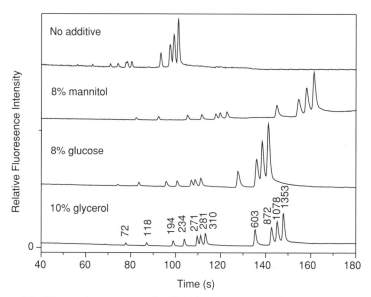

FIGURE 6.3. Electropherograms of ΦX174 Hae*III* restriction fragments on the Hitachi chip using 2% HPMC-50 / 1 × TBE solution without or with certain amounts of polyhydroxy additives. Numbers on electropherograms correspond to the size of DNA fragments in bp. Other conditions were the same as in Figure 6.2. Reproduced from[10], with permission.

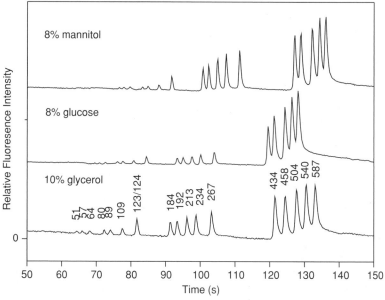

FIGURE 6.4. Electropherograms of pBR322 *Hae*III restriction fragments on the Hitachi microchip using 2% HPMC-50 / 1 × TBE solution enhanced by three polyhydroxy additives. Numbers on electropherograms correspond to the size of DNA fragments in bp. Applied field strength was 300 V/cm. Other conditions were the same as in Figure 6.2.

Previously, boric acid was thought to be an indispensable compound for the resolution enhancement by polyhydroxy compounds[15,17]. Whereas, our results using 1 × TE buffer with 2% HPMC-50 also show remarkably enhanced resolution by mannitol, glucose or glycerol, though boric acid is absent in the buffer (data not shown). At the optimal concentration of the additives of 6% mannitol and glucose and 8% glycerol in 2% HPMC-50 matrix, the $R_{s,271/281}$ values all increase to two times the $R_{s,271/281}$ in 3% HPMC-50 matrix.

The entanglement threshold concentration of HPMC-50 is about 3% for all polyhydroxy-enhanced matrices[10]. The additives do not evidently change the viscosity of the matrix due to the molecular-weights of the additives much lower than that of HPMC-50. So the 2% HPMC-50 solution, both with and without additives, belongs to the non-entanglement solution, and its viscosity is only about 40 cP at 25 °C. Compared to the conventionally used 1.2% HPMC-4000 solution whose viscosity is 335 cP, the additive-modified HPMC-50 matrix is easy for handling in future μ-TAS applications. At the same time, the plastic Hitachi chip could be repetitively used for dozens of times by simply replacing the polymer solution between runs, and was also disposable due to acceptable cost. The probable explanation to the resolution enhancement by additives in boric acid-deficiency buffer is that the remaining hydroxyl groups of HPMC-50 exhibit strong H-bonding to the additives which own many hydroxyl groups. These affinities increase the frictional characteristics of the dilute HPMC-50 matrix and the chance of transient coupling of DNA with polymer, which may shape more delicate networks in the solution suitable for the separation of DNA with different sizes. The results show that the presence of boric acid in the dilute HPMC-50 matrix is not a prerequisite for the separation.

3.2   Stepwise Gradient for High-Resolution Separation of DNA

Low polymer concentration is useful for the separation of large DNA fragments, while high concentration is helpful for the separation of small fragments. So the coexistence of a range of polymer concentrations is expected to be ideal for obtaining good separation of DNA with broad fragment distribution.

Figure 6.5 shows electropherograms of ΦX174 HaeIII in both isocratic and stepwise gradient modes. Baseline separation can be obtained for 1078/1353 bp in isocratic mode with 0.3% methylcellulose (MC), while the separation of 271/281 bp is very poor. The complete separation of 271/281 bp can be realized by increasing the MC concentration to 1.0%, but the resolution for 1078/1353 becomes worse. The stepwise gradient of these two matrices at an optimal volume ratio is expected to provide high resolution for all fragments simultaneously. As seen from Figure 6.5, better separation for both

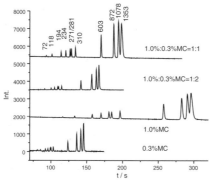

FIGURE 6.5. Electropherograms of ΦX174 HaeIII digest fragments in isocratic and gradient modes using the Hitachi chip. Numbers on electropherograms correspond to the size of DNA fragments in bp. Experimental conditions: 0.5 μg/ml EtBr in 50 mM Tris-boric acid buffer; applied field strength 117 V/cm; sample, 2 μg/mL ΦX174 HaeIII digest fragments. Reproduced from[9], with permission.

small and large fragments could be obtained with a stepwise matrix of 1.0% and 0.3% MC at the volume ratio of 1:1. However, the resolution of 1078/1353 bp could be further improved. With the increase of 0.3%MC in the sieving matrix to the volume ratio of 1.0% MC : 0.3% MC = 1:2, the optimal separation of all components was achieved.

In six consecutive injections into one channel, the RSD of migration times of ΦX174 HaeIII fragments is less than 0.7%, demonstrating the good reproducibility of microchip electrophoresis in stepwise gradient modes. A multiple stepwise gradient of matrices was also realized based on the changes of both the type and the concentration of polymers (Figure 6.6). Baseline separation was achieved for all fragments within 160 s using the stepwise separation, with higher resolution than any of the isocratic modes.

3.3 Determination of DNA Polymorphism on the Human Y-Chromosome

The reproducibility and accuracy of DNA analysis in an Agilent 2100 Bioanalyzer were checked by using DNA 500 Assay Kit for the separation of 25 bp to 500 bp fragments. All fragments ranging from 20 to 330 bp in 10 bp DNA ladder were successfully separated within 85 s except for the 10 bp coeluted with the lower marker (15 bp) (see Figure 6.7). Good reproducibility was obtained with RSD of fragment sizes less than 2.5%, which was contributed to the automatic calibration of migration times in each run based on the double internal markers added in each sample. The intraday quantitation validations of DNA concentration was performed by using low DNA mass ladder with known concentration of each fragment from the manufacturer

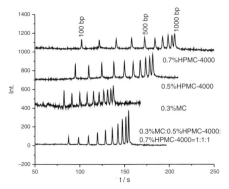

FIGURE 6.6. Electropherograms of 100 bp DNA ladder in isocratic and three-step gradient modes using the Hitachi chip. Experimental conditions as in Figure 5; sample: 1μg/mL 100 bp DNA ladder. Reproduced from[9], with permission.

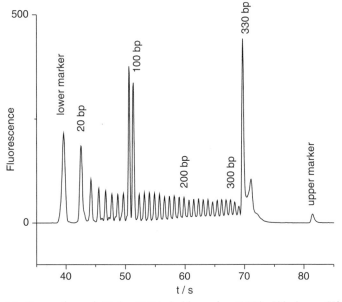

FIGURE 6.7. Separation of 10 bp DNA ladder using DNA 500 Assay Kit in the Agilent 2100 Bioanalyzer. Reproduced from[3], with permission.

and the DNA 7500 Assay Kit, which was suitable for fragment sizing from 100 to 7500 bp. The quantitation accuracy of fragment concentration was within 1.3%. The baseline separation of all components with fragment size ranging from 100 to 2000 bp could be realized within 80 s (data not shown).

The human Y-chromosome has unique characteristics in genetics, because it is a single haploid unit in the human genome that is only passed from father to son. The Y-chromosome represents the patrilineal contribution to the male genome[28]. DNA markers residing in the non-recombining portion of the human Y-chromosome have been shown to be useful for tracing male-specific gene flow and also in human evolution studies[30,31]. Hence, the development of high speed analytical methods for the polymorphisms on the Y-chromosome is necessary. In this report, three kinds of polymorphisms were analyzed by using Agilent microchip electrophoresis.

The first polymorphism was the YAP at locus DYS287, which is a simple polymorphism resulting from the insertion of *Alu* element on the long arm of the Y-chromosome and has proven to be useful for human population studies, since the distribution of the YAP differs among different populations[27,32]. During the PCR procedure for the sample preparation, the same primer sets have been used for YAP polymorphisms to obtain either YAP⁻ without Alu repeat (150 bp), or YAP⁺ with Alu repeat (455 bp) and the PCR products are further analyzed by microchip electrophoresis. YAP⁻ and YAP⁺ could be distinguished quite well within 90 s. The distributions of YAP⁻ and YAP⁺ were estimated by calculating their percentages in the 39 analyzed samples, which are 64% and 36%, respectively.

The second polymorphism sample was the probe 47z, which detects a polymorphism on the short arm of the Y-chromosome and long arm of the X chromosome[33]. The amplified genomic DNA (370 bp) could be distinguished by the digestion with *Stu*I. One of the alleles might be digested into three fragments, 370 bp, 270 bp and 100 bp, while the other allele can not be digested so that only one fragment of 370 bp could be obtained. Figure 6.8 shows that all the digested products could be well separated within 90 s. The reproducibility and accuracy of DNA sizing have proven quite well.

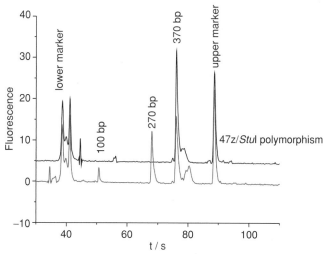

FIGURE 6.8. Separation of 47z/*Stu*I polymorphism with DNA 500 Assay Kit in the Agilent 2100 Bioanalyzer. Reproduced from[3], with permission.

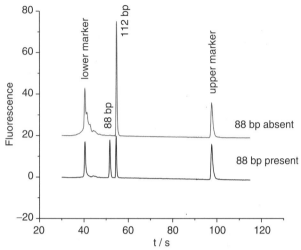

FIGURE 6.9. Separation of 12f2 polymorphism using DNA 500 Assay Kit in Agilent 2100 Bioanalyzer. Reproduced from[3], with permission.

The percentage of the allele that could be digested by *Stu*I is about 14% and the other keeping intact after digestion is 86%.

The 12f2 marker, the third analyzed sample, is located on the long arm of the Y-chromosome on the AZFa region and is also shown to be polymorphic among populations. During PCR process, two primer sets have been used, one to detect the absent or deletion of the 88 bp of the 12f2 marker, while the second set to amplify a Y specific marker, tat, with its 112 bp to serve as an internal control for each sample. As shown in Figure 6.9, the separation of the two alleles could be easily realized.

In order to examine the maximum resolution of the Agilent 2100 Bioanalyzer, a mixture of 9 DNA markers on the human Y-chromosome related to spermatogenic failure[34] was analyzed. From Figure 6.10, it could be seen that even fragments with 7 bp differences were successfully separated within 110 s, good reproducibility of DNA sizing could be obtained with RSD less than 1.2%.

4.   Conclusion Remarks

A dilute, low molecular-weight, and low viscosity sieving matrix has been explored for excellent separation of dsDNA mixtures on PMMA based microchips by introducing polyhydroxy additives (mannitol, glucose, and glycerol) into 2% HPMC-50 matrix. The additives at the optimal concentration will not increase the viscosity of HPMC-50, but will raise the separation resolution in dilute polymer solutions. In addition, separation using TE

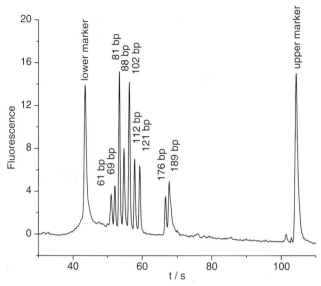

FIGURE 6.10. Separation of DNA markers on the human Y-chromosome in the Agilent 2100 Bioanalyzer. Reproduced from[3], with permission.

buffer in the absence of boric acid also exhibits a notably increased perform-ance, presumably due to formation of H-bonding interactions of polyhy-droxy additives with HPMC-50. Boric acid is not a prerequisite in polyhydroxy-enhanced HPMC-50 solution for DNA separation.

A stepwise gradient of linear polymeric matrices for microchip elec-trophoresis was utilized. The multiple steps of sieving matrices with different concentrations or different polymers in the separation channel provides higher resolution for complex DNA samples containing a wide range of frag-ment sizes than that achieved by isocratic systems. The technique allows the volume ratio of steps to be set optimally according to the distribution of DNA fragments.

The analysis of DNA polymorphisms on the human Y-chromosome was successfully achieved within 110 s. Through the analysis of 9 DNA markers in one run, high resolution with only 7 bp difference was obtained. All these results demonstrate that DNA polymorphisms can be readily and rapidly analyzed by microchip electrophoresis.

Acknowledgements. This study was partially supported by CREST of JST (Japan Science and Technology Corporation), a Grant-in-Aid from the New Energy and Industrial Technology Development Organization (NEDO) of the Ministry of Economy, Trade and Industry, Japan, a Grant-in-Aid for

Scientific Research from the Ministry of Health and Welfare, Japan, a Grant-in-Aid for Scientific Research from the Ministry of Education, Science and Technology, Japan, and a Grant-in-Aid from Shimadzu Corp., Japan. The authors would like to thank Dr. Lihua Zhang for providing some materials.

References

1. Wooley, A. T., Mathies, R. A., Proc. Natl. Acad. Sci. USA 1994, 91, 11348–11352.
2. Doyle, P. S., Bibette, J., Bancaud, A., Viovy, J.-L., Science 2002, 295, 2237.
3. Jabasini, M., Zhang, L., Dang, F., Xu, F., Almofli, M. R., Ewis, A. A., Lee, J., Nakahori, Y., Baba, Y., Electrophoresis 2002, 23, 1537–1542.
4. Wilding, P., Shoffner, M. A., Kricka, L. J., Clin. Chem. 1994, 40, 1815–1818.
5. Khandurina, J., McKnight, T. E., Jacobson, S. C., Waters, L. C., Foote, R. S., Ramsey, J. M., Anal. Chem. 2000, 72, 2995–3000.
6. Cohen, A. S., Najarian, D., Smith, J. A., Karger, B. L., J. Chromatogr. 1988, 458, 323–333.
7. Kheterpal, I., Mathies, R. A., Anal. Chem. 1999, 71, 31A–37A.
8. Wei, W., Yeung, E. S., Anal. Chem. 2001, 73, 1776–1783.
9. Zhang, L., Dang, F., Baba, Y., Electrophoresis 2002, 23, 2341–2346.
10. Xu, F., Jabasini, M., Baba, Y., Electrophoresis In press.
11. Jacobson, S. C., Ramsey, J. M., Anal. Chem. 1996, 68, 720–723.
12. Baba, Y., Ishimaru, N., Samata, K., Tsuhako, M., J. Chromatogr. A 1993, 653, 329–335.
13. Figeys, D., Arriaga, E., Renborg, A., Dovichi, N. J., J. Chromatogr. A 1994, 669, 205–216.
14. Vazquez, M., Schmalzing, D., Matsudaira, P., Ehrlich, D., McKinley, G., Anal. Chem. 2001, 73, 3035–3044.
15. Han, F., Huynh, B. H., Ma, Y., Lin, B., Anal. Chem. 1999, 71, 2385–2389.
16. Braun, B., Blanch, H. W., Prausnitz, J. M., Electrophoresis 1997, 19, 1994–1997.
17. Cheng, J., Mitchelson, K. R., Anal. Chem. 1994, 66, 4210–4214.
18. Fung, E. N., Yeung, E. S., Anal. Chem. 1995, 67, 1913–1919.
19. Salas-Solano, O., Carrilho, E., Kotler, L., Miller A. W., Goetzinger, W., Sosic, Z., Karger, B. L., Anal. Chem. 1998, 70, 3996–4003.
20. Luckey, J. A., Smith, L. M., Anal. Chem. 1993, 65, 2841–2850.
21. Endo, Y., Yoshida, C., Baba, Y., J. Biochem. Biophys. Methods 1999, 41, 133–141.
22. Chen, H.-S., Chang, H.-T., J. Chromatogr. A 1999, 853, 337–347.
23. Kheterpal, I., Mathies, R. A., Anal. Chem. 1999, 31A–37A.
24. Albarghouthi, M. N., Barron, A. E., Electrophresis 2000, 21, 4096–4111.
25. Piggee, C. A., Karger, B. L., Methods Mol. Biol. 2001, 163, 89–94.
26. Hammer A. F., Mol. Biol. Evol. 1994, 11, 749–761.
27. Hammer, M. F., Horai, S., Am. J. Hum. Genet. 1995, 56, 951–962.
28. Shinka, T., Tomita, K., Toda, T., Kotliarova, S. E., Lee, J., Kuroki, Y., Jin, D. K., Tokunaga, K., Nakamura, H., Nakahori, Y., J. Hum. Genet. 1999, 44, 240–245.
29. Blanco, P., Shlumukova, M., Sargent, C. A., Jobling, M. A., Affara, N., Hurles, M. E., J. Med. Genet. 2000, 37, 752–758.
30. Poobo, S., Science 1995, 268, 1141–1142.
31. Thomas, M. G., Ben-Ami, H., Skorecki, K., Parfitt, T., Bradman, N., Goldstein, D. B., Nature 1998, 394, 138–140.

32. Persichett, F., Blasi, P., Hammer, M., Malaspina, P., Iodice, C., Terranato, L., Novelletto, A., Ann. Hum. Genet. 1992, 56, 303–310.
33. Nakahori, Y., Tamura, T., Yamada, M., Nakagome, Y., Nucleic Acid Res. 1989, 17, 2152.
34. Nachamkin, I., Panaro N. J., Li, M., Ung H., Yuen, P. K., Kricka L. J., Wilding P., J. Clin. Microbiol. 2001, 39, 757–757.

7

Microfabrication and Application of Recessed Gold Electrodes in Microchip Electrophoresis System

CHONGGANG FU
Department of Chemistry, Liaocheng University, 252059, Liaocheng, P.R. China

Abstract: Based on photolithographic technique, a simple and novel way to construct micro recessed gold electrodes (μRGE) using recordable compact disk was described in this paper. μRGE were characterized by a remarkable versatility, great availability, good reproducibility, insensitive response to the fluctuation of flow speed, and very low price. The applicability of μRGE in microchip electrophoresis system was demonstrated for the anodic detection of dopamine and catechol. A high sensitivity and low noise was obtained, consequently a lower detection limit (0.31 for dopamine and 0.62 μ mol/L for catechol) was achieved, which was far lower than that reported in the literature.

Key words: Microfabrication, Microchip, electrophoresis, electrochemical detection.

1. Introduction

Microfabricated capillary electrophoresis(CE) chip have received great interests in recent years.[1,2]. Such miniaturized devices have several advantages over the benchtop separation systems. e.g. higher throughput could be achieved while consuming only picoliters of sample volume. Currently, such electrophoresis chip rely primarily on Laser-induced Fluorescence (LIF) to obtain high detection sensitivities. Yet, LIF detection typically requires derivatization of the analytes with a fluorophore. Moreover, only a select number of wavelengths can be used for excitation. Recently, Electrochemistry (EC), an alternative detection technique has witnessed a great success in microchip CE system because of its high sensitivity, tunable selectivity, independence of path length, and inherent miniaturization. In the chip-based CE-EC system, the configuration and position of the working electrode play an key role for achieving high detection sensitivity

132

and separation efficiency. Woolley[3] first reported on capillary electrophoresis chip with integrated electrochemical detection, based on the photolithographic placement of the working electrode positioned outside the exit of the electrophoretic separation channel. Wang et al[4] described an chip-formatted eletrophoretic system with electrochemical detector based on sputtering the working electrode directly onto the channel outlet. The above two method for electrode preparation are both involved the access to the complex and expensive equipments. Recently, Wang et al[5] introduced easily-performed electroless deposition procedure for fabricating gold electrode just outside the exit of separation channel to serve as a working electrode. Although the chip-integrated electrode eliminates the need for a elaborate channel-electrode alignment, electrode cleaning and replacement due to severe surface poisoning and damage become hardly possible. For this reason, the stand-alone electrode seems promising. Wang et al[6] described a thick-film electrode detector for eletrophoretic chip. This coupling obviates the need for permanent attachment of the electrode, allows a convenient surface modification, a fast replacement of passivated electrode, or the comparison and use of different electrode materials. In this paper, we described a convenient way of construction of micro recessed gold electrode using recordable compact disks, investigated its reproducibility and stability, and examined primarily its utility as the detector of electrophoretic chips.

2.   Experimental

2.1   Apparatus and Reagents

A home-built high-voltage power supply, with a adjustable voltage range between 0 and 2000V, was used for the electrophoretic separation. Amperometric detection was performed with an Electrochemical Analyzer 812 (CH Instruments, Austin, TX), which was connected to a Pentium 1.7G computer with 128M RAM. The glass microchannel chips, fabricated by combining photolithographic, wet-chemical etching and thermal bonding techniques, were made in the present laboratory using Au/Cr-coated glass slide.

All the solutions used were prepared with tri-distilled water, all the reagents were of analytical grade except the specially indicated. Dopamine, catechol and MES were obtained from Sigma, Potassium chloride, Potassium ferricynide was obtained from Beijing Chemicals Factory.

2.2   Preparation of μRGE

The electrodes used throughout this work were constructed with small ports of recordable compact disks (Kodak, bought from the local electric market).

The whole procedure is as follows: Firstly, a whole CD was immersed in concentrated nitric acid for 2 min, then took out and thoroughly rinsed with tap water to remove the plastic protective film from its surface. Secondly, the disk was cut into as many as rectangular slices (1.5×0.5 cm) as possible with a large scissors. Each slice was sequentially rinsed with isobutnol and de-ionized water and dried in oven at 110 °C for 30 min. Thirdly, a layer of negative photoresist is spin-coated on the upper and side face of a slice to completely cover the gold layer, then dried at 90 °C for 30 min, after that the photoresist in two circular regions was exposed for 80s under UV lamp. The exposed regions were defined by the printed pattern on a photomask made from a transparent plastic film. The smaller region (40μm in diameter) served for working electrode; the larger (2mm in diameter) for electrical contact. After development in petroleum ether, the slice was dried at 90 °C for 30 min. Electrical contact was performed with a copper wire using conductive silver paint. Thus a μRGE was obtained with 5μm recession.

3.   Results and Discussion

3.1   Electrochemical Properties of μRGE

Potassium ferricynide as a model compound was used for examining the electrochemical properties of a μRGE. Figure 7.1 shows a cyclic voltammogram of a μRGE in 5 mmol/L $K_4Fe(CN)_6$ and 0.5mol/L KCl. This diagram is of zigzag, which is typical property of a microelectrode. The dominating process

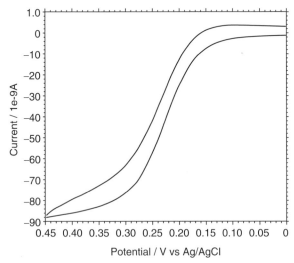

FIGURE 7.1. Cyclic voltammogram of a μRGE in 5 mmol/L $K_4Fe(CN)_6$ and 0.5mol/L KCl.

of mass transport in static solution is radial diffusion for microelectrode, whereas linear diffusion for macroelectrode. Generally, the rate of radial diffusion is far greater than that of linear diffusion. Therefore, the microelectrode is more prone to attain a steady state of mass transport; a larger limiting current density could be obtained. Moreover it is hardly affected by convection. So the microelectrode is particular suitable for the microfluidic system.

The effect of the potential scan rate on the electrode response was examined in the range of 10-500 mV/s. As shown in Figure 7.2, the limiting current doesn't significantly vary with increasing scan rate up to 100 mV/s. After that the limiting current rapidly increase with the scan rate, but the relationship doesn't obey Randles equation. This phenomenon could be explained as following: In case of lower scan rate, the diffusion layer is relatively thick, so the radial diffusion plays an important role. In this case, the limiting current i_l should obey the modified Bond[7] equation. That is:

$$i_l = \frac{AnFcD}{l + r} \tag{1}$$

where n represents number of electrons transferred, F the Faradic constant, D the diffusion coefficient, C the concentration of the electroactive species in the bulk of the solution. r and A the radius and surface area of the working electrode, l the depth of the recession. As shown in Equation(1), i_l is independent of the scan rate. When the scan rate further increase, the diffusion

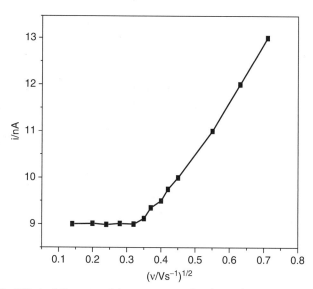

FIGURE 7.2. Effect of the potential scan rate on the electrode response.

layer is relatively thin, both radial and linear diffusion process determine the attitude of limiting current, representing a quasi-stable state of mass transport. Expectedly, As enough high scan rate attained, the electrode response would follow Randles equation[8]. i.e.

$$i_l = 2.69 \times 10^5 \, n^{3/2} \, A D^{1/2} \, v^{1/2} \, c \tag{2}$$

where v is the scan rate. But it is a pity that experiments using higher scan rate than 500mV/s were not able to be performed because of limitation of the instrumentation used (allowable maximum scan rate 500 mV/s).

3.2   Stability and Reproducibility of μRGE

μRGEs fabricated using photolithographic technique and versatile recordable gold CDs possess the advantages of great availability, very low price and easy-to-fabrication. The reproducibility of eleven parallelly made μRGEs were examined using cyclic Voltammetry. It was found that the variance coefficients of limiting current were within 5%, indicating that the fabrication method is highly reproducible.

The robustness and stability of a μRGE was studied by periodically determining the area-normalized capacitance (C) over several successive days. If there is no Faradic current, the value of C could be obtained according to the following equation:

$$C = \frac{I_c}{vA} \tag{3}$$

where A is the apparent area of working electrode. If the seal of a μRGE is perfect, that is the insulation layer of photoresist around the recession has no deficiency, the value of C would always remain constant. Experimental results show that the capacitance of the studied μRGE didn't significantly vary during seven successive days of immerse in 0.5 mol/L KCl solution. Longer time immerse results irreproducibility probably due to the damage of the insulation layer. Better results could be obtained if the μRGE was stored in dryness.

3.3   Application of μRGE in Electrophoresis Chip System

Figure 7.3 displays a electropherogram for an equimolar mixture of dopamine and catechol (each at 1×10^{-4} mol/L) using a μRGE as the detection electrode. The primary results show that the μRGE display well-defined concentration dependence. The calibration curve was linear with sensitivities of 0.15 and 0.08 nA/μM for dopamine and catechol. Based on three-time signal-to-noise ratio, detection limits of 0.31 and 0.62μM was obtained for dopamine and catechol respectively. Such values were lower than those reported in the literature[3]. So low detection limits are attributed to the insensitiveness

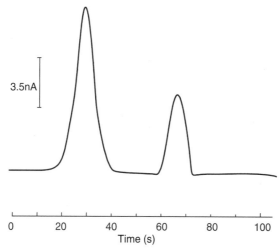

FIGURE 7.3. Electropherogram for an equimolar mixture of dopamine and catechol. Condition: 25mM MES buffer (pH6.8), separation voltage, 1800V; amperometric detection at 0.70V vs. Ag/AgCl.

to hydromechanical conditions for μRGEs. More extensive applications of μRGEs.in electrophoresis chip are in progress in our laboratory.

References

[1] V. Dolnik, S. Liu, S. Jovanovich, Electrophoresis, 2000,21,41-54
[2] N. A.Lacher, K.E. Garrison, R.S. Martin, S.M. Lunte, Electrophoresis, 2001,22,2526-2536
[3] A.T. Woolley, K. Lao, A.N. Glazer, R.A. Mathies, Anal. Chem. 1998,70,684-688
[4] J. Wang, B. Tian, E. Sahlin, Anal. Chem. 1999,71,3901-3904
[5] A. Hilmi, J.H.T. Luong, Anal. Chem. 2000,72,4677-4682
[6] J. Wang, B. Tian, E. Sahlin, Anal. Chem. 1999,71,5436-5440
[7] R. Lenigk, H. Zhu, T-C. Lo, R. Renneberg, Fresenius J. Anal. Chem. 1999,364,66-71
[8] R. Greef, R. Peat, L.M. Peter, D. Pletcher, J. Robinson, Instrumental Methods in Electrochemistry, Ellis Horwood Limited, 1985.

8

Fast Screening of Single-Nucleotide Polymorphisms Using Chip-Based Temperature Gradient Capillary Electrophoresis

PENG LIU[1,2*], WAN-LI XING[1,2*], DONG LIANG[1,2], GUO-LIANG HUANG[1,2], AND JING CHENG[1,2**]
[1] *National Engineering Research Center for Beijing Biochip Technology, Beijing, 100084, P. R. China*
[2] *Department of Biological Sciences & Biotechnology, Tsinghua University, Beijing, 100084, P. R. China*

Abstract: Recently, the analysis of single-nucleotide polymorphisms has attracted much attention. Although many techniques have been reported, new methods with high resolving power, low-cost and fast speed are still in demand. We present a fast SNP detection scheme using chip-based temperature gradient capillary electrophoresis to separate the homoduplex and heteroduplex PCR products which contain one or two SNP sites. The total time of a single run was only 8 minutes.

Key words: Single-nucleotide polymorphisms, temperature gradient, capillary electrophoresis chip.

1.   Introduction

As the most common type of human genetic variation, single-nucleotide polymorphisms (SNPs) have attracted much attention. It is estimated that there is almost 1 SNP/1000bp [1]. SNPs are important for understanding the relationship between genetic variants and diseases. It can also be used for identification purposes, such as forensics. Low-cost, reliable, fast speed and high-throughput methods for analyzing SNPs become increasingly more important [2].

Direct sequencing of a gene is the ultimate way of identifying the variants. However, this approach is not commonly used because of its high cost and long duration. For this reason, many other methods have been developed, such as DNA chip [3], mass spectroscopy [4]. The techniques based on

---

* Joint first authors with equal contributions; ** Corresponding author

conformational differences of DNA are very important in this field. They include Single-strand Conformational Polymorphism (SSCP) [5], Consistent Denaturant Gel Electrophoresis (CDGE) [6], Denaturant Gradient Gel Electrophoresis (DGGE) [7], Temperature Gradient Gel Electrophoresis (TGGE) [8], Denaturant High-performance Liquid Chromatography (DHPLC) [9], etc. In order to improve their speed and separation efficiency, these methods have been adapted to capillary electrophoresis, including Consistent Denaturant Capillary Electrophoresis (CDCE) [10, 11], Denaturant Gradient Capillary Electrophoresis (DGCE) [12], and Temperature Gradient Capillary Electrophoresis (TGCE) [13].

Chip-based capillary electrophoresis is a powerful separation technique and has become an attractive alternative to slab-gel electrophoresis and capillary electrophoresis in many fields [14]. Several groups have used this approach to detect SNPs, yet found that the separation efficiency and convenience in practice are still inadequate [11]. In the current study, we developed a method for conducting the SNPs detection by using a normal crossed-channel electrophoresis chip which is featured with temperature gradient programmed by a computer.

2.   Experimental Section

2.1   Chemical Reagent

The sieving matrix used was a 2.5% hydroxyethylcellulose (HEC, 200-300 cps, 2% in water, at 20 °C) (Tokyo Kasei, Tokyo, Japan) solution prepared in 1×TBE buffer (89mM Tris-boric acid, 2mM EDTA, pH=8.0) and was vacuumed to remove bubbles. The samples were labeled in the PCR procedure with Cy5-dCTP, which was purchased from Amersham Pharmacia (Piscataway, NJ).

2.2   Capillary Electrophoresis Chip

The crossed-channel electrophoresis chip is made of poly-(dimethylsiloxane) (PDMS). The cross-section dimension of the channel is 50×20 μm and the length of the effective separation channel is 50 mm (Figure 8.1).

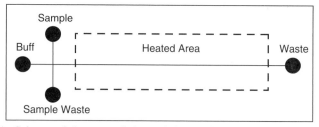

FIGURE 8.1. Scheme of the crossed-channel electrophoresis chip.

2.3   Experimental Setup

A laboratory-assembled chip-based CE system with laser-induced fluorescence (LIF) detection and heaters was used in our work. The design of the CE system was similar to that reported in previous papers [14]. Briefly, a 632 nm He-Ne laser was used as the excitation source. The fluorescence was collected by a PMT with a 670 nm band-pass filer and the frequency of sampling was 10 Hz.

To achieve the temperature gradient along the separation channel, three heaters (DN515, ThermOptics, Carson City, NV) were placed against the backside of the chip. The temperature of the heaters was controlled with a digital potentiometer tuned by a computer. All the parameters for generating the temperature gradient can be easily programmed via the computer. The precision for temperature gradient control reaches 0.1 °C.

2.4   DNA Samples

Two Cy5-labeled PCR products with the same length of 101bp containing one or two SNP sites were analyzed. The characteristics of the samples are listed in Table 8.1.

Before the analysis, heteroduplex PCR products were generated by heating the wild type and mutant PCR products at a ratio of 1:1 in the same test tube at 94 °C for 5 min and then by decreasing the temperature to 56 °C for 1 h to facilitate the reannealing of the DNA amplicons.

3.   Result and Discussion

3.1   Results

The results in Figure 8.2 and 8.3 show that a particular homoduplex DNA fraction was base-line resolved from its corresponding heteroduplex composites through the crossed-channel chip in only 8 minutes by applying a temperature gradient along the separation channel. The samples were also run separately through the same chip with the same temperature gradient. As a result only a single band was detected. The samples containing two SNP sites were tested under the temperature gradient of 62-67 °C for 240 s with a precision

TABLE 8.1. Characteristics of the DNA Samples

No.	Length (bp)	SNP type	SNP position	Sample source
1	101	C to T, G to A	50, 60	HLA_A1101
				HLA_A2501
2	101	C to T	53	HLA_B2703
				HLA_B2705

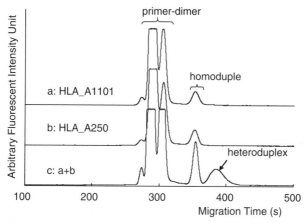

FIGURE 8.2. The electropherograms of DNA samples containing two SNP sites.

of 0.1 °C per step and the separation electric field of 150 V/cm. The temperature gradient for the samples containing one SNP site was 62.5-67.5 °C and the other conditions were the same as above. In our experiment, the sieving power and the heat durability of the separation media were satisfactory.

3.2   Different Temperature Gradient

We also analyzed the samples with two SNP sites under different temperature gradients. As shown in Figure 8.4, with only one degree C difference in the

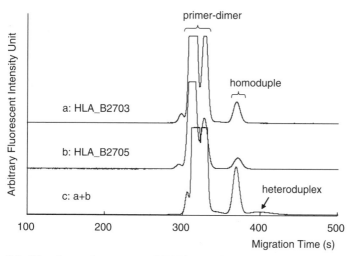

FIGURE 8.3. The electropherograms of DNA samples containing one SNP site.

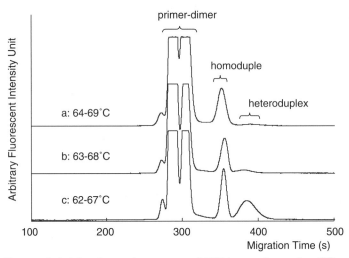

FIGURE 8.4. The electropherograms of DNA samples under different temperature gradients.

temperature gradient, the patterns of electropherograms changed dramatically. So in practice, we should choose the most suitable temperature gradient for each sample to obtain the best results.

3.3   Separation Efficiency

It is obvious that, in our experiment, only partial resolution was achieved, and two peaks, homoduplexes and heteroduplexes, were obtained. But this is good enough for us to differentiate homoduplexes from heteroduplexes. So we conclude that this method is very suitable to rapidly recognize the presence of a broad range of SNPs prior to further characterization.

4.   Conclusion

We have demonstrated a reliable chip-based temperature gradient capillary electrophoresis system for fast SNPs detection. Using this method, DNA samples can be analyzed for their SNPs in a single run. It can thus simplify the analysis process and shorten the total time needed for analysis.

Acknowledgements. This work was supported by the National Natural Science Foundation of China (No. 39825108 and No. 30000040).

References

1. Wang, D. G., Fan, J. B., Siao, C. J., et al., Large-scale identification, mapping, and genotyping of single-nucleotide polymorphisms in the human genome, Science, 1998, 280: 1077-1082.
2. Kristensen, V. N., Kelefiotis, D., Kristensen, T., Borresen-Dale, A., High-throughput methods for detection of genetic variation, Biotechniques, 2001, 30: 318-332.
3. Cheung, V. G., Spielman, R. S., The genetics of variation in gene expression, Nat. Genet., 2002, 32: 522-525.
4. Alper, J., Weighing DNA for fast genetic diagnosis, Science, 1998, 279: 2044-2045.
5. Orita, M., Iwahana, H., Kanazawa, H., Hayashi, K., Sekiya, T., Detection of polymorphisms of human DNA by gel electrophoresis as single-strand conformation polymorphisms, Proc. Natl. Acad. Sci. U.S.A., 1989, 86: 2766-2770.
6. Borresen, A., Hovig, E., Smith-Sorensen, B., Malkin, D., Lystad, S., Andersen, T. I., Nesland, J. M., Isselbacher, K. J., Friend, S. H., Constant denaturant gel electrophoresis as a rapid screening technique for p53 mutations, Proc. Natl. Acad. Sci. U.S.A., 1991, 88: 8405-8409.
7. Fischer, S. G., Lerman, L. S., DNA fragments differing by single base-pair substitutions are separated in denaturing gradient gels: correspondence with melting theory, Proc. Natl. Acad. Sci. U.S.A., 1983, 80: 1579-1583.
8. Toliat, M. R., Erdogan, F., Gewies, A., Fahsold, R., Buske, A., Tinschert, S., Nürnberg, P., Analysis of the NF1 gene by temperature gradient gel electrophoresis reveals a high incidence of mutations in exon 4b, Electrophoresis, 2000, 21: 541-544.
9. Liu, W. G., Smith, D. I., Rechtzigel, K. J., Thibodeau, S. N., James, C. D., Denaturing high performance liquid chromatography (DHPLC) used in the detection of germline and somatic mutations, Nucleic Acids Res., 1998, 26: 1396-1400.
10. Khrapko, K., Hanekamp, J. S., Thilly, W. G., Belenkii, A., Foret, F., Karger, B. L., Constant denaturant capillary electrophoresis (CDCE): a high resolution approach to mutational analysis, Nucleic Acids Res., 1994, 22: 364-369.
11. Tian, H. J., Brody, L. C., Landers, J. P., Rapid detection of deletion, insertion and substitution mutations via heteroduplex analysis using capillary- and microchip-based electrophoresis, Genome Res., 2000, 10: 1403-1413.
12. Gelfi, C., Righetti, S. C., Zunino, F., Della Torre, G., Pierotti, M. A., Righetti, P. G., Detection of p53 point mutations by double-gradient, denaturing gradient gel electrophoresis, Electrophoresis, 1997, 18: 2921-2927.
13. Gao, Q. F., Yeung, E. S., High-throughput detection of unknown mutations by using multiplexed capillary electrophoresis with poly(vinylpyrrolidone) solution, Anal. Chem., 2000, 72: 2499-2506.
14. Effenhauser, C. S., Bruin, G. J. M., Paulus, A., Integrated chip-based capillary electrophoresis, Electrophoresis, 1997, 18: 2203-2213.

9

Electro-Osmotic Flow Micro Pumps for Cell Positioning in Biochips

Rafael Taboryski, Jonatan Kutchinsky,
Ras Kaas Vestergaard, Simon Pedersen, Claus B. Sørensen,
Søren Friis, Karen-Margrethe Krzywkowski,
Nicholas Oswald, Rasmus Bjørn Jacobsen, Corey L. Tracy,
Margit Asmild, and Niels Willumsen
Sophion Bioscience A/S, Pederstrupvej 93, DK-2750, Denmark

Abstract: A feasible scheme for positioning of cells in a patch clamp bio-chip for high throughput drug screening comprises the application of suction by means of on-chip micro-pumps. A practical realization of such micro-pumps is based on Electro-Osmotic Flow (EOF). The principle of operation, and the design considerations for such pumps is described. Specific EOF pump geometries are demonstrated.

Key words: Electro-osmotic flow, micro-pump, patch clamping, electro-physiology, high throughput drug screening.

1. Background

In modern drug discovery carried out by the pharmaceutical industry, the primary screening of large compound libraries established through combinatorial chemistry, is typically based on methods targeting cloned ion channels expressed in mammalian cell lines Ref. 1. Today, the most commonly used such method for High Throughput Screening (HTS) is based on the fluorescence of compounds binding to intracellular calcium. However, despite the commercial breakthrough of the fluorescence based equipment for HTS, these techniques are only indirect markers of the ion channel response to applied drugs. The only direct and the most reliable method for studying the detailed function of ion channels is electrophysiology, and the state of the art is here the patch clamping technique (Hamill et al., 1981). This technique is however slow and labour-intensive and requires the presence of skilled operators. All these factors disqualify conventional patch clamping for HTS, where a single compound library may comprise hundreds of thousands of compounds to be screened, preferably at very low cost. Patch clamping is however commonly used by the pharmaceutical industry for lead optimisation and for safety pharmacology. The ideal HTS method should combine the high information content of conventional patch clamping with the high throughput and low cost of the existing fluorescence based methods. In addition this combination will allow the pharmaceutical industry to cut down on the development time of drugs, as high quality data will be available at an earlier stage of the drug discovery course.

Sophion Bioscience (www.sophion.dk) is developing automated patch clamping equipment (QPatch) with throughputs substantially higher than conventional manual patch clamping. An essential part of this equipment is a multi channel lab-on-a-chip measurement plate with high functionality. The plates are operated and handled by a screening station, comprising a plate handling robot, an electronics unit with patch clamping amplifiers and pump drivers, and a data acquisition unit. The automatic operation of the plates comprises automatic positioning of cells, establishment of the measurement configuration and carrying out measurements of ion-channel currents before and after application of compounds. The strategy behind QPatch is to allow users to obtain a high quality data set comprising a full patch clamp experiment for each drug dispensed onto the chip. As a consequence of this requirement an individual and independent pumping capability associated with every single channel is required. This can be achieved either by interfacing to a number of parallel external pumping lines, or by having on chip micro pumps located in conjunction with the individual channels. The main advantage associated with on chip pumps is the scalability with the number of channels per consumable. It is easy to imagine, that when the number of parallel measurement sites is increased, from 16 (first generation) to 96 (second generation) it will not be feasible to rely on a technology based on a massive parallel gas pressure interface to external pumping lines. The second option comprising on chip micro pumps is therefore chosen, despite the increased complexity of the consumable associated with this solution. Moreover, it appears that EOF pumps that generally have a relatively low volumetric flow rate, but can be designed to have a high stall pressure are particularly well suited to pump on loads representing a high flow resistance, like an orifice for patch clamping. Such holes are typically made on a planar Si membrane with standard Si processing technologies. The holes typically have a diameter of about 1 μm and flow conductance of order 1-20 pl/s/mbar. Thus the requirement for a pump will be its ability to exert a pressure of approximately 50-100 mbar in a time sufficiently long to position a cell on the orifice. Moreover, once the cell has been positioned, a pressure of approximately 300 mbar is required in order to rupture the cell membrane. In this paper we will present all the necessary considerations for designing and producing such micro pumps.

2.   The Patch Clamp Orifice

The patch clamping technique represents a major development in biology and medicine, since it enables measurement of ion flow through single ion channel proteins, and also enables the study of a single ion channel activity

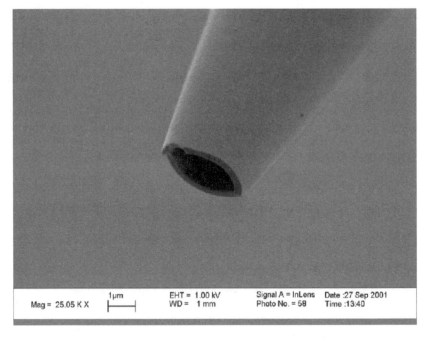

| Mag = 25.05 K X | 1 μm | EHT = 1.00 kV | Signal A = InLens | Date :27 Sep 2001 |
| | ├———┤ | WD = 1 mm | Photo No. = 58 | Time :13:40 |

FIGURE 9.1. Scanning electron micrograph of a micro pipette used for patch clamping.

in response to drug exposure Ref. 2. Briefly, in standard patch clamping, a thin (approx. 0.5-2 μm in diameter) glass pipette is used. The tip of this patch pipette is pressed against the surface of the cell membrane. The pipette tip seals tightly to the cell membrane and isolates a small population of ion channel proteins in the tiny patch of membrane limited by the pipette orifice (Figure 9.1).

The activity of ion channels can be measured individually ('single channel recording') or, alternatively, the patch can be ruptured, allowing measurements of the channel activity of the entire cell membrane ('whole-cell configuration'). High-conductance access to the cell interior for performing whole-cell measurements can be obtained by rupturing the membrane by applying negative pressure in the pipette. For patch clamping on planar substrates, the pipette tip is replaced by an orifice made on a Si membrane Figure 9.2. A typical data set showing the effect of an ion-channel blocker obtained by a patch clamp measurement on a chip is shown in Figure 9.3. From a micro fluidic point of view the patch clamp orifice represents a load to the pump quantified by a flow conductance. The flow conductance of a patch clamp orifice can be measured. A result of such a measurement is shown in Figure 9.4. This information is relevant when estimating the cell capture

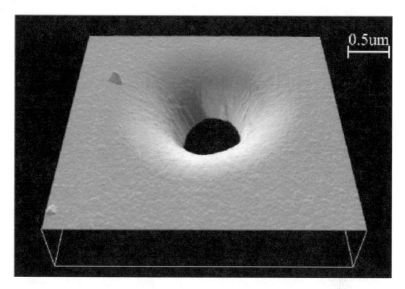

FIGURE 9.2. Atomic force micrograph of a patch clamp orifice on a planar Si substrate with silicon oxide surface coating.

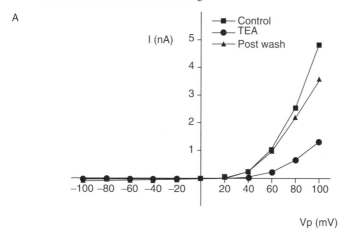

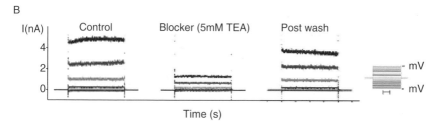

FIGURE 9.3. Effect of the K$^+$ channel blocker TetraEthylAmmonium chloride (TEA) measured on a chip. (A) I/V relations obtained under control conditions, after the addition of 5 mM of the K$^+$ channel blocker TEA, and after washout of the blocker. It is seen that the effect of TEA is reversible. (B) The graphs were constructed from 800 msec current sweeps recorded at voltages ranging from −100 mV to 80 mV. The applied voltage protocol is indicated to the right.

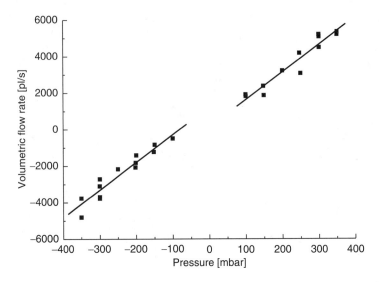

FIGURE 9.4. The flow rate of a physiological buffer solution through a patch clamp orifice was determined as a function of applied pressure resulting in a flow resistance of 15 pl/s/mbar for this particular orifice with a diameter of approximately 1.5 µm.

capability of a given pump. Any load of this dimension is likely to force the pump to operate near its stall pressure. An estimate of the load flow resistance will then provide information about the "capture radius" of the pump.

3.   Electro–Osmotic Flow

Electroosmotic flow (EOF) is generated by application of an electric field through a solution in a channel defined by insulating walls. The phenomenon depends on ionisation of sites on the surface, so that for electro-neutrality there is an excess mobile charge in the solution, predominantly located close to the walls within a thin screening layer given by the Debye length $\lambda_D \approx$ 1–10nm for the interface. An electric field applied across the solution acts on the excess charge in the solution causing the fluid to flow. The quantity and distribution of excess charge in the solution depends on the surface material (density of ionisable sites) and on the solution composition, especially pH and ionic concentration. The charge distribution is related to the zeta (ζ) potential, which is defined as the electric potential at the thin shear liquid plane exhibiting anomalous elastic properties due to its proximity to the surface. This potential can be related to electroosmotic flow. However, although values for the zeta potential are measured and published for material/solution combinations it is not really a readily controllable parameter, and as it arises from the ionisation of surface sites, ζ and EOF are very susceptible to changes in surface condition and contamination. A value of 75 mV for ζ is given in the literature for aqueous solutions of sodium and potassium at silica surfaces. For glass the values may be twice those for silica but for both the effects of pH and adsorbing species can in practice very significantly reduce the values. Such a value for ζ may be used in design calculations but it is wise to ensure that adequate performance is not dependant on it being achieved in practice. The direction of EOF is determined by the excess mobile charge in the solution generated by ionisation of the surface sites. As pKa for the ionisable groups on silica or silicate glass is ~2, then at neutral pH values the surface is negatively charged and EOF follows the mobile positive ions towards a negatively polarized electrode. The volumetric flow rate I_{vol}^{eof} associated with electroosmotic flow for a flow channel of length L, and constant cross sectional area A is given by the Helmholtz-Smoluchowski equation

$$I_{vol}^{eof} = \frac{A\varepsilon\zeta}{L\eta} U, \tag{1}$$

where ε is the permittivity and η the viscosity of the liquid, while ζ is the zeta potential of the interface between the liquid and the channel boundaries. U is the driving voltage applied across the ends of the channel with length L and constant cross sectional area A. **Eq.1** defines the maximum possible flow rate an EOF pump can deliver with no load connected. Similar expressions can also be derived for more complex pump geometry's, and here we will use the notation I_{max} to denote the maximum flow rate. The average velocity of the fluid particles in the channel is in general given by $u = I_{vol} / A$, and the electric field strength by $E = U / L$, allowing the definition of the electroosmotic mobility $\mu_{eof} = u / E = \varepsilon\zeta / \eta$ to be independent of any particular geometry of the flow channel containing the EOF pump, and solely to characterize the interface between the liquid and the walls. With a load connected to the pump, the EOF driving force will be accompanied by a pressure driven flow (Poiseuille flow). The volumetric flow associated with laminar Poiseuille flow is given by $I_{vol}^{Poiseuille} = K_{pump} \Delta p$, where Δp is the pressure difference across each end of the pump channel, and K_{pump} the flow conductance of the pump. The total flow rate is then given by

$$I_{vol} = K_{pump} \Delta p - I_{max}. \tag{2}$$

Note that the electroosmotic flow and the pressure driven flow are in the opposite directions. The pressure compliance or the stall pressure of the pump is found by putting $I_{vol} = 0$, and solving for Δp:

$$\Delta p_{max} = \frac{I_{max}}{K_{pump}}.$$

The overall performance of any particular EOF pump can be quantified by a quantity expressed in the unit of power (Watt) and given by the product

$\Delta p_{max} I_{max}$. The higher power, the better is the overall performance of the pump. If the pump is loaded with a flow conductance K_{load}, the pressure difference across the load is given by:

$$\Delta p_{load} = \frac{I_{max}}{K_{load} + K_{pump}}, \tag{4}$$

while the volumetric flow through the load is given by

$$I_{vol}^{load} = K_{load} \Delta p_{load}. \tag{5}$$

A specific choice of pump configuration will give rise to an electrical conductance of the pump channel G_{pump}. In response to the EOF driving voltage, the electrolyte inside the pump channel will carry the electrical current I_q. Design considerations associated with EOF pumps should comprise heat sinking due to the power dissipation in the pumps. Moreover, the location and design of electrodes should be considered. In devices to be used for biomedical purposes, the natural choice of electrode material is Ag/AgCl, with the process Ref. 3

$$AgCl(s) \xleftrightarrow{\pm e} Ag(s) + Cl^-(aq),$$

and hence the consumption of such electrodes when operating the pump should be considered. The rate of consumption of the electrode material expressed in volume per time unit is given by:

$$\Delta V_{\Delta t} = \frac{I_q m_{AgCl}}{e N_A \rho_{AgCl}}, \tag{6}$$

where $m_{AgCl} = 143.321$ g/mol and $\rho_{AgCl} = 5.589$ g/cm^3 is the molar mass and the mass density of AgCl, while $e = 1.602 \times 10^{-19}$ C and $N_A = 6.02 \times 10^{23}$ mol^{-1} is the elementary unit of charge and the Avogadro constant.

An alternative to the use of consumable electrodes involves the use of an external electrode linked to the chamber by an electrolyte bridge with high resistance to hydrodynamic flow. This might be a thin channel, similar to that providing the EOF pumping, but with a surface having low density of charged sites (low zeta potential) or where the surface has opposite polarity charge to the EOF pumping channel. In the latter case the low flow conductance channel to the counter electrode contributes towards the EOF pumping. Most wall materials tend, like glass or silica, to be negatively charged in contact with solutions at neutral pH. However it is possible to identify materials which bear positive charge. Alumina based ceramics may be suitable, especially if solutions are on the low pH side of neutral. Alternatively polymer or gel material, such as Agarose, polyacrylamide, Nafion, cellulose acetate, or other dialysis membrane-type materials may produce the bridge with high resistance to hydrodynamic flow. Preferably these should have low surface charge density or an opposite polarity to that of the EOF pumping channel.

4. The Corbino Geometry

A practical realization of an EOF pump is based on the so-called Corbino geometry known from conductivity measurements in the field of semiconductor physics Ref. 4. The Corbino geometry pump is comprised of plates with silica or glass surfaces separated by spacers, mounted in a laminated polymer holder. The Corbino configuration is particularly suitable for integration into a pipette well. The channel flow conductance for this geometry can be derived from a simple conservation law by exploiting a general analogy between fluid flow and current flow. In this geometry the fluid flows between the plates of annular shape and the flow is radial with a drain in the center of one of the plates. The principle is shown in Figure 9.5. The distance between the plates h has to be small compared with both the inner (r_{in}) and outer (r_{out}) radius of the annulus. The overall performance of any particular EOF pump is quantified by the stall pressure obtained when the pump is loaded with an infinite flow resistance, and the maximum volumetric flow obtained when the pump is free running (zero load). The flow properties for the Corbino geometry EOF pump were derived.

The flow conductance

$$K_{channel} = \frac{i\,\pi h^3}{6\eta \ln\left(\frac{r_{out}}{r_{in}}\right)} \tag{7}$$

The maximum volumetric flow rate

$$I_{max} = \frac{2\pi h}{\ln\left(\frac{r_{out}}{r_{in}}\right)} \mu_{eof} U, \tag{8}$$

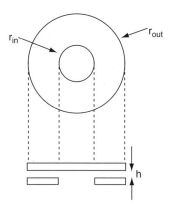

FIGURE 9.5. The Corbino geometry.

where η is the viscosity of the pumped liquid, and U the driving voltage. For the estimation of current flow in response to applied driving voltage, the electrical conductance for the pump channel is given. This can also be used to calculate the Joule heat dissipated in the pump.

$$G_{channel} = \frac{2\pi h}{\ln\left(\frac{r_{out}}{r_{in}}\right)} \sigma, \tag{9}$$

here σ is the electrical conductivity of the pumped liquid.

For any parallel plate pump configuration it can be shown that the flow rate (at zero pressure) is determined mainly by the x-y 2D layout while the stall pressure is given by h^2. Thus in particular, the Corbino geometry was shown to have an advantage over square channel parallel plate geometries of similar footprint in terms of max flow (Figure 9.6). For comparison, for a 10 × 10 mm square layout we found a max flow of only 0.5 nl/sec and stall pressure of 778 mbar. When comparing the benchmark numbers for max flow and stall pressure with the analytical model, the so-called zeta potential for the pump channel surface was found to be 17 mV. This zeta potential corresponds to an electro-osmotic mobility of the order of 1.3×10^{-4} cm^2/Vs.

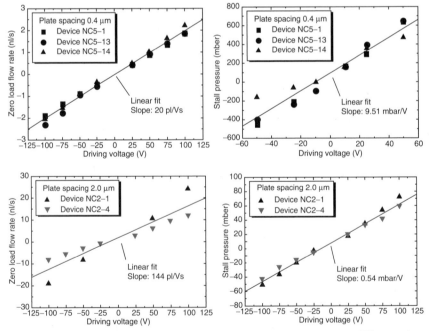

FIGURE 9.6. Benchmark data for Corbino geometry EOF pumps. Two different plate spacings were tested, 0.4 µm and 2.0 µm. The narrow channel clearly have the smallest flow rate and the highest stall pressure.

5.   Conclusion

It was found that EOF pumps are feasible for lab-on-a-chip applications, in particular for positioning of cells and for rupturing cell membranes in patch clamp applications. One particular novel pump geometry, the Corbino geometry was demonstrated to fulfill the requirements for pumping on loads associated with the small orifice used for patch clamping on planar substrates.

References

1. Xu J, Wang X, Ensign B, Li M,, Wu L Guia A, Xu J (2001) *Drug Discovery Today* **6**: 12781287
2. Hamill O, Marty A, Neher E, Sakmann B, Sigworth FJ (1981) *Pflügers Arch.* **3 91**:85-100.
3. Oldham, H.B, Myland, J.C., "Fundamentals of electrochemical science", Academic Press; ISBN: 0-12-525545-4
4. O.M. Corbino, Phys. Z. **12**, p561 (1911).

Part III

Surface Chemistry

10

The Application of Novel Multi-Functional Microarray Slides for Immobilization Biomolecules

YAPING ZONG, YOUXIANG WANG*, JENNIFER SHI, AND SHANNON ZHANG
Full Moon BioSystems, Inc., 845 W. Maude Ave, Sunnyvale, CA 94085, USA

Abstract: Microarray has revolutionized the study of molecular biology, especially the application in clinical diagnostics. When used in clinical diagnostics, microarray has to meet a high degree of reproducibility, reliability and quality in order to become a standard tool. Repeatability and reproducibility are essential for providing the best data and process control. The real challenge for microarray is, however, how to produce consistent and reliable data. The variance of microarray data is contributed by the quality of sample source, the quality of glass substrates, hybridization, and probe labeling and spotting. The quality of coated glass substrates is one of the main factors. This paper is focusing on discussing how to optimize coating conditions to improve the slide quality, consequently improve the data quality such as sensitivity and data reliability.

Key words: Microarray, slide, glass substrate, coating.

1. Introduction

Studying cellular processes and responses require monitoring expression levels for thousands of genes at a time. Time-consuming and labor-intensive traditional technologies such as Northern blot for the study of gene expression were significantly limited in both breadth and efficiency since these studies typically allowed investigators to study only one or a few genes at a time. However, the recently developed microarray technology has dramatically enhanced our ability to study biology and explore the molecular basis of disease. Microarray enables massively parallel molecular analyses to be carried out in a miniaturized format with a very high throughput. It allows mRNA expression to be assessed on a global scale and the parallel assessment of gene expression for hundreds or thousands of genes in a single experiment[1-3]. Besides its main application for detecting global patterns of gene expression, it has many other potential applications including identification of complex genetic diseases[4-5], mutation/polymorphism detection[6], and drug discovery and toxicology studies[7].

In general, there are two main types of technology for manufacturing DNA microarray. The first is on-chip synthesis of DNA molecules by either photolithographic synthesis or piezoelectric printing, often referred to as the Affymetrix method[8-11]. The second is immobilization of prefabricated DNA or oligonucelotides onto microchips, often referred to as Patrick Brown method[12-20]. The two methods enjoy different advantages over each other but they both suffer from experimental variance or inconsistencies, which ultimately affects the experimental reliability of microarray-based analyses[21]. The variance and inconsistency can be listed and contributed by the quality of sample source, the quality of coated glass substrates, hybridization, probe labeling, spotting, etc. Figure 10.1 is an example of hybridization results with the same probes on slides manufactured by two different companies (data provided by Dupont). One can tell from the images that data obtained with slides manufactured by Full Moon BioSystems has much better sensitivity, great spot morphology. In this paper, we are going to focus on discussing how to improve the quality of coated glass substrates.

A great quality of coated glass substrates should have following special features: very low background, great sensitivity for immobilization biomolecules, batch to batch consistency, great spot morphology, very stable in high humidity environment. These special features are mainly depended on the source of glass substrates, cleaning process, coating material and coating

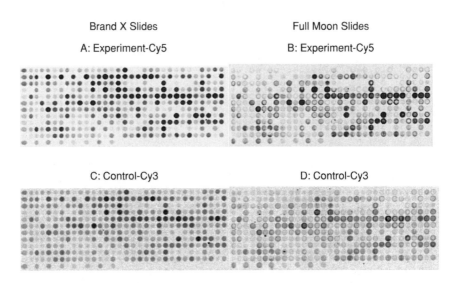

FIGURE 10.1. Images in the left side were obtained on Brand X slides. Images on right side were obtained on Full Moon slides. 100pmol was used for each probe, same probe on both slides. Brand X hybridization solution was used on Brand X slides. Full Moon slide hybridization solution was used on Full Moon Slides. Standard washing protocol was used. (Data provided by DuPont).

process. In order to improve the quality of coated glass substrates, Full Moon BioSystems has optimized the whole process.

2.   Experiment

All the chemicals were bought from Aldrich. 1 × 3 inch microscope slides are customized. It is 1 mm thickness. Each slide is handpicked before it is ready to be cleaned. The clean started with soaking the slides in commercially available ethanol at different temperature for different time courses to optimize the conditions. They were rinsed with running DI water and dried with nitrogen. They were then soaked in 1 N H_2SO_4 for different time at different temperature. They were washed with water and dried. The clean slides are ready to be coated with polymer materials. The 25% polymer is dissolved in different solvents for optimize the coating conditions. The slides were dipped for 2 mins and then rinsed with isopropanol and dried with nitrogen.

All the coated slides are evaluated with real hybridization. Total RNA was bought from Clontech. Probes were labeled with Clontech kit. After labeling, the probes were spun dried and dissolved in Full Moon BioSystems hybridization buffer. For cDNA slides, the targets were spotted in 50% DMSO with Amersham GenIII spotter. After spotting, the slides were UV cross-linked. Then they were pre-treated with 2 × SSC/0.1% SDS/1%BSA. After pre-treatment, they are ready to do hybridization with cover slip for 12 to 18 hours. They were finally washed and dried with nitrogen. For oligos, targets were spotted in 3 × SSC or 150 mM Sodium Phosphate buffer. The slides were then humidified at 65-75% humidity for 4 to 12 hours. After humidity, they were UV cross-linked at 600 mJ. They were then pre-treated to block the slides. The slides are ready to do the hybridization with probes dissolved in Full Moon BioSystems hybridization buffer. The slides were finally washed and nitrogen dried. The hybridized slides were scanned with Axon scanner and analyzed with their software.

To compare the slide quality manufactured by different companies, the slides were bought directly from different manufacture companies. The hybridization procedures were carried out based on protocols provided by manufacture companies.

3.   Results and Discussion

3.1   The Source of Glass Substrate

Currently, microscope glass slides are selected as substrates because of their favorable optical characteristics. Common glass is made from sand or silica (SiO_2), sodium carbonate (Na_2Co_3), limestone ($CaCO_3$), Magnesium Carbonate ($MgCO_3$), additives to improve the glass quality and to color the glass. Based on its composition, commercial microscope glass slides can be typed as vitreous silica, soda-lime-silicate glass, borosilicate glass-pyrex, aluminosilicate glass, lead "crystal" glass, etc. Full Moon BioSystems has decided to use borosilicate glass as substrate. It has following special features: low expansion, 3-5 ppm/ °C, softening point ~700-800 °C, high chemical durability[22]. These special features fit the microarray application the best. The commercial borosilicate glass has following composition: $80SiO_2$ + $~10B_2O_3$ + $~2Al_2O_3$ + $~5Na_2O$. Remain 3% are impurities. The impurities affect the glass surface property dramatically, such as pH and background. Consequently it causes slide to slide variation and batch to batch variation if one does not control the impurity carefully. Full Moon BioSystems has worked with glass manufacturing company to customize the glass substrates. Special additives were also added. Customized glass substrates have much lower background and secured batch to batch consistency.

3.2 Cleaning Process

Slide surface is usually dirty and contaminated with stains and has scratch marks. Microscope slides can not be used for microarray application without pre-clean treatment. Each slide has to be hand-selected. Slides with scratch marks will be trashed. Slides with stains that can not be cleaned will be trashed. After pre-selection, slides are ready to be cleaned. There are a few steps to clean slides. Most slides are first treated with detergent or organic solvent such as acetone or ethanol. Detergent can get rid of dust, dirt and residues on the surface. Acetone and ethanol can clean oils and organic residues that appear on glass surfaces. One can also sonicate slides in detergent and organic solvents. The detergent and organic solvents can be pre-warmed. For deep clean, the second step is using inorganic acid such as HCl, Nitric acid, Aqua Regia. HCl can be used to clean slides because it can mildly etching the surface of the glass. Nitric acid can be used to clean slides because it can leech the ions from within the surface of the glass. Aqua Regia is the strongest acid. It can etch many metals, including gold. Extra care is needed while using Aqua Regia due to that it is the strongest corrosive reagent. One can also use ammonium water or NaOH and KOH to clean slides. The slides can be cleaned in warm acid solution or base solution for a few minutes or a few hours. After acid or base clean, the last step is to rinse the slides with DI water and then blow dry.

We have found out that hybridization results can be very different by slightly changing the clean conditions such as reaction time and temperature. Surface properties can be different by using different clean conditions. It consequently affects coating and hybridization results. Without controlling the clean process consistently, it is impossible to obtain coated slides with batch to batch consistency. Full Moon BioSystems has developed its own unique clean process. The clean process has been dramatically simplified. It starts to clean the slides with commercially available ethanol in warm conditions for a few hours. The slides are then rinsed with water and dried with nitrogen. The slides are further soaked in 1.0 N H_2SO_4 for a few hours at warm temperature. Then it is rinsed with water and dried with nitrogen. Both temperature and reaction time have been optimized. The resulted clean slides give much lower background and are very consistent from batch to batch examined with scanner.

3.3   Coating Material

The quality and surface coating of the glass substrate are fundamental factors for a successful DNA microarray system. A poor quality substrate will lead to a low DNA binding efficiency and hence, to a poorly defined fluorescent image. The most popular coating materials used for coating microarray slides are 3-aminopropyl-triethoxysilane (amino-slides), 3-glycidoxypropyltrimethoxysilane (epoxy slides), polylysine (polysine slides), telephthaldicarboxaldehyde (aldehyde slides), hydrogel coated slides, polymer coated active ester slides[22-25]. The resulting coated slides may have combined following drawbacks: 1) mono-layer with less binding capacity such as amino slides, epoxy slides, aldehyde slides; 2) mono-functional reactive group on the surface with less binding capacity; 3) high background due to intrinsic fluorescence of the material itself. In ontrast, the coating material developed by Full Moon BioSystems is a polymer. The polymer has multifunctional reactive groups on its branch. Each of the functional groups takes particular role in improving the coating quality. One of the functional groups helps to concentrate the DNA onto the glass surface. Therefore hybridization efficiency has been improved dramatically. There is a functional group to increase the slide stability.

Therefore the slides are very stable in high humidity environment. It can be boiled in hot water for 20 minutes. There is a functional group to protect dye degradation, especially for cy5. Therefore the slides have much better sensitivity for both cy3 and cy5, especially for cy5. Figure 10.2 shows the

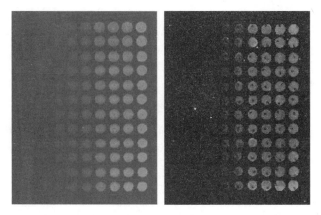

FIGURE 10.2. Image in the left side was obtained with Full Moon slide. Image in the right side was obtained with Vendor slide. The same target was spotted on both Full Moon and vendor slides with dilution series. Each dilution has twelve repeats. Probe labeled with cy5 was hybridized.

hybridization results with cy5 labeled probes onto two different slides. Fluorescent signal intensity on Full Moon slides is much stronger with better sensitivity. Furthermore morphology is much better. There is a functional group to lower background. Cy3 channel has background lower than 200 counts with axon scanner scanning at 700 PMT. Cy5 channel has background below 100 counts with axon scanner scanning at 700 PMT. Therefore the state of art novel coating material with multi-functional reactive groups developed by Full Moon BioSystems promises to deliver the best quality of coated glass substrates for immobilization biomolecules.

3.4   cDNA Slides

The resulted cDNA slides have following special features:

a) slide surface contains multifunctional reactive groups.
b) Greatly improved DNA attachment efficiency (Figure 10.3).
c) Enhanced sensitivity for both Cy5 and Cy3 (Figure 10.2).
d) Excellent stability with self-life for half-years. No degradation was observed boiled in 95 °C water.
e) Excellent spot morphology.
f) Very consistent from batch to batch.

Figure 10.3 shows the hybridization results compared to all other leading commercial slides. Full Moon Slides have the best signal intensity for both cy3 and cy5. Figure 10.4 shows the spot morphology. The signal intensity is uniformly distributed within a single spot.

3.5   Oligo Slides

The resulted oligo slides have following special features:

a) Slides are coated with multifunctional reactive groups.
b) Much high binding efficiency for DNA.
c) Small spot size (100 μm) is very good for manufacturing high-density arrays.
d) Very stable in high humidity environment with self-life about half years.
e) Very low background with much high sensitivity.
f) It is efficient for immobilization both modified (react with 5′-amine) and unmodified (react with 5′-OH) oligos.

Figure 10.5 shows the signal intensity comparing with all other leading commercial slides available on the market. Full Moon slides have much better signal intensity for both modified and unmodified oligos.

Figure 10.6 shows the hybridization efficiency for different length oligos. Longer oligos have better signal intensity. 30 mer oligos already have enough

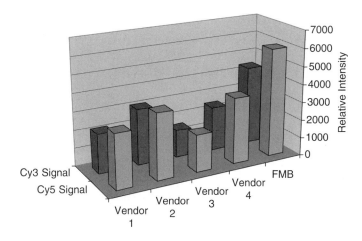

FIGURE 10.3. Signal intensity of the hybridization results obtained with different vendor slides. Target and probe are the same. Target was dissolved in 50% DMSO. Hybridization was carried out based protocols provided by each vendor.

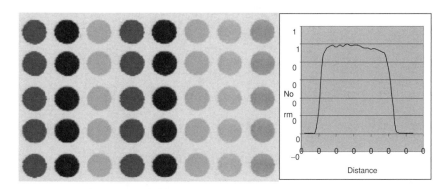

FIGURE 10.4. Target was dissolved and spotted in 50% DMSO. The complementary strand labeled with cy5 and cy3 was mixed in different concentration ratio. The signal intensity distribution was analyzed with ImageQuant.

hybridization efficiency for detection. Oligos spotted in 3 × SSC have better sensitivity than that spotted in 150 mM Sodium phosphate buffer.

4.   Immobilization Proteins

The oligo slides can also be used to immobilize proteins. Figure 10.7 shows an example of immobilization of FluoroLinks cy3 labeled goat anti-rabbit IgG (H+L). The reagent was bought from Amersham Biosciences. The slides can

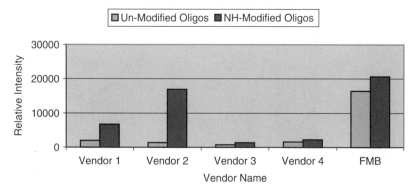

FIGURE 10.5. 43-mer oligos in 150 mM phosphate buffer, pH = 8.0, with a final concentration of 5 nmol/mL is spotted. Complementary strand labeled with cy3 and cy5 was hybridized on each vendor slide based on protocols provided by each vendor. Images were obtained with Axon scanner. Data was analyzed with ImageQuant software.

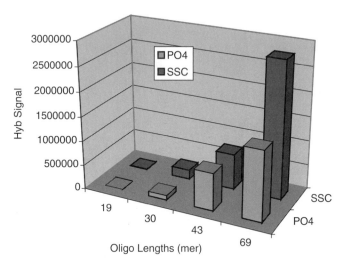

FIGURE 10.6. Different length of oligos were spotted in 3 × SSC and 150 mM Sodium Phosphate buffer (pH = 8.5). Complementary strand labeled with cy3 was hybridized.

attach proteins with great signal intensity and morphology. There is a company who has used oligo slides successfully for studying cytokines.

5.   Conclusions

The state of art novel coating chemistry developed by Full Moon BioSystems promises to deliver the best-coated glass substrate for immobilization biomolecules. A thin layer of synthetic engineered materials with

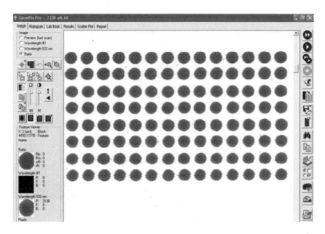

FIGURE 10.7. FluoroLinks cy3 labed goat anti-rabbit IgG (H+L) was spotted in 3 × SSPE. It was then washed with standard wash conditions.

Multi-functional reactive groups is uniformly and consistently coated on the glass surface, which ensures results with the highest reproducibility and quality. The substrates are ideal for attaching a variety of biomolecules, including cDNA, modified and un-modified oligonucleotides, and proteins.

References

1. Brown, P.O., Botstein, D., Nat. Genet., 1999, 21:33-37.
2. Young, R. A., Cell, 2000, 102:9-15.
3. Lockhart, D. J., Winzeler, E. A., Nature, 2000, 405:827-836.
4. Dill, K., Stanker, L. H., and Young, C. D., J. Biochem. Biophys. Methods, 1999, 41:61-67.
5. Struelens, M. J., De Gheldre, Y., Deplano, A., Infect. Control Hosp. Epidemiol., 1998, 19: 565-569.
6. Wodicka, L., Dong, H. L., Mittmann, M., Ho, M. H., Lockhart, D. J., Saccharomyces Cerevisiae. Nat. Biotechnol., 1997, 15:1359-1367.
7. Cheung, V. G., Gregg, J. P., Gogolin-Ewens, K. J., Bandong, J., Stanley, C. A., Baker, L., Higgins, M. J., Nowak, N. J., Shows, T. B., Ewens, W. J., Spielman, R. S., Nature Genet., 1998, 18:225-230.
8. Lipshutz, R. J., Fodor, S. P., Gingeras, T. R., Lockhardt, D. J., Nature Genet., 1999, 21:20-24.
9. McGall, G., Labadie, J., Brock, P., Wallraff, G., Nguyen, T., Hinsberg, W., Proc. Natl Acad Sci. USA, 1996, 93:13555-13560.
10. Singh-Gasson, S., Green, R.D., Yeu, Y., Nelson, C., Blattner, F., Sussman, M. R., Cerrina, F., Nature Biotechnol., 1999, 17:974-978.
11. Milner, N., Mir, K. U., Southern, E. M., Nature Biotechnol., 1997, 15:537-541.
12. Proudnikov, D, Timofeev, E., Mirzabekov, A., Anal Biochem., 1998, 259:34-41.
13. Chrisey, L. A., Lee, G. U., O'Ferrall, C. E., Nucleic Acids Res., 1996, 24:3031-3039.
14. Rasmussen, S. R., Larsen, M. R., Rasmussen, S. E., Anal. Biochem., 1991, 198: 138-142.

15. Salo, H., Virta, P., Hakala, H., Prakash, T. P., Kawasaki, A. M., Manoharan, M., Lonnberg, H., Bioconjug. Chem., 1999, 10:815-823.
16. Rogers, Y. H., Jiang-Baucom, P., Huang, Z. J., Bogdanov, V., Anderson, S., Boyce-Jacino, M. T., Anal. Biochem., 1999, 266:23-30.
17. Cohen, G., Deutsch, J., Fineberg, J., Levine, A., Nucelic Acids Res., 1997, 25, 911-912.
18. Beier, M., Hoheisel, J. D., Nucelic Acids Res., 1999, 27:1970-1977.
19. Cheung, V. G., Morley, M., Aguilar, F., Massimi, A., Kucherlapati, R., Childs, G., Nature Genet., 1999, 21:15-19.
20. Morozov, V. N., Morozova, T. Y. A., Anal. Chem., 1999, 71:3110-3117.
21. Kothapalli, R., Yoder, S. J., Mane, S., Loughran, T. P., BMC Bioinformatics, 2002, 3, 22.
22. Hill, J. H., Petrucci, R. H., General Chemistry, Prentice Hall, Inc. ISBN 0-02-354481-3. 1996.
23. Kumar, A., Liang, Z., Nucelic Acids Res., 2001, 29:e2.
24. Diehl, F., Grahlmann, S., Beier, M., Hoheisel, J. D., Nucleic Acids Res. 2001, 29:e38.
25. Halliwell, C. M., Cass, A. E. G., Anal. Chem., 2001, 73:2476-2483.

11

Novel Surface Technologies for Genomics, Proteomics, and Drug Discovery

Ye Fang, Anthony G. Frutos, Joydeep Lahiri,
Dana Bookbinder, Darrick Chow, Allison Tanner,[1]
Qin Zong, Ann M. Ferrie, Yijia P. Bao, Fang Lai, Xinying Xie,
Brian Webb, Margaret Faber, Santona Pal, Ollie Lachance[1],
Paul Gagnon[1], Megan Wang[1], Marie Bryhan,
Lyn Greenspan-Gallo, Greg Martin[1], Larry Vaughan[2],
Camilo Canel,[1] Kim Titus,[1] Debra S. Hoover,[1] John Ryan,[2]
Uwe R. Muller, James B. Stamatoff,[1] Laurent Picard,
Anis H. Khimani,[2] and Jeffrey L. Mooney
Biochemical Technologies, Science & Technology Division, Corning Incorporated, Corning, NY 14831, USA, and Life Sciences Division, Corning Incorporated, [1]Kennebunk, ME, and [2]Acton, MA 01720, USA

Abstract: Following the recent progress in functional genomics and proteomics, and high-throughput screening (HTS) in drug discovery, evolving technologies over the last decade have offered a tremendous leap over the caveats of traditional techniques. In response to this metamorphosis of technologies through different platforms, Corning has introduced a suite of surface technologies with applications in microarray printing, enhanced attachment, and consumables in drug discovery. Microarrays generated on an ultra-flat glass substrate with GAPS coating exhibiting a robust chemistry and low surface background have led to higher sensitivity and reproducibility for the expression assay. Recent introduction of UltraGAPS™ surface enables oligo attachment for use in differential gene expression analysis. Various attachment surfaces to meet the needs of the applications in genomics, proteomics and drug discovery will be discussed.

Key words: Surface technologies, GPCR, Microarrays, Genomics, Proteomics, Drug Discovery, Cell Culture.

Introduction

The genome sequencing projects have resulted in a plethora of sequence information primarily for the human genome, and various other organisms. Complete knowledge of the primary sequence has initiated the study of the functionality of each gene and its role in the complex intricacies of metabolism.

The increased understanding of the interactive pathways and its components will lead to the identification and qualification of reliable targets for therapeutic intervention. Ongoing work in functional genomics and proteomics will potentially lead to achieving these goals, and various contemporary and emerging technological tools will enable this endeavor. The successful tools must offer the ease and rapidity in identification, isolation and characterization of biomolecules, enable their production, immobilization, labeling, detection, analysis, data management, and archival for referencing, mapping and interactive modeling. Each one of these technological development areas invite challenges or opportunities in material sciences, surface technologies, biochemical and physical analyses, synthesis, scale-up technologies, reagents development, detection technologies, and instrumentation. The leading candidates in this technological development will be the ones that promise to offer efficiency, rapidity, sensitivity, reliability, reproducibility, higher throughput, and applicable breadth or vision.

One of the historical and continued needs in the development areas listed above has been in immobilization of biomolecules, preferably in their native or active state, followed by detection and interaction with their counterparts in a particular assay. Surface technologies have largely contributed in achieving this goal where molecules may be attached to specific chemical moieties on a specialized surface. A variety of surface chemistries have enabled the immobilization of different biomolecules for binding, detection, quantification, and competitive inhibition and analysis of complementary targets. Corning's expertise in surface and materials science has contributed to the development of a suite of surface chemistries for the attachment of a wide variety of molecules. A list of various substrates currently available is provided in Table 11.1. Selected candidates from this list are described in further detail in this review focusing on specific functionalities and applications.

I.   Non-Binding Surface (NBS™)

Non-specific binding of proteins and nucleic acids to polymer surfaces is a widely appreciated problem in homogeneous assays involving these biomolecules. Corning has developed a novel surface that significantly minimizes non-specific binding of biomolecules to polystyrene and polypropylene materials used in the manufacture of consumables for assays. To demonstrate the performance of this novel surface, radioassays were performed that used radiolabeled protein and DNA. ^{125}I-labeled IgG, BSA, and ^{32}P-labeled DNA (20-mer oligo and HindIII-λDNA) were used in binding assays in NBS-coated and non-coated microplates. The data in Table 11.2 clearly indicates that the NBS significantly reduces protein and nucleic acid binding to polymers, and consequently increases signal-to-noise. The trend towards lower assay volumes (leading to higher surface to volume ratios inside the assay wells) heightens the benefits of using NBS to reduce non-specific binding in protein- or DNA-based homogeneous assays, such as Scintillation Proximity

TABLE 11.1. List of Surfaces for Biomolecule Attachment and Applications

Platform	Surfaces	Application	Binding
Assay	Medium bind	EIA/RIA	250 ng IgG/cm^2
	High bind	EIA/RIA	500 ng IgG/cm^2
	Non bind (NBS)	Homogeneous assays: fluorescence, luminescence, absorbance detection modes	>95% binding reduction vs polystyrene
	Sulfhydryl bind & Carbo bind	Assays requiring site directed orientation of the biomolecule	Covalent, site-specific
	Universal bind	Immobilization of a mixture of biomolecules, dsDNA, antigens of unknown structure	Covalent, non-specific
Genomics & Proteomics	DNA bind™	Immobilization of aminated biomolecules: nucleic acid hybridization, solid phase PCR	Covalent, site-specific
	GAPS II (Gamma Amino Propyl Silane)	High quality DNA binding: printing microarrays; cDNAs	Ionic via free amino groups
	UltraGAPS™	High quality DNA binding: printing microarrays; cDNAs, long oligos (75 mers & higher)	Ionic via free amino groups
	Streptavidin surface	Binding assays of biotinylated ssDNA, dsDNA, peptides, proteins, and small organic molecules	> 25 pmoles of d-biotin/well
Tissue Culture	Standard tissue culture surfaces	Cell/Tissue culture	Treated for growth of anchorage dependent (AD) cells
	Poly-D-Lysine coated	Cell/Tissue culture: enhanced attachment of cells	Coated surface
	Ultra Low Bind Surface	Inhibits attachment of cells: e.g., maintenance of stem cells in an undifferentiated state	Neutrally charged, hydrophilic layer
	CellBIND™ Active Surface	Enhanced attachment of AD cells that are resistant to attachment on standard surfaces	Non-coated, treated surface, highly hydrophilic; long-term stability & storage of the surface

TABLE 11.2. Protein and Nucleic Acid Binding

	Binding in ng/cm2				
	^{125}I-IgG	^{125}I-BSA	^{125}I-Insulin	^{32}P-Oligo DNA	^{32}P-λphage DNA
Polystyrene (PS)	400	450	310	22	6
Polypropylene (PP)	380	440	370	3	<2
NBS on PS	<2.5	<2.5	5	<2	<2

Assay (SPA), enzyme kinetics, fluorescence polarization (FP), and other assays in high throughput screening.

II.   GAPS (Gamma Amino Propyl Silane) Surface

Numerous surfaces have been developed to facilitate attachment of DNA for the generation of DNA microarrays. One of the first surfaces used was Poly-lysine developed by the pioneers in this field (1,2) Schena et al., 1995; DeRisi et al., 1996). Alternatively, amino silane-based surfaces were developed where a coating of the silane renders the slide hydrophobic with active amine groups for the attachment of negatively charged DNA molecules. In addition, numer-ous surface chemistries (e.g., aldehyde, epoxy, maleimide, etc.) have emerged since to facilitate immobilization of DNA. Important considerations in select-ing glass-based substrate for microarray printing are the quality and flatness of the glass (Figure 11.1) and the attachment chemistry. The GAPS and UltraGAPS™ chemistry from Corning is available on ultraflat glass that enables consistency in printing and DNA retention, uniform spot morphol-ogy, low backgrounds, and high signal-to-noise (Figure 11.2, UltraGAPS™).

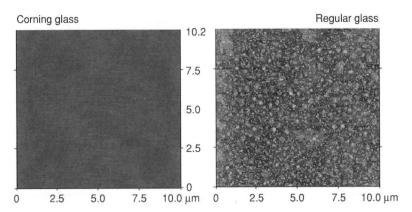

FIGURE 11.1. Atomic force micrographs of aminosilane-coated glass slides. Surface of the regular sodium-rich glass coated with aminosilane reveals the presence of globu-lar structures and crystalline bodies that were not observed on GAPS-coated slides.

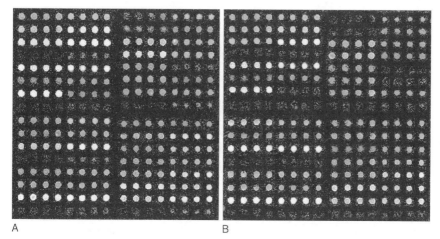

FIGURE 11.2. Panels show microarrays printed on UltraGAPS™ slides. In each slide a total of 32 human genes have been printed across 4 subgrids. Within each subgrid, each row of eight spots represents one gene. The first 4 spots within each row are 75 mer oligonucleotides and the last four represent dsDNA PCR product. Cy3- and Cy5-labeled probes generated from Testis and Placental RNA, respectively, were hybridized with the arrays (A). In another set of hybrdizations labeled probes were generated by inverse labeling (Placental-Cy3 and Testis-Cy5) and hybridized with the arrays (B).

III.   G Protein-Coupled Receptor (GPCR) Arrays on GAPS

Microarray technology enables probing the genome or proteome in a way that is naturally both systematic and comprehensive. The power and versatility of DNA microarrays is currently being realized; protein arrays have, however, lagged behind in development because of issues associated with maintaining the correctly folded conformations of proteins on solid substrates. The fabrication of arrays of membrane proteins including GPCRs is particularly challenging because these proteins typically need to be embedded in a membrane environment in order to maintain their native conformations. Membrane-bound proteins represent the single most important drug targets, as approximately 50% of current drug targets are membrane bound and 20% of the top 200 best selling drugs target GPCRs. G protein-coupled receptors (GPCRs), a large family of membrane-bound proteins that compose seven transmembrane domains (Figure 11.3) that mediate a great number of key cellular processes through the binding of ligands to the extracellular side of the receptor. The ligand binding leads to activation of the receptor and G-proteins associated with the receptor on the cytosolic side. There are an estimated ~ 400-700 GPCRs, approximately

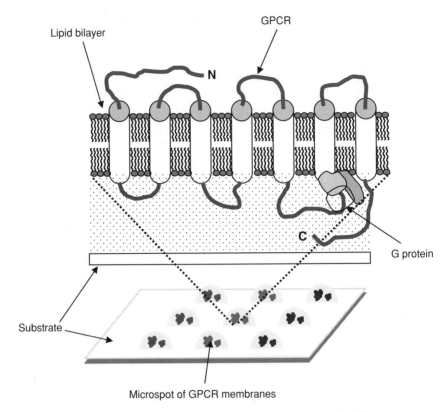

FIGURE 11.3. A 3 × 3 G protein-coupled receptor microarray on GAPS.

190 of which have known ligands. GPCRs for which ligands are unidenti-
fied are termed as orphan GPCRs, many of which are presumed to be
important drug targets. Given the large number of GPCRs as drug targets,
many of which bind identical or analogous ligands and yet carry out distinct tissue-specific functions, the implementation of microarray methods is
particularly relevant for the development of high-specificity ligands during
the process of drug discovery.

We recently described the fabrication of microarrays of GPCRs on surfaces coated with GAPS and described their use for ligand screening
assays (3-5). Surface chemistry plays a major role in the array performance. We were interested in surfaces on which membranes exhibited two
contradictory properties: high mechanical stability while preserving long-
range lateral fluidity. GAPS coated surfaces provide an optimal balance-
supported membranes exhibit long range lateral fluidity, and are also
resistant to desorption exposed to air, independent of the phase (gel or
fluid phase) of the lipid.

The GPCR microarrays can be used for a number of pharmacological applications:

(1) Determination of ligand affinity to the receptors: Arrays of human neurotensin receptor subtype 1 were used as a model system to investigate the feasibility of estimating the affinities of ligands using GPCR arrays (3). A modified saturation assay was developed and the binding of fluorescently labeled neurotensin (Bodipy-TMR-neurotensin, BT-NT) to the NTR1 arrays was examined. Results revealed that the binding of BT-NT to the NTR1 arrays was specific with a binding constant of 2 nM, consistent with those reported in literature. This result also suggested that receptors arrayed on GAPS retain their binding affinity and specificity for ligands.

(2) Determination of compound affinity to the receptors: In drug discovery and development against a GPCR, one common approach used involves competitive binding of a library of compounds of interest against a known, labeled ligand. This type of competitive binding assay was adopted to estimate compound affinity to GPCRs in the arrays. Arrays of human $\beta1$ adrenergic receptor and neurotensin receptor were treated with a solution containing a fixed concentration of fluorescently labeled ligands (1 nM BT-CGP 12177 for $\beta1$ or 2 nM BT-NT for NTR1) and different concentrations of a given compound of interest (3-5). From a plot of the inhibition profile, a Ki value for the given compound was calculated and found to be in agreement with the literature, suggesting that GPCR arrays can be used to obtain compound affinities. Coupled with the ability to test multiple GPCRs in parallel, these experiments also suggest the viability of using GPCR microarrays for highly multiplexed studies of compound potency.

(3) Compound screening: The screening of compounds using GPCR microarrays can be carried out in many different formats and with different types of GPCR "probes". For example, compounds could be tested against an array consisting of one member of each GPCR family or against an array consisting of all of the GPCRs within a family (e.g. the adrenergic receptors), or against a full index GPCR array. Ligand fishing for orphan GPCRs could also be carried out using GPCR microarrays. One example is presented in Figure 11.4. Arrays of the adrenergic receptor– each array contained the $\beta1$, the $\beta2$ and the $\alpha2A$ subtypes were used to screen highly selective compound against a subfamily member using a competitive binding assay (3-5). Fluorescently labeled CGP 12177 (Bodipy TMR CGP 12177, BT-CGP 12177), a β-selective ligand, was used as a probe. A small set of compounds including betaxolol, xamoterol, and ICI 118551 at different concentrations were used to examine the ability to inhibit the binding of the probe to the receptors in the arrays. Results shown that the differential affinity of ICI118551 for the $\beta1$ and $\beta2$ receptors was observed at 10 nM; while negligible inhibition is observed for the $\beta1$ receptor, approximately 74% inhibition is observed for the

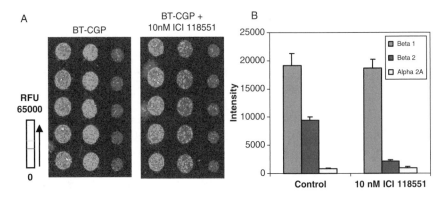

FIGURE 11.4. Demonstration of the use of GPCR microarrays for determining the selectivity of compounds among the different subtypes of a receptor. Each microarray consists of three columns; each column contains, from left to right, five replicate microspots of the b1, b2 and a2A adrenergic receptors, respectively.

(A) Fluorescence false-color images (from left to right) of the array incubated with solutions containing BT CGP12177 (5 nM), and mixtures of BT-CGP 12177 (5nM) and ICI 118551 (10 nM) to the array.

(B) Histogram analysis of the images in (A) showing the relative fluorescence intensities (RFU) of the arrays incubated with BT CGP 2177 in the absence and presence of ICI 118551.

(C) Table showing the amounts of inhibition and the Ki values for the inhibitors used in the experiment.

β2 receptor (Figure 4). These experiments clearly demonstrate that GPCR microarrays are a powerful tool to test compound selectivity across closely related GPCRs.

IV.   Novel Tissue Culture Surfaces

A.   Ultra Low Bind

Ultra low bind is a hydrophilic and neutrally charged surface that inhibits non-specific and hydrophobic interaction of proteins thus inhibiting subsequent cell attachment (6-9). This property has been found to be useful in maintaining cells in suspension, e.g., in the growth of stem cells in suspension and undifferentiated. The ultra low bind surface has been shown to successfully inhibit attachment of anchorage-dependent MDCK, VERO, and C6 cells grown to relative confluence.

B.   CellBIND™ Active Surface

Corning has recently introduced a unique surface for the enhanced attachment of fastidious cells that normally do not attach efficiently to standard tissue culture-treated surface or require protein coating of the latter for attachment and growth. CellBIND surface is a non-proteinaceous surface that has demonstrated efficient attachment and growth of a number of cell lines and primary cells, and for these cells eliminates the need for tedious, time-consuming, high cost and low stability biological coatings. The surface enables process economy and minimizes bottlenecks in cell culture scale-up, cells grown for reagent and assay development, and isolation, growth, and expansion of primary cells. Furthermore, since the surface is not coated it does not require special storage conditions and offers prolonged stability and storage, comparable to a standard tissue culture-treated surface.

Acknowledgements. We would like to thank Dr. Linda Belkowski for reading the manuscript and providing comments.

References

1. Schena, M., Shalon, D., Davis, R.W., and Brown, P.O. 1995. Quantitative monitoring of gene expression patterns with a complementary DNA microarray. Science 270, 467-470.
2. DeRisi, J., Penland, L., Brown, P.O., Bittner, M.L., Metlzer, P.S., Ray, M., Chen, Y., Su, Y.A., and Trent, J.M. 1996. Use of a cDNA microarray to analyze gene expression patterns in human cancer. Nat. Genet. 14, 457-460.
3. Fang, Y., Frutos, A.G., and Lahiri, J. 2002. Membrane protein microarrays. J. Am.Chem.Soc. 124, 2394-2395.
4. Fang, Y., Frutos, A.G., and Lahiri, J. 2002. G-protein-coupled receptor microarrays. Chem. Bio. Chem. 3, 987-991.
5. Fang, Y., Frutos, A.G., Webb, B., Hong, Y., Ferrie, A., Lai, F., and Lahiri, J. 2002. Membrane biochips. Biotechniques, In Press.
6. Andrade J.D., Ed. 1985. Surface and interfacial aspects of biomedical polymers. Vol. 2, Plenum Press, New York.
7. Martlin, K.S., and Valentich, J.D., Ed. 1989. Functional epithelial cells in culture. Alan R. Liss, Inc., New York.
8. Ratner, B.D., Ed. 1988. Surface characterization of biomaterials. Vol. 6, Elsevier, New York.
9. Shalaby, S.W., Hoffman, A.S., Ratner, B.D., and Horbett, T.A., Ed. 1984. Polymers as biomaterials. Plenum Press, New York.

12

Photoactivatable Silanes: Synthesis and Uses in Biopolymer Array Fabrication on Glass Substrates

HANDONG LI AND GLENN MCGALL

Chemistry, Affymetrix, Inc. 3380 Central Expressway, Santa Clara, CA 95051, USA

Abstract: We wish to report a fast and easy way to build hydrophobic layers and hydrophilic functional polymer layers thereafter onto glass surfaces. A benzophenone-based silane was synthesized and used to prepare stable, hydrophobic, photo-activatable coatings on glass supports. Hydrophilic polymers were then applied to the substrate, and photochemically cross-linked to the underlying silane. The resulting substrates had not only increased functional density due to plurality of functional groups on the polymer, but also enhanced stability against solvents and displacement reagents (such as water and phosphate salts). The substrates were suitable for fabricating oligonucleotide probe arrays either by in situ synthesis or immobilization methods.

1. Introduction

Non-porous flat solid substrates (e.g. glass substrates) have provided a format for manufacturing microarrays, which have revolutionized biological analysis in many ways such as miniaturization and parallel analysis. Biological materials, such as genes and antibodies, can be deposited in a precisely defined location. Small sample volume, high sample concentration and rapid hybridization or binding kinetics are possible to be achieved on such chip formats [1].

However, data quality and reproducibility will all rely on a stable and functional substrate. Many silylating agents produce coatings with undesirable properties including instability to hydrolysis and the inadequate ability to mask the silica surface which may contain residual acidic silanols. Methods have been developed for stabilizing surface bonded silicon compounds. For example, hydrophobic and sterically hindered silylating agents were described by Kirkland [2] and Schneider [3]. However, the use of these surface bonded silylating agents is disadvantageous, because they typically require very forcing conditions to achieve bonding to the glass, since their hindered nature makes them less reactive with the substrate.

Previously, high-capacity flat glass surfaces were prepared either by etching or coating with colloidal silica to increase the surface area and capacity[4-6].

In this case, the chemical nature of the glass surface and silane coating is the same as in the current flat glass supports. In another process, reactive polymer brushes were built on glass substrates using surface-initiated polymerization[7-9]. While this approach offers great potential, it is somewhat complicated and difficult to control. Recently, a method has been reported for grafting polystyrene films to glass surfaces using a photo-activatable silane[10]. Here we have adapted a similar approach for the attachment of polymers containing reactive functional groups which provide supports suitable for the synthesis and immobilization nucleic acids and other biomolecules. This paper describes the synthesis of the photoactivatable silanes, silanation on the glass surfaces, polymer coating and subsequent photoattachemt onto the surfaces. We also present photolithographic oligonucleotide synthesis on the surfaces and the hybridization of oligonucelotide target molecules. The polymer coated surfaces prepared by this method have the following advantages:

1. An initial hydrophobic silane coating offers protection from unwanted hydrolysis, and thus increased coating stability.
2. Polymer grafting provides multiple points of attachment to the substrate, which further increases the stability of the coating.
3. Plurality of functional groups on the polymer provides increased capacity for subsequent attachment of nucleic acid probes or other biomolecules, compared to conventional silanated flat substrates.
4. The polymer composition can be controlled in order to optimize properties for a given application (porosity, functional group content, intermolecular spacing, etc.).
5. Photolithographic oligonucleotide synthesis using MeNPOC chemistry [9] proceeds with substantially higher yield on these supports.

2.   Materials and Experiments

GC-MS analyses were performed on Agilent 6890 GC System with 5973 MD detector. UV-Vis data were acquired on a Varian Cary 3E spectrophotometer. Proton NMR was recorded on a Varian Gemini-400 spectrometer. All reagents and anhydrous solvents were purchased from Sigma-Aldrich and used without further treatment, except for the following: MeNPOC polyethyleneglycol phosphoramidite, Pierce Biochemical (Milwaukee, WI); dimethyl-N,N-diisopropylphosphoramidite, 2-[2-(4,4'-dimethoxytrityloxy) ethylsulfony]ethyl-(2-cyanoethyl)-(N,N-diisopropyl) phosphoramidite (5'-phosphate-ON reagent), ChemGenes (Waltham, MA); fluorescein phosphoramidite, 5'-DMT-dT 3'phosphoramidite, 5'-carboxyfluorescein phosphoramidite, 5'-MeNPOC 2'deoxynucleoside 3' phosphoramidites, Amersham Pharmacia Biotech (Piscataway, NJ); C3 spacer phosphoramidite, DNA synthesis reagents and anhydrous acetonitrile,

Glen Research (Sterling, VA); Silica gel (60 Å pore size, 230-400 mesh) from E. Merck.

2.1   Synthesis of 4-(3'-Triethoxysily) Propylamidobenzophenone (APTSBP)

2.1.1   Preparation of Benzoylbenzoyl Chloride (BPCOCl)

Thionyl chloride (50 ml) was introduced into a three neck round bottom flask equipped with a condenser, drying tube and a gas bubbler under Ar. Benzoylbenzoic acid (BP-COOH) (13.8 g, 0.061 moles) was added to the flask. The suspension was stirred for half hour at room temperature and then slowly heat to reflux. Control the heating and watch for gas release. The BP-COOH was totally dissolved after 20 minutes of reflux. No more gas was generating after 40 minutes, indicating the reaction was complete. The mixture was stirred for one more hour after no more gas is generating. The excess thionyl chloride was removed by distillation. The minor left over was thoroughly removed under vacuum. Use two dry ice cold trap and NaOH solid trap to avoid damage to the pump. An oiless, Teflon paraphram pump would be recommended for this usage. A light yellowish solid (14.6 g) was obtained. Yield: 97%

2.1.2   Preparation of APTSBP

Aminopropyltriethoxysilane (13.2 g, 14 ml, 0.06 moles), triethylamine (6.06 g, 8.3 ml, 0.06 moles), and anhydrous tetrahydrofuran (20 ml) were introduced into a three neck round bottom flask under Ar. The mixture was cooled in an ice bath. BPCOCl (14.6 g, 0.061 moles) dissolved in 50 ml of anhydrous THF was added drop wise with good stirring. Fume of Et3N-HCl could be observed. White precipitate formed after 10 ml was added. The ice bath was removed after addition. The final mixture was further stirred for one hour, at the time the mixture was warmed up to room temperature. Prepare filtration equipment in glove box under dry nitrogen. The solid was filtered off. Wash with dry THF (10 ml). The filtrate was passed through silica gel (30 g, in a Buchner funnel) under vacuum. Wash the silica with THF (10 mL). The final filtrate was put on a rotavap to remove THF. A light yellowish solid was obtained and dried under vacuum. The process has a yield of 24 g at 92%.

GC-MS: >95% purity, 429 (M+).
[1]   HNMR: (CDCl3, 400MHz), 0.75 (t, 2H,), 1.2 (t, 9H), 1.8 (quintet, 2H), 3.5 (quartet,2H), 3.8 (quartet, 6H), 6.8 (br, 1H), 7.5 (t,1H), 7.6 (d, 2H), 7.7 (d,2H),7.8 (d, 2H),7.9(d,2H).

2.2 Preparation of Hydrophobic and Photoactivable Layers on Glass Surfaces

Glass slides were cleaned by soaking successfully in Nanostrip (Cyantek, Fremont, CA) for 15 min, 10% aqueous NaOH/70 0C for 3 min, and then 1% aqueous 1% HCl fro 1 min, rinsing thoroughly with deionized water between each step, and then spin drying for 5 min. under a stream of nitrogen at 35 0C. The slides were then silanated for 1 hr in a gently agitating 1% solution of APTSBP in toluene, rinsed thoroughlywith toluene, then isopropanol, and finally dried under a stream of nitrogen.

2.3 Preparation of Functional Polymer Layers

A thin layer of polymer coat on glass surface was performed by spin-casting solutions of polymers at a spin speed of 2000 rpm for 20 second. Typical polymer solution were made in deionized water at 1% concentration by weight. The spin coated slides were dried in a 50 0C degree oven for 30 min and cooled down to room temperature before illumination. The illumination was carried out in a BioLink UV box at 254 nm, for 15 min. with a total energy of 2.2 joules. The slides were sonicated in water bath for 2 minute, then soaked in water at room temperature overnight. We rinsed slides with water and iso-propanol, then dried with a stream of nitrogen. The thickness of the resulting polymer layers were determined by ellipsometry.

2.4 Substrate Stability Test by Surface Fluorescence

A substrate slide was mounted onto a flow cell connected with an Affymetrix Arrray Synthesizer, following a standard synthesis cycle as described by McGall et al [11]. First the MeNPOC polyethyleneglycol phos-phoramidite was coupled on the substrate and capped with dimethyl-N,N-diisopropylphosphoramidite. The substrate was washed with acetonitrile and dried with Ar. A strip patterned mask was used to mask the substrate and a 365 nm light source was applied (total 6 Joules) for the photolysis. A mixture of fluorescein phosphoramidite (0.5 mM) and DMT-dT phospho-ramidite (49.5 mM) was introduced to react with the hydroxyls released from photolysis. The fluorescein was deprotected by ethylenediamine:ethanol (1:1) at room temperature for one hour. The substrate surface fluorescence intensity was acquired using a scanning confocal fluorescence microscope with photon-counting electronics. The intensity values are proportional to the amount of surface-bound fluorescein. The functional group density could be determined by direct comparison of the observed surface fluores-cence intensity. Surface fluorescence within the nonilluminated regions of the substrate was taken as nonspecific background. To test the stability of the substrate, the scanned slides were soaked in a 0.5 M sodium chloride

0.1 M sodium phosphate, 10 mM EDTA, 0.01% Triton, pH 7.8 buffer (SSPE) for 17 hours at 45 °C. The slides were rinsed with water and re-scanned.

2.5   Hydroxyl Site Density Measurement

Fluorescein tagged molecules were synthesized on the glass surface on an Affymetrix Array Synthesizer using phosphoramidite chemistry. A cleavable linker (5'-phosphate-ON reagent) was synthesized on the surface, followed by a spacer (C3 spacer phosphoramidite) and 5'carboxyfluorescein phosphoramidite). The surface was then diced into about 1 cm2 pieces, weighed, placed in a glass vial and cleaved into 1.0 ml of ethylenediamine:water (1:1) containing internal standard at 50 °C for 4 hours. The released 3'pC3-fluorescein was analyzed by a Beckman System Gold HPLC with an ion-exchange column and fluorescence detector. The internal standard 3'pC3C3-fluorescein was made separately on an ABI synthesizer and quantified by UV-Vis.

2.6   Photolithographic Synthesis of Oligonucleotides on Prepared Substrates

Oligonucleotides were synthesized on an Affymetrix Array Synthesizer using standard DNA synthesis cycles as described by McGall et al [11]. Photodeprotection was performed with an open square mask using 6 joules of 365 nm irradiation. All phosphoramidites were used at a concentration of 50 mM in anhydrous acetonitrile. The sequence of coupling is the following: a spacer, the MeNPOC polyethyleneglycol phosphoramidite; unreacted hydroxyls capping by dimethyl-N,N-diisopropylphosphoramidite; a cleavable linker, 2-[2-(4,4'-dimethoxytrityloxy)ethylsulfony]ethyl-(2-cyanoethyl)-(N,N-diisopropyl)phosphoramidite; a fluorescent label, 5-carboxyfluorescein phosphoramidite; and then a sequence of 5'-MeNPOC 2'-deoxyribonucleoside phosphoramidites. After synthesis, the entire synthesis area (about 1 cm²) was diced, placed in a glass vial, and cleaved into 1.0 ml of ethylenediamine:water(1:1) containing internal standard at 50 °C for 4 hours. The cleaved products were analyzed by a Beckman System Gold HPLC with an ion-exchange column and fluorescence detector. Elution had a linear gradient of 0.4 M $NaClO_4$ in 20 mM Tris pH 8.0 buffer at a flow rate of 1.0 ml/min.

To determine the efficiency of oligonucleotide synthesis, a relative synthesis yield was calculated by dividing the integrated area of the full length oligo peak by the total area of all products cleaved from the surface.

2.7   Target Hybridization to Probes

Oligonicleotide probes (20mer) were synthesized in a checkerboard pattern on Affymetrix Array Synthesizer using photolithography and phosphoramidite chemistry as described by McGall et al [11]. The removal of protecting groups was performed by soaking the slides in ethylenediamine:ethanol (1:1 by vol) solution for alt least 4 hrs at room temperature. A typical probe sequence (from the surface) was 3'-GACTTGCCATCGTAGAACTG-5'. The 5'-fluorescein-labeled oligo-nucleotide target had a concentration of 10 nM in a hybridization buffer containing 0.1M MES (2-[N-morpholino]ethanesulfonic acid, 0.89M NaCl, and 30 mM NaOH, pH 6.8). This target concentration is to saturate surface probes. The hybridization was carried out by soaking the slides in the target solution at 45 °C for 17 hrs with gentle agitation. The slides were scanned on a confocal fluorescence microscope after extensive washes of the slides with fresh hybridization buffer.

To evaluate hybridization discrimination, probe sequences on the surface were designed to include perfect matches, a single-base mismatches at base 10 (A substituted for T) and two-base mismatches at both bases 10 and 12 (A substituted or G). The discrimination ratio was defined in the following equation in which the Fpm is the fluorescence signal intensity for perfect match regions and Fmm is the signal intensity for mismatch regions after background correction).

$$Rd = (Fpm - Fmm) / (Fpm + Fmm)$$

3. Results and Discussion

3.1 Photoactivatable Silane

Shown in Scheme 12.1, the benzophenone moiety was coupled to a typical silane compound with high purity and high yields. A quick filtration through

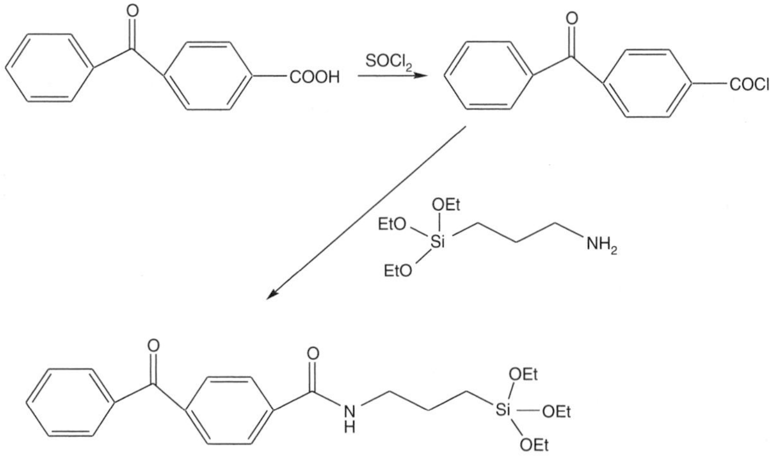

SCHEME 12.1. Synthesis of (4-(3'-triethoxysily) propylamidobenzophenone) (APTSBP).

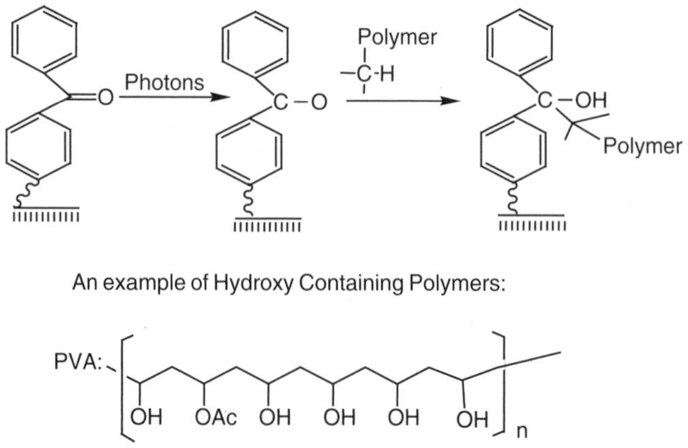

An example of Hydroxy Containing Polymers:

FIGURE 12.1. Photocrosslinking of polymers onto the benzophenone layers.

a short dry silica gel column provided a sufficient purification on the product. The purity and structure identity were confirmed by GC-MS and proton NMR. This synthetic strategy could be readily extended to make a variety of photoactivatable silanes, including benzophenone and azido compounds.

The photochemistry of benzophenone is depicted in Figure 12.1. The benzophenone structure was chosen for several advantages. It is not only chemically inert and stable as well under ambient light. It has well-characterized photochemistry and the light generated intermediates non-selectively insert to C-H bonds in a wide range of different chemical environments. Water molecules do not affect its photoactivity due to the reversibility of the mechanism in contrast to the water quenching problem in nitrene (azido) system [12].

The process is depicted in Figure 12.2. The benzophenone layer was first immobilized on glass surfaces using standard silanation. The molar concentration of the silane is aout 20 mM in anhydrous toluene. A base catalyst triethylamine is also added to a final concentration of 20 mM. This anhydrous procedure gives monolayers of the benzophenone silane on the glass surface [13].

Prior to photocrosslinking, the polymer was spin-coated on the glass surface. The polymer concentration of 1% by weight in water gave a thin layer coating on the top of benzophenone silanes. In the spin-coating process, it was found the polymers need to have surfactant effect, so that the polymer solutions had low surface tension to be able to spread on the hydrophobic benzophenone silane surfaces. Polyvinyl alcohols and polyacrylamide derivatives all were found to have those required properties. The polymer and benzophenone silane layers were then illuminated with UV light at 254 nm. Benzophenone has a maximum absorption around 260 nm. The intermediate is actually a biradical triplet. It either inserts to C-H bond in its vicinity or returns to the ground state which can be reactivated by light. This nature

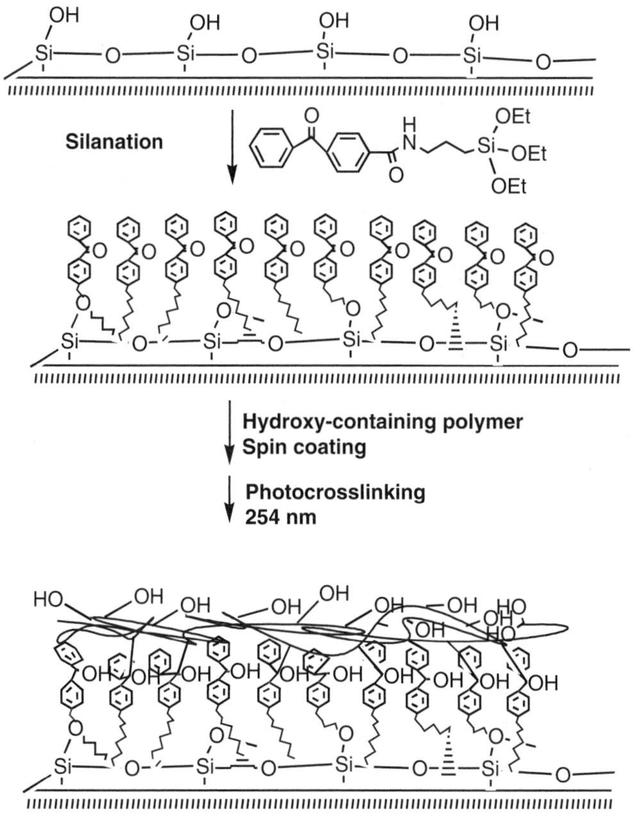

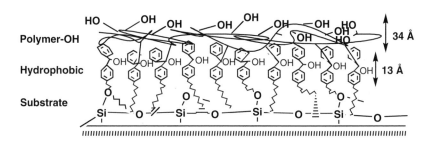

FIGURE 12.2. Process of polymeric surface on glass substrate

allows benzophenone to have higher photoattachment efficiency. The thickness of the hydrophobic layer and the polymer layer was measured at 13 Å and 34 Å respectively, each corresponding to a monolayer of molecules (Figure 12.3).

FIGURE 12.3. Thickness of the multilayered polymer surface.

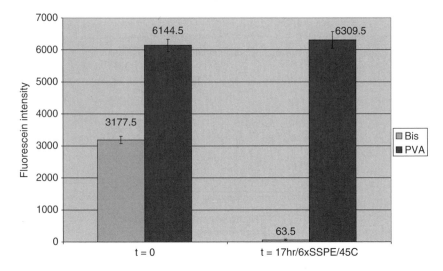

FIGURE 12.4. Substrate stability comparison by surface fluorescence analysis. Bis suface is a standard silence surface. PVA is the polymeric surface.

3.2 Stability of the Polymeric Surfaces

Surface fluorescence analysis was employed to test the substrate stability. The stability comparison is shown in Figure 12.4. The polymeric surfaces are stable under harsh hydrolysis conditions. The instability and loss of functional groups on the substrate are due to hydrolysis of siloxy bonds formed on the surfaces. Again the hydrolysis released silanol can go back to the surface to form siloxy bond. There exists an equilibrium of siloxy bond hydrolysis and reformation on the surface. The polymeric substrate prepared in this paper has the following characteristics. Multiple attachment points better prevented the loss of functional groups. On the other hand, the hydrophobic nature of silane layers efficiently protected the surface from water attack.

3.3 Hydroxyl Site Density and Oligonucleotide Synthesis Efficiency

Figure 12.5 shows the procedures for quantification of surface hydroxyl density and analysis of oligo synthesis efficiency. The cleaved products are quantified by HPLC. Comparisons of site density and synthesis efficiency between standard silane surface and the polymeric surface are summarized in Table 12.1.

Chromatograms of T6mer and mixed base oligo16mer synthesis were shown in Figures 12.6 and 12.7. The 16mer oligo sequence is GAATGA-CATTTACAGC. The hydroxyl site density on PVA surface is 30% higher than that on standard silane surface. T6mer synthesis efficiency improved by 55% and mixed base oligo16mer synthesis efficiency improved 62%. These results

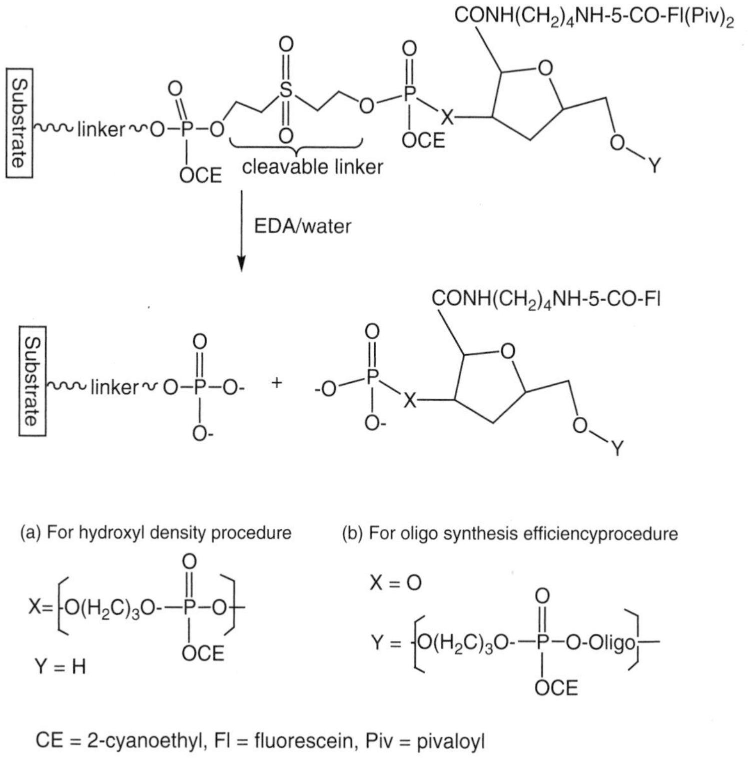

FIGURE 12.5. Procedures for surface hydroxyl density quantification and oligo synthesis efficiency analysis.

TABLE 12.1. Comparison of Site Density and Efficiency

Glass substrate	Hydroxyl density pmoles/cm^2	Relative T6 synthesis yield (%)	Relative oligo16mer synthesis yield (%)
Standard	141.2	31.3	6.3
PVA	182.8	48.5	10.2

clearly indicate the hydroxyl groups on the polymer are more accessible for extended oligonucleotide synthesis off the glass surface. It is expected that longer oligonucleotide synthesis will have bigger improvement on PVA surface versus standard silane surface.

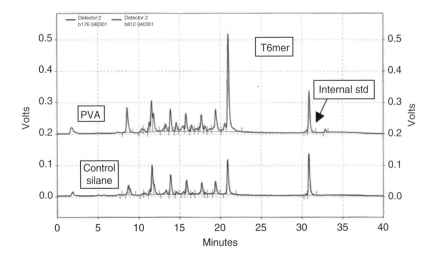

FIGURE 12.6. Synthesis of Oligonucleotides (T6mer) Comparison between standard silane and polymer surface.

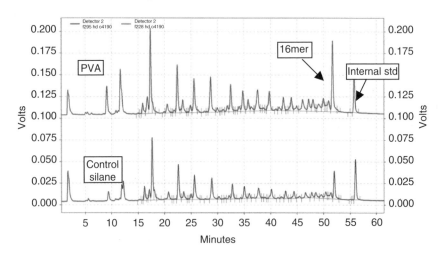

FIGURE 12.7. Synthesis of Oligonucleotides (16mer GAATGACATTTACAGC) Comparison between standard silane and polymer surface.

3.4   Hybridization and Discrimination

PVA surface gave hybridization signal more than four times higher than the standard silane surface after background correction. Both PVA and standard silane surfaces have similar discrimination ratio. The hybridization signal and discrimination ratio were shown respectively in Figure 12.8 and 12.9. Fpm, Fmm10, Fmm12, Fmm1012 are hybridization signal at perfect match, single

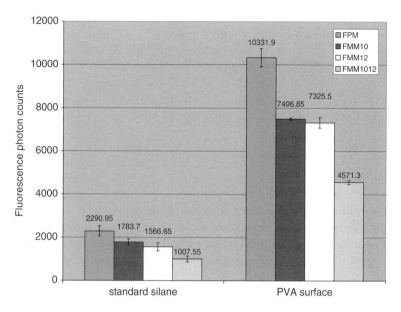

FIGURE 12.8. (also see the text).Hybridization (10 nM target in MES buffer at 45 0C for 17hrs).
Probe: 5′-GTCAAGATGCTACCGTTCAG-3′.
Target: 3′-CAGTTCTACGATGGCAAGTC-fluorescein.

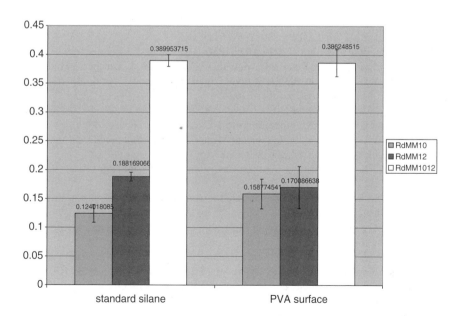

FIGURE 12.9. Discrimination ratio.

mismatch at base 10, single mismatch at base 12 and two mismatches at base 10 and 12. Rdmm10, Rdmm12 and Rdmm1012 are discrimination ratio for single mismatch at base 10, single mismatch at base 12 and two mismatches at base 10 and 12.

4. Conclusions

A photoactivatable silane was synthesized, characterized and used to photocross link functional polymers onto glass substrate. The layer of silane was hydrophobic thus to protect the surface from water and salt attack. Multiple crosslinking points of polymer macromolecule also enhance molecular attachment on the surface. They together provided a superior surface stability. The functional groups like hydroxyls on the surface bound polymer were more accessible to in situ synthesis of oligonucleotides with a 50-60% improvement on relative synthesis yield. The 20mer oligonucleotide probes synthesized on the polymeric surface were able to hybridize the complimentary target oligoncleotide with a similar specificity to what obtained on standard silane treated surface.

Acknowledgement. We thank Dr. Marc Glazer at Stanford University for his assistance with Elliposometry.

References

1. Schena M., DNA microarrays, a practical approach, Oxford University press (1999).
2. Kirkland et al., Anal. Chem., 1989, 61:2-11.
3. Schneider et al., Synthesis, 1990, 1027-1031.
4. Glazer M., Franck C., Vinci, RP., McGall, G.H., Fidanza, J., Beecher, J. High surface area substrates for DNA arrays, Organic/Inorganic Hybrid Materials II, Material Research, 1999, 576:371-376.
5. Glazer M. et al. Porous silica substrates for polymer synthesis and assays, Submitted as conference proceedings to Nucleoside and Nucleotides.
6. Hodges JC., et al., Free living radical polymerization of functional monomers on solid support, J. Comb. Chem., 2000, 2:80-88.
7. Currie EPK, et al., Stuffed brushes: theory and experiment, Pure Appl. Chem., 1999, 71:1227-1241.
8. Huang X., Wirth M. J., Surface-initiated radical polymerization on porous silica, Anal. Chem., 1997, 69:4577.
9. McGall G., et al., Macromolecular arrays on polymeric brushes and methods for preparing the same Patent Application– case #3254.1
10. Frank C. W., et al., Photochemical attachment of polymer film to solid surfaces via monolayers of benzophenone derivatives, J. Am. Chem. Soc., 1999, 121:8766.
11. McGall et al., The efficiency of light-directed synthesis of DNA arrays on glass substrates, J. Am. Chem. Soc., 1997, 119:5081-5090.
12. Elender, G., et al., Biosens. Bioelectron., 1996, 11:565.
13. Leyden, D. E., Silanes surfaces and interfaces, Gordon and Breach Science Publishers (1986).

Part IV

Bioinformatics and Drug Discovery

13

An Integrated Biochemoinformatics System for Drug Discovery
Managing and Mining Chemical Structure Information, Biological Activity Fingerprints, and Gene Expression Profiling Patterns

LEMING SHI, ZHENQIANG SU, AIHUA XIE, CHENZHONG LIAO, WEI QIAO, DAJIE ZHANG, SONG SHAN, DESI PAN, ZIBIN LI, ZHIQIANG NING, WEIMING HU, AND XIANPING LU
Chipscreen Biosciences, Ltd.

Abstract: Chipscreen Biosciences, Ltd. (www.chipscreen.com) is a drug discovery company specialized in novel small molecule therapeutics. Chipscreen has developed a proprietary chemical genomics approach to accelerate the discovery of new medicines from its collection of natural products, traditional Chinese medicines, and synthetic chemical libraries. Central to its drug discovery platform is Chipscreen's capability of integrating *in silico* drug design, chemical synthesis, unique parallel multi-target high throughput screening, global gene expression profiling, and informatics to rapidly and effectively advance the drug discovery process. To fulfill Chipscreen's drug discovery needs, we have developed an integrated biochemoinformatics system to efficiently manage and mine various types of experimental data, including chemical structure information, biological activity fingerprints, and gene expression profiling patterns. Well-informed decision on which drug candidates should be advanced into preclinical and clinical development can be made by maximizing the utilities of experimental data stored in the database, thereby lowering the risk and increasing the success rate of the drug discovery and development process.

Key words: Biochemoinformatics, bioinformatics, chemoinformatics, data mining, data visualization, computer-aided drug design, medicinal chemistry, high throughput screening, DNA microarray, gene expression profiling.

1. Introduction

The worldwide pharmaceutical industry is a huge business with annual sales of about 400 billion U.S. dollars (Class, 2002). This industry is very profitable and risky. The process of discovering and developing a new drug is very

time-consuming and expensive (Drews, 2000; Smith, 1992; also see information at the Pharmaceutical Research and Manufacturers of America
(PhRMA)'s web site: www.phrma.org), as is shown in Figure 13.1. On average, it costs a pharmaceutical company about 800 million U.S. dollars and 10
to 15 years to move a new chemical entity (NCE) through the Food and Drug
Administration (FDA) approval process. Most of the compounds synthesized by chemists fail at different stages in the process and on average, only
one out of every 5,000 to 10,000 compounds synthesized by chemists reaches
the market.

To ensure the safety and efficacy of approved drugs, the pharmaceutical
companies and the FDA take extraordinary measures to conduct and regu-
late the clinical trials for drug candidates after enough animal testing data
have been collected and evaluated. It usually takes a pharmaceutical company six and a half years of exploratory research and preclinical animal testing before an INDA (Investigational New Drug Application) is filed with the
FDA. Clinical trials cannot be initiated until the INDA is approved.

In Phase I toxicity testing which usually takes one and a half years, the
drug candidate is evaluated for safety in healthy volunteers in a population of
20 to 80 people per trial. A trial is conducted in a single dose of the drug candidate, beginning with small doses. If the compound is shown to be safe, multiple doses are evaluated in other clinical trials with escalating dosages. In
addition to safety, researchers also try to determine the most frequent side
effects and to learn how the compound is metabolized and excreted. If the

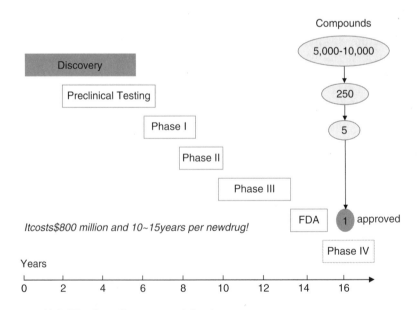

FIGURE 13.1. The drug discovery and development process.

compound is shown to be safe in Phase I, it is advanced to Phase II. Generally, only two out of five compounds tested in Phase I emerge as acceptable and progress to Phase II.

The main purpose of Phase II which takes about two years, is to determine the efficacy of the drug candidate in a population of 100 to 500 patient volunteers although safety is also studied. Data are collected and analyzed to determine whether the compound is efficacious and safe for the intended patient population. If so, the drug is advanced further to Phase III clinical trials. Only half of the compounds tested in Phase II progress to Phase III.

Under Phase III trials which takes about three and a half years, safety and efficacy are further evaluated in a much larger population of as many as 1,000 to 5,000 patient volunteers including the elderly, patients with multiple diseases, patients who take other drugs, and patients whose organs are impaired. Most drug candidates that are successful in Phase II also are found to be acceptable in Phase III.

If a drug candidate shows superior efficacy and safety profiles over available medicine approved by the FDA for the same disease, the pharmaceutical company can file with the FDA an NDA (New Drug Application) for an approval of marketing this drug. The NDA includes all animal and human testing data, analyses of the data, and information on how the drug is manufactured. The FDA makes a decision on 90% NDAs within 10 months. On average, only one out of five drug candidates in clinical trials is eventually approved by the FDA.

In some cases, drugs may be approved for sale before the FDA is totally satisfied with their safety data. But the pharmaceutical company is required to continue conducting additional safety studies and tracking drug safety among the treated patients. This process is called Phase IV, or post-market surveillance. New uses of the drug, new patient populations, and the long-term effects of the drug may be explored at this time.

The pharmaceutical company is under enormous pressure to fill its pipeline with novel drug molecules in order to maintain a competitive edge on the market (Class, 2002). On average, a research-based pharmaceutical company invests 10 to 20% of its annual revenue in research and development. Competition in the pharmaceutical industry forces the company to adopt innovative technologies that promise to bring a molecule to the marketplace in a shorter period of time. The key is how to shorten the period required for discovery and preclinical testing since the period of time required for clinical trials cannot be changed.

Advanced technologies, such as functional genomics, computer-aided drug design, automated synthesis, high throughput screening, global gene expression profiling, and informatics, have been increasingly used in the drug discovery and development process in the hope of increasing productivity and minimizing failure rate. Functional genomics promises to identify more biological targets that can be used for drug development. Computer-aided drug design, or rational drug design, eliminates the need of synthesizing unlikely drug candidates. Automated synthesis, or combinatorial synthesis, makes it possible for chemists to synthesize a large number of drug candidates in a short period of time that are to be evaluated in a high throughput screening manner. Promising candidates are further evaluated in preclinical testing before they can be tested in humans in various stages of clinical trials.

A challenge that comes with the increasing use of cutting-edge drug discovery technologies is the huge amount of data being accumulated. The efficient management and interpretation of these data present a new burden to informatics support within a pharmaceutical company.

2.   An Integrated Drug Discovery Platform

Chipscreen Biosciences, Ltd. (www.chipscreen.com) is a drug discovery company specialized in the discovery and development of novel small molecule therapeutics against type 2 diabetes, osteoporosis, benign prostate hyperplasia, and cancer. Chipscreen has developed a proprietary chemical genomics approach (Seghal, 2002) to accelerate the discovery of new medicines from its collection of natural products, traditional Chinese medicines, and synthetic chemical libraries. Central to its drug discovery platform is Chipscreen's capability of integrating *in silico* drug design, chemical synthesis, unique parallel multi-target high throughput screening, global gene expression profiling, and informatics to rapidly and effectively advance the drug discovery process.

Figure 13.2 shows the flowchart of Chipscreen's integrated drug discovery platform. Briefly, we start with computer-aided drug design, in which many computational approaches are being used. The compounds through computer-aided design are synthesized by our chemistry department using traditional medicinal synthesis and small-scale parallel synthesis. Along with our collections of traditional Chinese medicines and natural products, these samples are screened in multi-target assay models. Promising candidates from the screening are chosen to treat particular cell lines and their impact on gene expression profiling of those treated cell lines is determined by DNA microarray and compared with those of reference compounds. It helps us to select preclinical candidates rationally for animal studies or later on in clinical trials. We currently have several compounds in preclinical studies and hope that some of them will progress to clinical trials within the next two years.

2.1   Computer-Aided Drug Design

Traditional drug synthesis used to be solely based on a chemist's intuition, i.e., the chemist decides which compound to synthesize. This process is largely irrational and highly depends on a chemist's previous experience. However, with the rapid advances in functional genomics, structural genomics, and more in-depth understanding of human diseases on a molecular basis, many disease-related biological targets have been identified and shown to be draggable

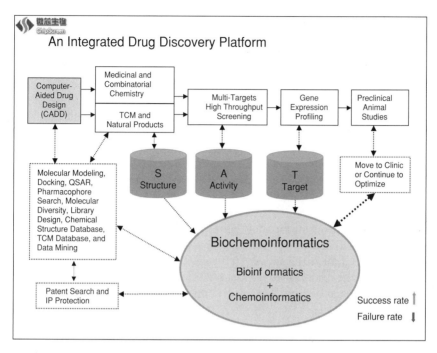

FIGURE 13.2. An integrated drug discovery platform.

targets (Drews, 2000; Hopkins and Groom, 2002; Knowles and Gromo, 2002). In many cases, the 3D structures of these targets have been determined either through X-ray crystallography or nuclear magnetic resonance spectroscopy. These advances have resulted in the increasingly wide uses of computer-aided drug design (CADD) methodologies for the rational design of drug molecules (Chen et al., 2000; Codding, 1998; Taylor et al., 2002).

CADD has become an essential component of a pharmaceutical company's drug discovery program. The goal of CADD is to increase the success rate of the drug discovery and development process by eliminating those candidates that are less likely to be developed as drug candidates due to, e.g. low activity or high toxicity. Depending on whether the 3D structure of the drug target is available and used in the CADD process or not, CADD methodologies can be divided into two major categories: structure-based drug design (SBDD) and ligand-based drug design (LBDD).

In SBDD, the 3D structure of a drug target, usually a protein, is used to search for small drug-like molecules that form a favorable match to the active site of a drug target, resulting in a therapeutic effect via a mechanism of inhibition of an undesirable enzymatic activity or the induction of a protein function through the activation of a nuclear receptor, etc. Molecular docking is the method of choice for evaluating the likelihood of binding between a drug target and a small molecule. A scoring function that evaluates the goodness of fit between drug targets and small molecules is employed to rank-order a library of small molecules, which may be the in-house compound repository of a pharmaceutical company or compounds enumerated virtually by a computer program through a combinatorial algorithm.

In situations where the 3D structure of a drug target is unavailable and no homology model is appropriate, LBDD methods can be used to assist the drug design process. A prerequisite for using LBDD method is the availability of biological activity for a set of known compounds. With modeling packages, each compound can be represented by a set of molecular descriptors that characterize different aspects of molecular features. These descriptors are treated as independent variables (x) and correlated with experimentally determined biological activity (y) through an array of mathematical and machine learning methods. The function, f, or a QSAR (quantitative structure-activity relationship) model that relates chemical structure information to biological activity, $y = f(x)$, can be applied with reasonable accuracy to the prediction of biological activity (efficacy or toxicity) of compounds not included in the training or model-building process. Therefore, synthetic efforts can be focused on those compounds that are predicted to be more active and less toxic.

From a practical point of view, CADD methods are used for two main purposes: lead identification and lead optimization. QSAR, pharmacophore identification and 3D database searching, molecular diversity analysis, drug-likeness analysis, and related virtual screening methods have been fulfilling CADD needs. CADD methods can point to the right direction for structural modification for enhancing biological activity and minimizing undesirable properties like toxicity.

A computer database of natural products has been used for virtual screening against drug targets used in Chipscreen's drug discovery programs.

2.2   Medicinal Chemistry and Traditional Chinese Medicines

Compounds designed through CADD methods are synthesized via traditional synthetic approach and/or combinatorial synthesis. For a small number of molecules, traditional, one-compound-at-a-time method suffices. On average a medicinal chemist can synthesize about 100 unique compounds per year using traditional medicinal chemistry approach. However, during lead optimization process a much larger number of compounds are needed to be synthesized in order to identify the best candidates. The use of high throughput screening techniques also makes it possible to screen a large number of compounds. Combinatorial or automated synthesis (Agrafiotis et al., 2002) is gaining popularity for being able to synthesize a large number of compounds in a parallel fashion in a short period of time and promises to make the drug discovery process more productive.

Let us use the reaction A + B + C => D as an example to illustrate the principle of combinatorial synthesis. If we have 10 options for reagents A, B, and C, respectively, we can generate a library of $10 \times 10 \times 10 = 1,000$ potential products D by making all the combinations. Special automated synthesizers and robotic systems have been made available for performing the synthesis, separation, and purification. Computational analysis of the diversity of reagents helps to choose which reagents to be used for the synthesis in order to maximize the diversity of the combinatorial library.

Combinatorial synthesis is divided into solid phase synthesis and liquid phase synthesis. Solid phase synthesis is a methodology whereby one of the reactant molecules is attached to a resin referred to as the solid support. After the completion of a reaction, the products are cleaved from the resin. On solid phase synthesis, the separation of products is achieved simply by filtration, and millions of compounds can be prepared at a time. The disadvantage of this solid method is that it can be only applied to a limited number of reactions, e.g. peptide synthesis. Liquid phase synthesis is a methodology whereby all reactions are carried out in their own separate reaction vessels. It is the typical automated parallel synthesis. It is suitable for many kinds of reactions in a relatively large scale. The disadvantage is that the number of compounds prepared at a time is smaller.

One challenge for combinatorial synthesis is to adapt the chemical reactions into an automated way and to be able to control reaction conditions in individual reaction vessels since the combinations of reagents are different and generally require different handling conditions. Another challenge is the limited structural diversity of the synthesized library. It comes without surprise because all compounds generated in a particular library from the same reaction share a core structural frame or scaffold. Therefore, combinatorial synthesis is useful for lead optimization after a lead is available. Finding a lead compound is more efficient from a collection of compounds with much more structural diversity.

Mother nature has been a rich source for therapeutic molecules. It has been estimated that more than half of the drugs has a nature's origin. Nature continues to be an important source for drugs. An important feature of molecules from the nature is their unparallel structural diversity and complexity compared to molecules synthesized from a chemist's lab. We have been collecting samples of natural products from extreme conditions in the hope to identify novel therapeutic molecules. Traditional Chinese Medicines (TCM) with proven clinical efficacy have also been under intensive study at Chipscreen in order to identify compounds or components that are responsible for efficacy.

Synthetic compounds and natural products are all undergone various screening assays for activity and toxicity evaluation.

2.3   Multi-Target High Throughput Screening

With the rapid advances in functional genomics, structural genomics, and more in-depth understanding of human diseases on a molecular basis, many biological targets related to diseases have been identified and are being used as targets for drug discovery. On the other hand, with the increasing use of combinatorial synthesis and the accumulation of a large number of compound samples from different sources, the number of compounds that need to be screened for a particular activity has been increased dramatically. How to rapidly identify the small subset of compounds that are active against a particular target is a critical step in the drug discovery process. This difficult task is analogous to finding a needle from a haystack.

High throughput screening (HTS) enables the screening of thousands to hundreds of thousands of samples per day (Bajorath, 2002). HTS technology was developed at the end of the 1990's and its core components include an *in vitro* assay model at a molecular or cellular level, a robotic liquid handling system, a sensitive biological signal detection system, and a computerized data acquisition and management system.

Each assay model detects only one aspect of the biological functions of a compound. However, the biological system is complex in which many biological molecules function in a well orchestrated way. Disruption of the functions of one drug target by a drug molecule is likely to cause a cascade of other functional changes related to such a disruption. Therefore, it is highly desirable to detect the biological effects of a drug molecule on different biological targets in order to obtain an overall survey of the activity pattern of a drug molecule.

At Chipscreen, we have developed a battery of assay models for the high throughput screening of drug candidates for treating type 2 diabetes, osteoporosis, benign prostate hyperplasia, and cancer. These models collectively detect the biological efficacy, toxicity, and metabolism of drug candidates. The resulting activity fingerprint or profile is used for selecting a subset of lead compounds for further evaluation, including their impact on gene expression of *in vitro* cell models.

2.4   Microarray Gene Expression Profiling

It is understood that thousands of genes and their products (i.e., RNAs and proteins) in a given living organism function in a complicated and orchestrated way that creates the mystery of life. However, traditional methods in molecular biology generally work on a "one gene in one experiment" basis, indicating that the throughput is very limited and the "global picture" of gene functions is hard to obtain. In the past several years, a new technology, called microarrays or gene chips (www.gene-chips.com), has attracted tremendous interests among researchers (Lockhart and Winzeler 2000; Schena, 1999, 2002; Shi et al., 2003). This technology promises to monitor the whole genome on a single chip so that researchers can have a much broader and better view of the interactions among thousands of genes simultaneously.

The fundamental concept of microarray technology is to miniaturize traditional bioanalytical detection system so that hundreds or even thousands of biomolecules with unique identity can be detected simultaneously in one single experiment by using a tiny amount of test sample. Therefore, it is essential to achieve high sensitivity for a tiny amount of analyte in test samples. High sensitivity is usually achieved by fluorescence or radioactivity.

The pharmaceutical industry has been a major driving force for the widespread utility of microarray technology. Identifying novel drug targets can give a pharmaceutical company many advantages in developing new therapeutics. For example, drugs against a novel target may be able to avoid the problem of drug resistance encountered with previous drugs. Such benefits have been seen in the development of new AIDS drugs against different targets. There are only several hundreds of targets for currently available drugs (Drews, 2000), but the number of druggable targets is estimated to be at least in the thousands (Drews, 2000; Hopkins and Groom, 2002; Knowles and Gromo, 2002).

Although gene expression analysis for drug discovery is still a very important area for microarray applications and will remain so for many more years to come, researchers are combining their microarray studies with other data, then trying to map all those data to biological pathways and systems. Meanwhile, an increasing number of microarray studies are being done at the early development stage, e.g. in lead optimization and preclinical evaluation, specifically, in toxicological studies.

At Chipscreen, an important aspect of applying microarray technology in our drug discovery and development process is gene expression profiling for candidate evaluation based on the principle of toxicogenomics (Nuwaysir et al., 1999; Hamadeh et al., 2002; Lakkis et al., 2002; Ulrich R and Friend SH, 2002), a hybrid of functional genomics and molecular toxicology. The goal of toxicogenomics is to find correlations between toxic responses to toxicants and changes in the genetic profiles of the objects exposed to such toxicants.

Gene expression profiles of our own lead compounds and controls, i.e. drugs on the market, candidates from our competitors, and similar drugs with adverse effects, are compared. The rationale is that for a candidate to move forward in the R&D pipeline, it should not show gene expression profile similar to drugs causing severe adverse effects. In addition, specific genes related to the mechanism of toxicological effects are being investigated in great detail.

2.5   Integration via Biochemoinformatics

Bioinformatics can be described as any use of computers to handle biological information. Similarly, chemoinformatics can be described as any use of computers to handle chemical information. Historically, these two branches of science have been evolved independently. In the practice of today's drug discovery and development, there is an increasing need of managing both biological and chemical information simultaneously. Therefore, we coined the term "biochemoinformatics" and define it as the integration of bioinformatics and chemoinformatics.

As we described above, drug discovery in the post-genomic era is featured by the application and integration of novel technologies of genomics, gene and protein expression, combinatorial chemistry, and high throughput screening and biological experimentation. The consequence is that huge amounts of data are being generated and accumulated at an unprecedented speed. A bottleneck problem in drug discovery is how to effectively store, manage, analysis, and interpret those data and turn them into knowledge, and ultimately to be able to guide the drug discovery process.

In the process of drug discovery and development, we are facing with the accumulation of both biological and chemical information. It is necessary for us to integrate bioinformatics and chemoinformatics in order to make maximal use of the data for drug discovery. The approach that is being taken at Chipscreen is to integrate the power of bioinformatics with that of chemoinformatics by developing an integrated biochemoinformatics system.

3.   Methods and Materials

Molecular modeling study is performed on SGI workstations or PCs using commonly available modeling packages. Databases are run under Oracle 9i on a Sun Enterprise 3500 server. Microsoft Visual C++ is used for software development.

4.   Results and Discussion

It becomes obvious from the above discussions that huge amount of data is being generated in the drug discovery process, including chemical structure information from chemistry, biological activity fingerprints from HTS, and molecular targets or ADMETox-related (absorption, distribution, metabolism, excretion, and toxicity) information from gene expression profiling patterns. Software packages for managing one or two types of these data have been made available. However, no software is commercially available for handling all three types of diverse data.

To effectively handle these diverse types of data and accelerate our drug discovery process, we have developed a biochemoinformatics software system for our internal uses. The system includes two main components: a relational database based on Oracle and an application system for data mining and visualization. It is expected that our drug discovery platform through biochemoinformatics integration will allow us to increase success rate and decrease failure rate.

4.1   Relational Database for Managing Chemical Structure Information, Biological Activity Fingerprints, and Gene Expression Profiling Patterns

Integrating and mining diverse data on chemical structures information (S), biological activity fingerprints (A), and biological target information from gene expression profiling patterns (T) are essential for making well-informed decisions during the process of drug discovery and development. Our database systems are built on the most widely used relational database platform, i.e. the Oracle platform, which has been considered as the industry standard.

Chemical structure information is usually handled by using a specially designed commercial chemical information management system. The most widely used system is MDL's ISIS system (www.mdli.com), which uses a flat or hierarchical database structure to store chemical structure information and related property data. One disadvantage of the ISIS system is the inconvenience in communicating with other types of data that are usually stored in a relational database system, e.g. an Oracle database. The difficulty in chemical information management is the efficient handling of chemical structure information, which requires special mechanism for information storage and retrieval. Generally, atom and bond information of a molecule can be represented in a tree structure in which the vertices represent atoms and the edges represent chemical bonds. More importantly, there must be intelligent algorithms for performing complicated, time-consuming chemical structure search, usually through a graph matching algorithm. Major chemical software vendors like MDL, Accelrys (www.accelrys.com), and Tripos (www.tripos.com) have all claimed to have implemented the Oracle version of their respective chemical structure database system. However, performance of the Oracle-based system is not satisfactory and the price tag is very high.

We have defined a unique database format to store chemical structure information in Oracle in a concise and searchable way. The storage, search, and display of chemical structures have been effectively implemented. In addition, chemical structure information has been fully integrated with other types of data, e.g. physicochemical properties, biological activity fingerprints, and gene expression profiling patterns. At present, we have built up an Oracle-based chemical structure database of about 1 million compounds, each of which is accompanied with an array of molecular descriptors. These descriptors can be used for CADD purposes.

Biological activity data are also stored in the Oracle database in a straightforward way. Assay information and assay results are specified and stored. Dose-response curve-fitting can be calculated on-the-fly using our in-house applications. For each compound an activity fingerprint is composed of the activities (e.g. IC_{50} for enzymatic inhibition or GI_{50} for cytotoxicity) across dozens of assay models. Such activity fingerprints are being used for the selection of lead compounds for further investigation.

The massively parallel nature of DNA microarray technology enables researchers to investigate the biological effects of drug candidates at a genome-wide scale. Making sense of the huge amount of gene expression data has become one of the most challenging tasks of bioinformaticians (Quackenbush, 2001). Of particular importance is the design of a database system to effectively store and manage the huge amount of gene expression data from DNA microarray experiments. The main components of our "microarray informatics" system are shown in Figure 13.3.

At the center of the microarray informatics system is a gene expression database, which has been designed to be in compliance with the MIAME (Minimum Information About a Microarray Experiment) guidelines (Brazma et al., 2001). The Image Analysis module locates microarray spots and quantifies their intensities, which are the real data required for the Data Mining and KDD module and the Visualization module (to be discussed in more detail below). Information on a selected subset of interesting genes is gathered through the use of an Intelligent Internet Robot by searching the WWW and publicly available databases like GenBank and PubMed. Information from microarray analysis can be used to annotate the functions of genes.

It should be pointed out again that the gene expression database is linked to both the chemical structure database and the biological activity database through common database fields, e.g. compound identification number, thereby allowing for seamlessly examination and analysis of different types of data.

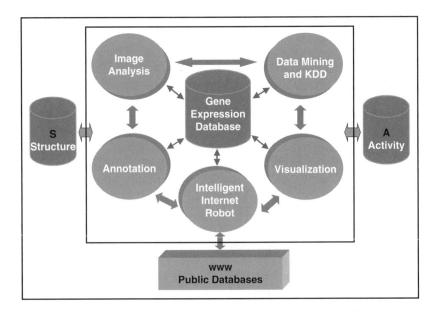

FIGURE 13.3. Structure of the microarray informatics system.

4.2   Data Mining, KDD, and Visualization

Data mining and knowledge discovery in databases (KDD) are new terms for describing research efforts of turning raw data into useful knowledge for decision-making (Frawley et al., 1992; Han and Kamber, 2001; Shi et al., 2000; Shi, 2001, 2002). Usually, these two terms are used interchangeably although data mining focuses on algorithms and KDD deals with the whole process that includes data storage, retrieval, pre-processing, and analysis. Data mining or KDD is defined as "the nontrivial extraction of implicit, previously unknown, and potentially useful information from data" (Frawley et al., 1992). "Nontrivial" means that data mining is not a simple task. To obtain hidden, previously unknown information from data requires special expertise and analysis tools. Figure 13.4 shows the graphical user interface of Chipscreen's biochemoinformatics system and sample output.

There is an ODBC connection between the data analysis system and the Oracle database. The dataset retrieved from the Oracle database is imported into a built-in spreadsheet for further analysis and visualization. Many exploratory analysis methods have been implemented in the system, including 2D and 3D scatter plots, histograms, pie charts, etc. More advanced data analysis and visualization methods are also implemented, include simple and multiple linear regression, principal component analysis, hierarchical cluster analysis, K-means clustering, correlation analysis, K-nearest neighbors, back propagation artificial neural networks, self-organizing maps, genetic algorithms, two-way clustered image maps, etc. These data analysis methods can be categorized into several classes according to their intended application: dimension reduction, clustering, and classification (with prediction).

This system has been used routinely used to analyze HTS data and gene expression data, e.g. finding the similarity or dissimilarity of our drug candidates to commercial products in terms of gene expression profiling or activity fingerprints. Our experience is that there is no universal best method that is suitable for all problems. Instead, each method has its own advantages and disadvantages. For a particular problem one method may be better than the other. It is up to the researcher to identify the most appropriate method(s) for a particular problem, usually by exploring all or most of the methods available with the same dataset (Shi et al., 2000; Shi, 2002).

5.   Conclusions

Chipscreen Biosciences has developed an integrated drug discovery platform that includes computer-aided drug design, chemical synthesis, unique parallel multi-target high throughput screening, global gene expression profiling, and informatics to rapidly and effectively advance the drug discovery process. To effectively manage and mine the huge amount of diverse data on chemical structures, biological activity fingerprints, and gene expression profiling

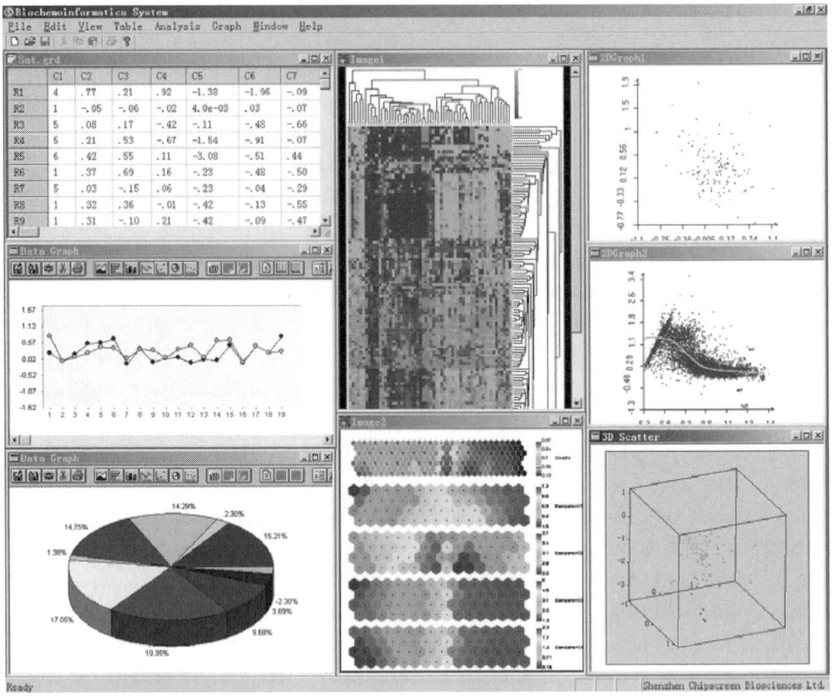

FIGURE 13.4. Graphical user interface of biochemoinformatics system.

patterns, we have developed an integrated biochemoinformatics system. We have implemented the database system on Oracle. Many data analysis and visualization applications have been implemented in Microsoft Visual C++. The whole system has been used on a daily basis in our internal drug discovery process and has been found very user-friendly and useful.

Acknowledgements. This work was supported in part by the National Hi-Tech ("863") Project of China. We are grateful to Dr. Weida Tong of the U.S. Food and Drug Administration, Dr. Lingwen Zeng of Capital Biochip Corporation, and Prof. Li Guo of the Chinese Academy of Sciences for helpful discussions during the development of Chipscreen's biochemoinformatics system and the preparation of this chapter.

Notes. ADMETox: absorption, distribution, metabolism, excretion, and toxicity; CADD: computer-aided drug design; FDA: Food and Drug Administration of the United States of America; HTS: high throughput screening; INDA: investigational new drug application; KDD: knowledge discovery in databases; LBDD: ligand-based drug design; MIAME: Minimum

Information About a Microarray Experiment; NCE: new chemical entity; NDA: new drug application; QSAR: quantitative structure-activity relationship; SBDD: structure-based drug design; TCM: traditional Chinese medicine.

References

1. Agrafiotis DK, Lobanov VS, Salemme FR (2002). Combinatorial informatics in the post-genomics era. Nat Rev Drug Disc 1:337-3346.
2. Bajorath J (2002). Integration of virtual and high-throughput screening. Nat Rev Drug Discov 1:882-894
3. Brazma A, Hingamp P, Quackenbush J, Sherlock G, Spellman P, Stoeckert C, Aach J, Ansorge W, Ball CA, Causton HC, Gaasterland T, Glenisson P, Holstege FC, Kim IF, Markowitz V, Matese JC, Parkinson H, Robinson A, Sarkans U, Schulze-Kremer S, Stewart J, Taylor R, Vilo J, and Vingron M (2001). Minimum information about a microarray experiment (MIAME) -toward standards for microarray data. Nat Genet 29:365-371.
4. Chen KX, Jiang HL, Ji RY (2000). Computer-aided drug design: principles, methodologies, and applications. Shanghai: Shanghai Science and Technology Press, pp 1-487.
5. Class S (2002). Pharma review. Chem & Eng News Dec 2:39-49.
6. Codding PW (1998). Structure-based drug design: experimental and computational approaches (NATO Asi Series. Series E, Applied Sciences, No. 352), Kluwer Academic Publishers.
7. Drews J (2000). Drug discovery: a historical perspective. Science 287:1960-1964.
8. W. Frawley and G. Piatetsky-Shapiro and C. Matheus, Knowledge Discovery in Databases: An Overview. AI Magazine, 1992, 13(3):57-70
9. Hamadeh HK, Bushel PR, Jayadev S, DiSorbo O, Bennett L, Li L, Tennant R, Stoll R, Barrett JC, Paules RS, Blanchard K, and Afshari CA (2002). Prediction of compound signature using high density gene expression profiling. Toxicol Sci 67:232-240.
10. Han J and Kamber M (2001). Data mining: concepts and techniques. San Francisco: Morgan Kaufmann Publishers, pp 1-550.
11. Hopkins AL and Groom CR (2002). The druggable genome. Nat Rev Drug Discov 1:727-730.
12. Knowles J, Gromo G (2002). Target selection in drug discovery. Nat Rev Drug Discov 1:63-69.
13. Lockhart DJ and Winzeler EA (2000). Genomics, gene expression and DNA arrays. Nature 405:827-836.
14. Nuwaysir EF, Bittner M, Trent J, Barrett, JC, and Afshari, CA (1999). Microarray and toxicology: the advent of toxicogenomics. Mol Carcinog 24:153-159.
15. Quackenbush J (2001). Computational analysis of microarray data. Nat Rev Genet 2:418-427.
16. Schena M (1999). DNA microarrays: a practical approach. New York: Oxford University Press, pp 1-210.
17. Schena M (2002). Microarray analysis. New York: Wiley.
18. Sehgal A (2002). Drug discovery and development using chemical genomics. Curr Opin Drug Discov Devel 5:526-531.

19. Shi LM, Fan Y, Lee JK, Waltham M, Andrews DT, Scherf U, Paull KD, and Weinstein JN (2000). Mining and visualizing large anticancer drug discovery databases. J Chem Inf Comput Sci 40:367-379.
20. Shi L (2001). Arrays, molecular diagnostics, personalized therapy and informatics. Expert Rev Mol Diagn 1:363-365.
21. Shi L (2002). Data mining: an integrated approach for drug discovery. In: W.L Xing and J. Cheng, ed. Biochip Technology, : Springer-Verlag Press, in press.
22. Shi LM, Hu WM, Su ZQ, Lu XP, Tong WD (2003). Microarrays: technologies and applications. in Fungal Genomics, Volume 3 of Applied Mycology and Biotechnology, D. K. Arora and G. G. Khachatourians (Editors), Elsevier Science, in press.
23. Smith CG (1992). The process of new drug discovery and development. Boca Raton: CRC Press.
24. Taylor RD, Jewsbury PJ, Essex JW (2002). A review of protein-small molecule docking methods. J Comput Aided Mol Des 16:151-166.
25. Ulrich R, Friend SH (2002). Toxicogenomics and drug discovery: will new technologies help us produce better drugs? Nat Rev Drug Discov 1:84-88.

Part V

New Technologies

14

Development of the MGX 4D Array System Utilizing Flow-Thru Chip Technology
Performance Evaluation of the MGX 4D Array System for Gene Expression Analysis

HELEN SCHILTZ, ADAM STEEL, BRADY CHEEK, ZIVANA TEZAK, DAVID COSSABOON, KATE SIMON, GANG DONG, MATT CHORLEY, PHIL BECKER, JINGYI LO, HARRY YANG, AND ANDREW O'BEIRNE
MetriGenix Inc., 708 Quince Orchard Road, Gaithersburg, Maryland 20878, USA

Abstract: A novel automated microarray platform, the MGX 4D™ Array System, for gene expression analysis based on the patented Flow-thru Chip™ (FTC) technology has been developed. FTC technology uses a well-defined porous substrate constructed of microchannels that connect the upper and lower faces of the array. The porous structure of the FTC enables fluid to flow through the channels and is amenable to automated assay procedures. The probe is bound along the inner surfaces of the channels giving a large surface area-to-volume ratio when compared to flat surface arrays. This increases the probe binding capacity by approximately 100 fold. The MGX *4D* Array System is a result of combining microfludics with the FTC technology. This integrated system consists of three main components: (1) The *4D* Array; (2) the MGX 2000, a microfluidic controller on which hybridization assays are performed; and (3) the MGX 1200CL, a microfludic detection station using a CCD camera for image capture and the MetriSoft™ software package for data analysis.

Key words: Microarray, Flow-thru Chip, microfluidic, chemiluminescence, gene expression profiling, and automation.

1. Introduction

Microarray technology has become a standard technique that is widely used in today's functional genomic research field to assign biological function of genes based on their expression profiles (Schena et al, 1998; Brown and

---

[1] Flow-thru Chip™, 4D™, and MetriSoft™ are the trademarks of MetriGenix, Inc.

Botstein, 1999; and Lockhart and Winzeler, 2000). High-density oligonucleotide microarrays and cDNA microarrays, whether commercially available or homemade, are the common vehicles for studying transcription profiles of thousands of genes from a single biological sample (for example, Blanchard et al, 1996; Lockhart et al, 1996; and Shalon et al, 1996). Once genes related to a specific biological process are discovered from these large-scale microarray experiments, the research focus is then directed to an in-depth study of a subset of genes, generally from tens or hundreds. Detailed analysis provides for characterization of gene involvement and relation in biological processes such as disease or metabolic pathways, with the aim of applying the knowledge to drug discovery, drug development, or diseases diagnostics (Afshari et al, 1999; Marton et al, 1998; and Wallace, 1997). As a result, the demands for sensitive, affordable, easy to use, and scalable microarray platforms are ever increasing. Flow-thru Chip™ (FTC) technology is an advanced microarray platform (Beattie, 1996) that meets these requirements with high performance, amenability for automation, and flexibility for low to high throughput applications.

Unlike conventional two dimensional planar surface microarray technologies, the FTC employs a three-dimensional porous substrate constructed with well-defined microchannels connecting the upper and lower faces of the substrate such that fluid can flow through the chip (Figure 14.1). The microchannel structure results in a much larger surface area to volume ratio than is attained with flat surface platforms, on the order of 100 fold, which yields performance enhancements in capacity, kinetics, and uniformity of biological assays (Steel et al, 2000; Steel et al, 2001; and Cheek et al, 2001). Previous studies demonstrate that probes, either as nucleic acids or proteins, can be deposited into a discrete bundle of microchannels on the chip to detect multiple analytes from a single test (Steel et al, 2001; Benoit et al, 2001; and Cheek et al, 2001). The "probes" referred to here are the immobilized biomolecules attached to the surface of the microchannels that specifically interact with the "target molecules". The target molecules are prepared from biological samples and labeled, for example with fluorophores or antigens, for their detection.

Microfluidic based assay miniaturization and automation has revolutionized the design of biosensors. Agilent Lab-on-a-Chip technology, for instance, uses semiconductor microfabrication techniques to translate complex experimental and analytical protocols, developed in software, into chip architectures consisting of interconnected fluid reservoirs and pathways (Woolley and Mathies, 1994, Kuschel et al, 2001). Liquid movement on the chip, in as small as a few picoliters, is driven by a pressure gradient or electrokinetic force. The electrode-pump and valve systems are capable of performing reagent dispensing and mixing, incubation and reaction, as well as sample partition and analyte detection. Microfluidics undoubtedly provides a means to total automation and assay miniaturization, thereby improving assay accuracy and reliability by reducing human intervention while also reducing sample size and lab space requirements.

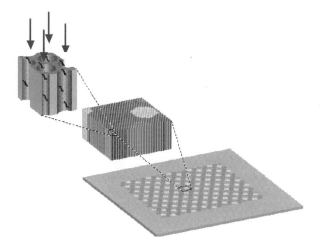

FIGURE 14.1. Conceptual schematics of the Flow-Thru chip.

The MGX *4D* Array System is the result of combining the unique properties of the three dimensional FTC platform with the precise fluid control of
the microfluidic technology. The integrated MGX *4D* Array System consists
of three main components: (1) The *4D* Array that utilizes the FTC technology; (2) the MGX2000, a microfluidic assay controller on which the automated array processing is performed; and (3) the MGX1200CL, a
microfluidic detection station using a CCD camera for image capture and the
Metrisoft software package for data analysis. This chapter provides a descrip-
tion of the principles and mechanisms of the MGX *4D* system and a performance evaluation of the *4D* system using Inflammation *4D* arrays
developed for gene expression measurement in a biological context.

2.   The MGX *4D* Array

The MGX *4D* Array is the central component of the MGX *4D* System. The
4D Array uses a porous silicon substrate that is 500 μm thick with 10 μm
microchannels and an open area ratio of approximately 50% (Figure 14.2).
There are approximately one million microchannels within one centimeter
square area of this substrate. Prepared by a proprietary etching process in a
semiconductor fabrication facility, the microchannels of the silicon substrate
are highly uniform, resulting in high uniformity of spot morphology
(Lehmann, 1999; and Steel et al, 2001). The pristine, well-ordered surface of
the microchannels also accounts for homogeneity of the surface fabrication
and biomolecular reactions in the downstream processes. The microchannels
of the FTC are in fact capillaries, which yield slower evaporation of

deposited probe solutions, resulting in a smaller spot footprint, better spot uniformity, and higher probe density deposited on the chip (Steel et al, 2001). Furthermore, the opaque structure of the silicon FTC eliminates cross talk of the emitted light from channel to channel. The FTC structure provides light guiding properties that enable efficient transmission of photons from inside the microchannels to the CCD detector (Cheek et al, 2001).

Silicon wafers require surface modification to form an intermediate film layer for enhanced biomolecule immobilization. This intermediate layer is produced by immersion of the wafers in a solution of 2% glycidoxylpropy-ltrimethoxysilane in methanol for one hour. Silanized wafers are then dried and cleaned using sonication, and finally dried in a drying oven at 85 °C. The wafers are divided into chip sections in a photolithographic step during pore formation process and then are easily broken into individual chips prior to the printing process.

The bioreactive elements of the *4D* arrays are composed of oligonucleotide probes of 60 to 70-mers with 5′ amine modification. The 5′ amine modification allows for covalent attachment of the probes to the epoxy monolayer on the chip. Design of probe sequences is accomplished using a proprietary in-house procedure based on the selection criteria of probe length, location relative to the 3′ untranslated region of the gene, secondary structure, Tm, and specificity

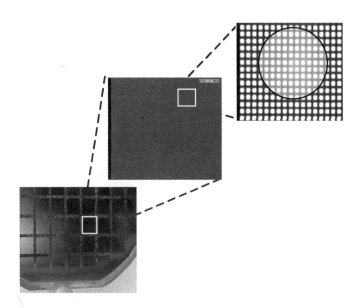

FIGURE 14.2. MGX *4D* Array under different magnitude scales. A picture of the silicon wafer used to make the array is shown at the bottom. The wafer is precut into individual chips of 1 cm² in size (middle picture) that have approximately 1 million of microchannels per chip. Each array spot occupies approximately 100 microchannels as depicted in the top picture.

in detecting the target gene. Probes are further screened based on functional testing using biological model systems. Oligonucleotide probes are qualified using standard OD and capillary electrophoresis methods. A BioChip™ piezo-electric spotter (PE Life Sciences) is used to dispense 5nL of 30 µM probe solution in 1x SSC onto the silanized chips, with center-to-center spacing of 350 µm. Spotted chips are incubated overnight, blocked in a solution containing 0.1% polyvinylpyrrolidone and 0.1% Ficoll, and then baked at 85 °C. Post spotting treatment reduces the incidence of nonspecific binding of target and reagent molecules onto the non-spotted area of the chip while enhancing the immobilization of the DNA probe on the chip. The 4D Arrays prepared with this method generate uniformly round spots of approximately 100 µm diameter, and a probe density of approximately 10^{13} probes per cm², or 1.6×10^9 probes per microchannel (Steel et al, 1998; and Cheek et al, 2001).

The MGX 4D Inflammation Array described in this study contains a total of 192 elements. Eighty inflammation related genes, with a broad range of biological functions, were selected based on their biological importance, expression signature and fold change between the normal and inflammatory tissues. The array also includes 16 assay controls, comprised of three hybridization controls, one negative control, eight housekeeping genes, three staining controls, and a sample preparation control for signal normalization and quality control purposes. All probes on the 4D Arrays are printed in duplicate adjacent to each other on the chip.

The spotted 4D array is assembled into a molded plastic cartridge, which is sealed with adhesive using a customized cartridge press. The 4D array cartridge also contains three reservoirs for holding samples and reagents (Figure 14.3). For automated microfluidic processing, the cartridge's reservoir ports precisely align with the fluid line ports on both the MGX 2000 Hybridization Station and the MGX1200CL Detection Station.

3.   MGX2000 Hybridization Station

Like any conventional nucleic acid hybridization or immunoassay, multiple fluid transfer steps, such as blocking, sample addition, hybridization, staining, and washing steps, are required to complete a microarray assay. Conventional microarray technologies that utilize flat surface platforms, either as glass slide or nylon membrane, generally require manual manipulation to perform an assay. As such, microarray assays are labor intensive and technique dependent, and results are thus generally less reproducible due to high assay variation. Reducing assay hands-on time and improving assay reproducibility and accuracy through automation are necessities of any microarray platform with high performance and the potential utility for in-vitro diagnostics.

The MGX 2000 Hybridization Station was designed specifically for automated processing of the 4D arrays based on the FTC technology. The

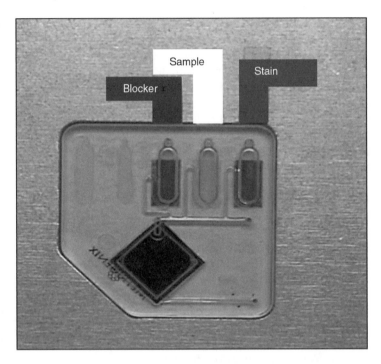

FIGURE 14.3. MGX *4D* Array Cartridge. The cartridge contains a chip chamber and reservoirs. Shown is the three-reservoir cartridge for holding test sample and reagents as illustrated.

FIGURE 14.4. MGX 2000 Hybridization Station.

instrument is an integrated microfluidic device with a small footprint of $26.5 \times 26.5 \times 12$ cm (W×D×H) (Figure 14.4). As illustrated in the schematic drawing in Figure 14.5, the MGX2000 is composed of a stepper motor syringe pump and solenoid valve system. The pump generates a pressure gradient across the chip, driving fluid flow through the microchannels and controls the speed of the fluid movement. The sequential movements of the valves control the direction of fluid movement through individual reservoirs and the chip chamber on the cartridge. The instrument is capable of processing two cartridges per run time, common delivery of two buffers,

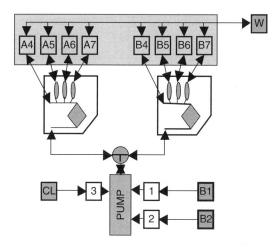

FIGURE 14.5. Schematic Representation of the MGX2000 Hybridization Station. A single pump is connected to multiple valves 1 to 3, A4 to A7, and B4 to B7. Valves 1 to 3 connect to buffer B1, buffer B2, and a cleaner "CL", respectively. Letter "W" denotes to the waste collector.

and has a dedicated cleaning solution. The entire hybridization and staining reactions are fully automated with user defined assay parameters, which are stored and easily accessed on the machine through the touch pad.

When designing the instrument and the chip cartridge, the unique geometry of the FTC was taken into consideration to achieve uniform fluid distribution. Since target molecule cannot laterally penetrate into other microchannels on the chip until it exits a microchannel, the instrument employs oscillation to mix sample or reagents back and forth through the microchannels during the hybridization and staining reactions. The residence time of targets within the microchannels is long relative to the time needed for target to migrate the short distance to the microchannel wall via diffusion. Confining the probes and targets inside the 10 μm microchannels enhances the rate of probe and target molecule interaction, a step generally constrained by the diffusional transport of target molecule to the immobilized probes in flat system geometries.

As mentioned above, the MGX 2000 is operated via a touch pad LCD interface that provides ten programmable assays and additional cleaning and preparatory routines. The instrument provides the user with a set of optimal assay parameters but also allows end-users to further modify the routine parameters such as oscillation flow rate, oscillation volume, reagent flow rate, hybridization time and temperature for specific assay requirements. In general, hybridization of the *4D* array can be conducted from one minute to several hours, and from room temperature up to 70 °C. The self-contained *4D* array cartridge, once preloaded with all necessary sample and reagents, is placed against the adapter plate on the MGX2000 and compressed with a polycarbonate cover plate to provide an airtight seal. The ports of both the cartridge and the adapter plate are sealed via O-rings on the adapter plate, permitting fluid flow through the chip chamber during an assay.

4.   MGX1200CL Detection Station

While conventional microarray platforms use fluorescence for detection, the MGX *4D* Array takes advantage of chemiluminescence (CL) and the FTC geometry to create a simple array readout device, the MGX 1200CL (Figure 14.6). *4D* arrays are imaged on the MGX1200CL after hybridization and staining reactions are completed on the MGX2000. The MGX1200CL Detection Station has a small footprint of 25×42×28 cm (W×D×H).

As an alternative to fluorescent detection method, CL offers the attractive features of inherently low background and simple instrumentation requirements. Fluorescence requires an excitation light source that can produce nonspecific radiation such as reflection. Consequently, some amount of fluorescent signals can be generated in areas of the microarray where the fluorophore is not present, resulting in higher background in fluorescence methods. In contrast, nonspecific radiation is significantly reduced in CL imaging. Warm-up and drift of the light source and interference from light scattering present in fluorescence methods are absent in CL, making the background level much lower. CL imaging systems generally include only an imaging device such as a charge coupled device (CCD) and lack the excitation light source and wavelength selection optics of fluorescence systems. Although CL detection technology has been widely applied to commercial immunoassay and nucleic acid detection methods in the microplate and membrane assay platforms, using CL imaging methods in microarray platforms is unprecedented.

The main concerns when designing a CCD based CL imaging device, MGX1200CL, include the selection of the CL substrate and the resolution of the CCD in order to provide adequate signal intensity and image resolution. For the CL substrate, the MGX1200CL uses a "flash" CL substrate, which emits light on the order of seconds, over a "glow" substrate. "Flash" substrate provides better spatial resolution due to a minimum diffusion of light-emitting molecules during the reaction. Since the detection reaction of the *4D* array occurs within the microchannels of the *4D* array, the diffusion of the light-emitting species is limited, conserving the image spatial resolution. Based on this criterion, an enzymatic reaction of the SuperSignal® West Femto Maximum Sensitivity Substrates Luminol and Hydrogen Peroxide (Pierce) catalyzed by horseradish peroxidase was validated for use with the MGX *4D* Arrays (Cheek et al, 2001).

The MGX1200CL Detection Station, like the MGX2000 Hybridization Station, takes advantage of the FTC geometry and microfluidics. Once the cartridge processing is complete on the MGX2000 station, the cartridge is then placed against the adapter plate on the MGX1200CL and compressed with a cover plate that is both air and light tight. The MGX 1200CL combines stepper

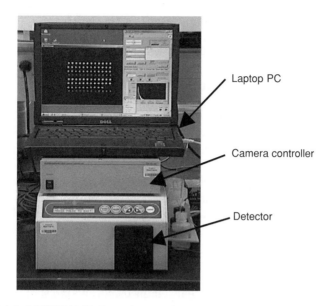

FIGURE 14.6. MGX1200CL Detection Station.

motor driven syringe pumps and solenoid valves for CL substrate delivery in a one-directional fashion and cleaning routines (Figure 14.7). While capturing the image, CL substrate is continuously pumped through the chip to ensure an enzyme-limited reaction, which results in higher signal intensities per unit time compared to static CL detection methods used in typical membrane-based microarrays (Cheek et al, 2001). The FTC geometry combined with flush CL substrates allows for dynamic CL measurement of the *4D* arrays.

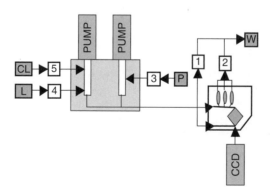

FIGURE 14.7. Schematic Representation of the MGX1200CL Detection Station. P, L and W represent the peroxide, luminal and waster containers, respectively. Numbers 1 to 5 represent valves used to control fluid delivery throughout the system. CCD denotes the CCD camera.

A 12-bit cooled CCD camera is used for the MGX1200CL Detection Station. A CCD is a collection of light sensitive diodes (or pixels) with dimensions of approximately 14 μm, which convert photons of light into electrons (i.e. electrical charge). An analog-to-digital converter (ADC) converts each pixel value, based on the amount of electrical charge it collects, into a digital value. The amount of electrical charge a pixel receives is limited, and a 12-bit ADC can convert the analog information into 2^{12}, or 4096. Therefore, the 12-bit camera has a dynamic range of 0-4095, or over 3 logs; however, noise limits the useful dynamic range to roughly 1-800, or close to 3 logs. The MGX 1200CXL uses an automatic exposure routine to take advantage of the maximum dynamic range of the detection system.

Image capture and analysis are carried out via *MetriSoft* software, preinstalled on a Pentium® (Intel) laptop computer. *MetriSoft* controls both the fluidics and the data capture routines. The software has an easy to use tab format, guiding the user through the data entry, image capture, and image analysis, and data query processes. The software allows users to perform auto exposure and manual exposure routines during the image capture step, for up to 10 seconds of exposure time. It also implements an auto spot finding routine to assist in image analysis. Total intensity, that is, the net intensity of the spot after subtracting the local background, and the normalized signals relative to user-selectable reference gene(s) are generated from captured images. Data are exported into an Excel workbook in tabular form (see example in Figure 14.8). Data queries can be performed to retrieve previous data for assay-to-assay comparison, for up to 100 assays at a time. Results of a data query are exported into tables and graphs in Excel as well.

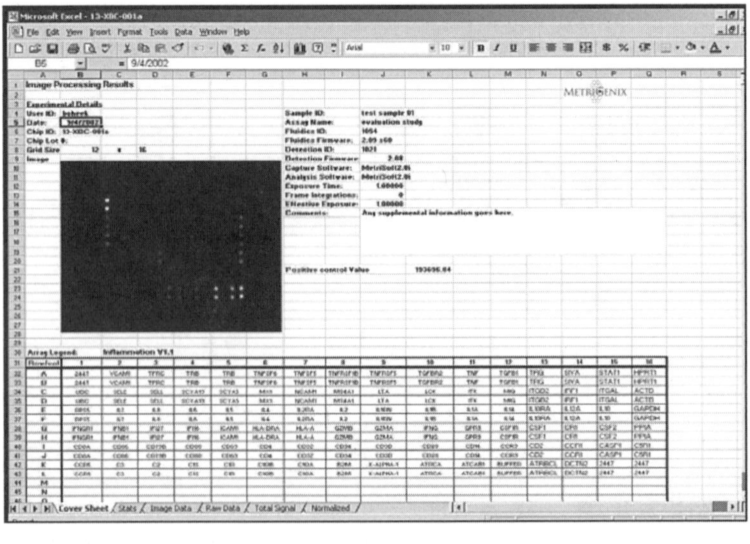

FIGURE 14.8. Example of the *MetriSoft* Data Output of the *4D* Array Assay.

5.   MGX *4D* Array Assay Protocols

5.1   Labeled RNA Sample Preparation

The MGX *4D* Array detects biotin labeled nucleic acid fragments derived from total RNA or mRNA of the biological specimens. The labeled RNA samples used in this study were prepared by the widely used in-vitro transcription (IVT) method (Lockhart et al, 1996). Briefly, total RNA samples were purchased from Ambion or Clontech. First strand cDNA was synthesized from the total RNA using SuperScript II™ (Invitogen) reverse transcriptase, and a poly (T) T7 primer (GGCCAGTGAATTGTAATACGA CTCACTATAGGGAGGCGGT24). The second strand cDNA was synthesized using *E. coli* DNA Ligase and DNA polymerase I (Invitrogen). Both reactions were performed following the manufacturer's protocols. The cDNA sample cleanup was performed using the phenol/chloroform extraction method. The cDNAs were precipitated and resuspended in DEPC water. An IVT reaction was performed with the purified cDNA as template, using a MEGAscript™ T7 High Yield Transcription Kit (Ambion). Biotin-11-CTP and Biotin-16-UTP (Perkin Elmer) were added to the IVT reaction to label the RNA transcripts, following Ambion's recommended protocols. The cRNA products were purified with a Qiagen RNeasy Mini Kit, and their integrity and purity were determined by OD ratio and by running a RNA 6000 Nano Assay on Agilent 2100 Bioanalyzer. Sample aliquots were prepared and stored at −80 °C for use during this study.

5.2   MGX 4D Array Assay Procedure

The minimum assay manipulation required by the MGX *4D* Array system includes: 1) sample dilution, 2) loading the sample and assay reagents on the cartridge, and 3) transferring the cartridge to the MGX 2000 station and the MGX 1200CL station.

The *4D* Array assay requires a blocking reagent and a staining reagent in addition to the labeled target RNA sample. Blocking reagent is a 16% goat serum buffered solution, whereas the staining reagent is a streptavidin-horseradish peroxidase conjugate at 10 μg/mL in a stabilizer mixture. The cartridge used in this study takes 65±2 μL of the sample or reagent volume in the reservoir. For the studies reported here, we diluted a 10 μg cRNA sample aliquot into a hybridization buffer containing MES, NaCl, EDTA, sarcosine, herring sperm DNA and 30% formamide. An additional 2 μL of hybridization spikein control DNA, a mixture of three synthetic biotinylated oligonucleotides at various concentrations, were added into the diluted cRNA sample, for a total of a 70-uL volume. A five-minute incubation at 90±5 °C was performed to denature the diluted sample, followed by 2-5 minutes of chilling on ice and a quick centrifugation to collect the condensation. The blocking reagent, denatured sample, and the staining reagent were loaded into their respective reservoirs

TABLE 14.1. *4D* Array Assay Parameters

Steps	Duration	Volume (μL)	Flow (μL/min)	Buffer Type	Temperature
Conditioning		1250	500	Buffer 1	RT
Blocker 1	6 min		10		RT
Flush 1		1250	500	Buffer 2	RT
Hybridization	2 hours		10		37 °C
Flush 2		2000	500	Buffer 1	RT
Blocker 2	6 min		20		RT
Flush 3		1000	500	Buffer 1	RT
Staining	16 min		10		RT
Flush 4		2000	500	Buffer 1	RT

on the cartridge. The loaded cartridge was mounted on the MGX 2000 Hybridization Station for a two-hour hybridization reaction at 37 °C unless otherwise specified.

A typical *4D* assay program, as illustrated in Table 14.1, can be programmed and saved using the LCD touch pad. During the assay, the instrument utilizes the MGX Buffer 1 at various time points for cartridge conditioning, flushing, and delivery of the blocking and staining reagents to the chip. MGX Buffer 2, the hybridization buffer, is pumped through the sample reservoir to deliver the biotinylated cRNA targets into the chip chamber, where hybridization on the array takes place. A SA-HRP staining solution is subsequently delivered to the chip, where streptavidin molecules bind to the captured biotin labeled cRNA targets. Unbound materials are removed through flushing cycles.

After the final flush, the cartridge is transferred to the MGX 1200CL Detection Station. Chemiluminescent substrates luminol and peroxide are pumped into the instrument and equally mixed, and the substrate mixture is then delivered into the chip chamber of the cartridge. In the presence of hydrogen peroxide, HRP enzymes catalyze the oxidation of luminol, emitting light that is captured by the CCD camera. In summary, the entire procedure includes a sample preparation step, a hybridization preparation step, a hybridization step, and a detection step. Excluding the front-end sample isolation and labeling, the total assay time with the *4D* Array System is approximately 4 hours with less than 15 minutes hands-on time (Figure 14.9).

6.   Performance Characteristics of the MGX *4D* Array

6.1   Image Uniformity

The image uniformity of the *4D* Array is the compounded effect of multiple factors, including uniformity of the array printing quality, homogeneity of the assay conditions such as fluid distribution, as well as the uniformity of the image detection device. In the context of array spot uniformity, printing

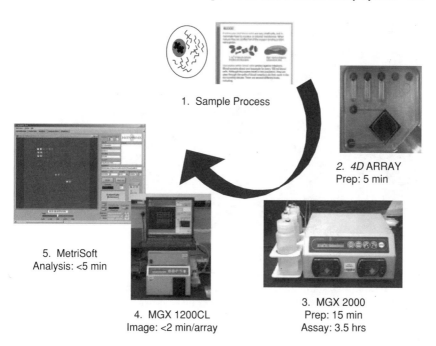

1. Sample Process

2. *4D* ARRAY
Prep: 5 min

5. MetriSoft
Analysis: <5 min

4. MGX 1200CL
Image: <2 min/array

3. MGX 2000
Prep: 15 min
Assay: 3.5 hrs

FIGURE 14.9. *4D* Array Assay Procedure Overview.

conditions such as probe concentration, printing buffer composition and printing environment (i.e. humidity and temperature), and surface modification of the substrates are all critical for any microarray platform. Likewise, the uniformity of the *4D* array prepared under the conditions described here has been evaluated as a key performance characteristic.

Image uniformity is measured by the variability of the signal intensity of the replicate spots across the *4D* array and expressed as a percent coefficient of variation (%CV). A 21-mer oligonucleotide probe modified with a 5′ C6 amine group and a 3′ biotin (T-amine-CCCAGGGAGACCAAAAGC-biotin) was spotted on the *4D* chip in an 8 by 8 array for a total of 64 spots. Chip cartridges were assayed on the MGX2000 and imaged on the MGX1200CL. The intra-chip %CV calculated from a total of 64 replicate spots on the chip was less than 15%, with probe concentrations ranging from 6.25 to 200 nM tested. An example of these chip images and signal intensity for 200 nM probe concentration are given in Figure 14.9 where intra-chip %CV was 8.1%. The local background intensities across the same array are also highly uniform with a CV of 0.9%. Visual examination of these images showed uniform spot morphology as well. These results illustrate that the *4D* arrays prepared using the current procedures described here are highly uniform in terms of the signal and background intensity generated.

The signal to noise (S/N) ratio at full exposure of the detection station was 38.9 for arrays spotted with the 6.25 nM probe solution in the above study.

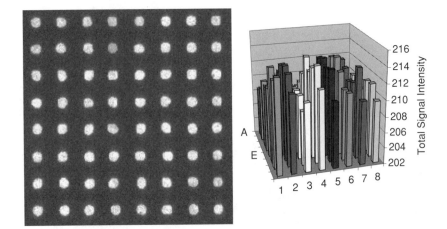

FIGURE 14.10. Signal Uniformity of *4D* Array. A single biotin labeled oligonu-cleotide probe of 23mers was spotted in an 8 by 8 array on the Flow-thru chip. The chip cartridge was stained on the MGX 2000 hybridization station and imaged on the MGX 1200CL detection station. Shown at left is the array image taken from *MetriSoft*, and at right is the spot signal intensity of the 8 x 8 array at one-second exposure.

Provided that probe attachment efficiency is 10% and that biotin coupling efficiency of the oligo probes is 100%, an array spot deposited with 5nL of 6.25 nM of probe solution would have 1.9×10^6 biotin molecules. Therefore, if the sensitivity of detection is set as S/N of 3, the lower limit of detection of the MGX1200CL is approximately 1.9×10^5 biotin molecules. This esti-mate agrees with the previous determination of the detection limit of the pro-totype *4D* system of 5×10^4 HRP molecules given that streptavidin-HRP conjugate has a streptavidin to HRP ratio of 2 and that biotin to streptavidin ratio is 4 in detection (Cheek et al, 2001) (Figure 14.10).

6.2   Intra-Chip Variability

The uniformity of the *4D* arrays was further examined in a hybridization detection using labeled RNA samples prepared from total RNA on the MGX Inflammation *4D* Arrays. The Inflammation *4D* array contains 192 duplicate probe spots. As illustrated in Figure 14.11, spleen and thymus labeled RNAs were tested on Inflammation *4D* arrays, with the positions of the key control probes indicated as well. Duplicate spots were printed in adjacent position in the same column.

Intra-chip variability of replicate spots measures the precision of the analysis on the same array. In one experiment, four Inflammation *4D* arrays were hybridized to spleen cRNA samples on two instruments on the same day. All of the genes that were above the assay cutoff, that is, three times the signal intensity of the negative controls, were included in the analyses. The mean intra-chip %CV from the duplicate spots from these arrays was 9.6 ± 15.6%, with a median %CV value of 5.2. A large study of testing a total of 36 chips from two different production lots of the Inflammation *4D* arrays using identical thymus and spleen sample aliquots showed similar results, with the mean intra-chip %CV being less than 15. These results indicate that the intra-chip variability of the *4D* array is comparable to or better than that of other DNA microarray platforms (for example, Li et al, 2002).

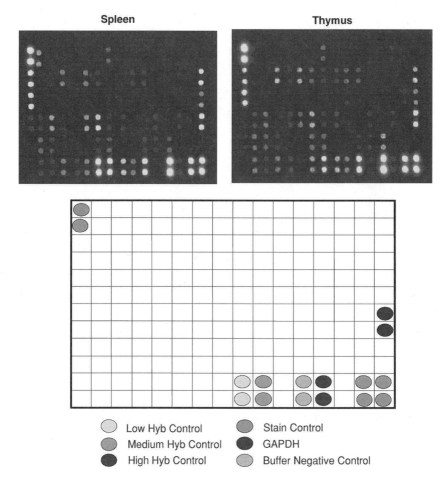

FIGURE 14.11. Hybridization Uniformity of the Inflammation *4D* Arrays. The Inflammation *4D* Arrays were hybridized respectively to spleen and thymus cRNA on the MGX2000 Hybridization Station and imaged on the MGX 1200CL. Examples of the images are shown. The schematic diagram of the array is given at the lower panel to indicate the control spots on the chip.

6.3   Dynamic Range

The Inflammation *4D* array detects a range of inflammation genes from the RNA samples tested. Relative abundance of the detected genes on the Inflammation array are reflected by the relative intensities of the duplicate spots on the chip. The relative expression of each gene between samples is determined via normalization to that of the GAPDH gene. From the same experiments in Figure 14.11, the normalized signals of the positive genes averaged from four arrays were plotted in Figure 14.12. As indicated in the graph, the relative expression levels of the genes on the chip ranged over 2.5 logs. The results agree with previous determination of the dynamic range of the prototype *4D* system in a different experimental design. We note that analyses were performed based on a single exposure time of the images, suggesting that the biological dynamic range is larger than the dynamic range of the device, as is seen with most microarray systems (Stoss et al, 2002).

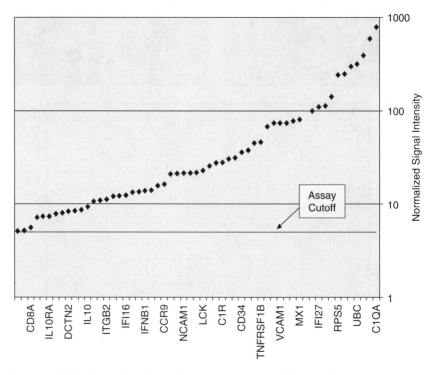

FIGURE 14.12. Detection Dynamic Range of the Inflammation *4D* Array. The Inflammation *4D* array was hybridized with a spleen cRNA sample on the MGX2000 and imaged on the MGX1200CL. The assay cutoff was set at 3 times the signal intensity of the buffer negative spot. Signal intensities of genes that were above the assay cutoff were normalized to that of the GAPDH gene and depicted in this graph, with GAPDH value being 100.

6.4   Shorter Reaction Rate

A typical glass-slide or membrane based microarray assay requires long hybridization incubation (such as over night) to achieve the acceptable results in terms of sensitivity and reproducibility. Enhanced hybridization kinetics is achieved with the FTC technology due to increased surface area, dynamic fluidic delivery of the *4D* instrument, and the optimized probe density on the *4D* array.

To determine the assay kinetics of the *4D* array, thymus cRNA samples were hybridized to the Inflammation *4D* arrays for various time intervals at 37 °C on the MGX 2000 station using standard *4D* assay parameters and procedures. The average normalized intensities of six genes from at least two duplicate chips were plotted as a function of hybridization time in Figure 14.13. These genes represent high (PPIA and GAPD), medium (CD2, CD3d, and CD3z), and low (CD4) abundance genes on the chip. As demonstrated, hybridization of the highest abundance genes reaches saturation at the three-hour time point, and low abundance genes approach saturation at the

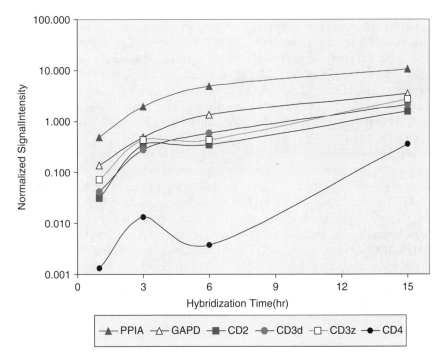

FIGURE 14.13. Hybridization Kinetics of the *4D* Array. Thymus cRNA samples were hybridized to Inflammation *4D* Arrays on the MGX2000 hybridization for 1, 3, 6, and 15 hours, respectively. Normalized signal intensities of high (PPIA and GAPD), medium (CD2, CD3d and CD3z), and low (CD4) abundance genes were plotted as a function of hybridization time.

fifteen-hour time point, significantly shorter than the overnight requirement of the conventional two dimensional microarray platforms (Shalon et al, 1996). These results indicate that depending on the assay sensitivity requirement, a *4D* assay could be performed with a hybridization time of less than one hour using the current instruments and reagents for high abundance targets. Alternatively, a 4D array could be performed for 'standard' microarray hybridization times to detect low abundance genes.

6.5   Chip-to-Chip Variability

A key performance characteristic of the *4D* system that we evaluated is the chip-to-chip reproducibility. Inter-chip %CV is the measure of the system variability on a repetitive testing basis. Analogous to the intra-chip variability, factors contributing to the uniformity of the signal and background distribution also attribute to the inter-chip %CV. The chemiluminescent single channel detection of the MGX *4D* array eliminates the common reproducibility problems encountered when using two-channel fluorescence methods, and simplifies the statistical methods required to interpret the array results. The difference in labeling efficiency of cy3 and cy5 when performing dual-dye labeling, for instance, not only increases the complexity of the experiment but also yields high chip-to-chip variability with that method. Using the *4D* array for comparison of differential expression of different biological samples, signal normalization is only pertinent to normalization of sample preparation and hybridization efficiency using the same external controls on the chip. Therefore, chip-to-chip reproducibility of the *4D* system is only constrained by the reproducibility of the signal intensity generated from each chip.

Inter-chip variability of the *4D* array system may arise from variations in chip-to-chip printing quality or from instrument related performance variation. Instrument variability could result from variations between the left and right cartridge heads of the MGX2000 or from variation between different MGX2000 instruments. In one study, the same thymus RNA sample aliquots were hybridized to Inflammation *4D* arrays over three days on three different MGX2000 stations. A total of 18 chips were assayed in 9 runs on the left and right cartridge heads. Linear regression analyses of the normalized signal intensities of genes from the entire detection range were performed. The median normalized signal intensities of genes from the left cartridge head was compared to that from the right cartridge head, as shown in Figure 14.14, based on these 9 pairs of thymus samples (n = 18). The slope of the linear regression equation is 0.98, with 1 being identical between the median values of the left and right heads. Similar liner regression analyses were performed on normalized signal intensities of individual arrays tested on the left cartridge head to the average value of all arrays obtained on the left cartridge head. As a result, the average slope value of the individual chips on the left cartridge head yielded 0.98 ± 0.098. Similarly, the average slope value of the

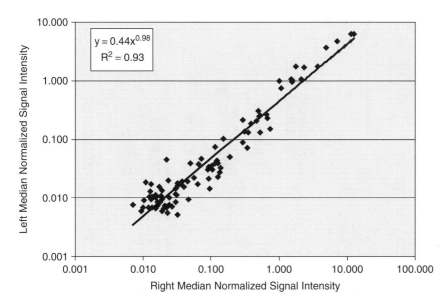

FIGURE 14.14. Left to right cartridge head variation of the *4D* array system.

individual chips tested on the right cartridge head yielded 0.96 ± 0.083 when compared to the average of all chips obtained on the right cartridge head. These results suggest that no systematic variation exists between the left and the right cartridge heads, a conclusion that was further confirmed by performing paired t-test on the data set. In addition, no significant difference was found between day-to-day or instrument-to-instrument as well (data not shown).

6.6   Fold Change Correlation

The immediate application of the *4D* Array System is to provide a useful tool to researchers to investigate differential gene expression of defined gene sets. The methods for gene content development *in silico* and for probe validation using biological model systems and RT-PCR are out of the scope of this report and thus are not discussed in depth here. Probe selection conforming to specificity and efficiency of detection is nonetheless crucial to the performance of the *4D* system. We evaluated the detection accuracy of the MGX Inflammation *4D* Array using thymus and spleen tissues and assessed the fold changes of genes that have known differential expression between these two tissues.

In one such experiment, 9 pairs of thymus and spleen samples were tested on three MGX2000 station over three days. Signals generated from each chip were normalized to the hybridization control on the chip, and expressed as relative signal. The averaged relative signals of 7 genes were measured and are

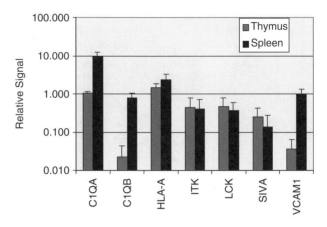

FIGURE 14.15. Relative Signal Intensities of Selected Genes on Inflammation *4D* Arrays. Thymus and Spleen cRNA sample aliquots were hybridized to Inflammation *4D* arrays on MGX2000 stations. Average relative signal intensities of these seven genes from 9 replicate chips are plotted here.

depicted in Figure 14.15. These 7 genes had distinct expression patterns in spleen and thymus tissues based on results from a high-density glass surface microarray platform. The difference in expression of these genes is expressed as fold change of spleen to thymus. For a control experiment, we hybridized the *4D* arrays by rotating them overnight in a rotisserie oven at 37 °C. After the hybridization, the cartridges were returned to the fluid delivery system IBBU (Steel et al, 2001) for staining and washing. Overnight hybridizations were performed with no flow through control of fluid, in a manner similar to performing a flat surface microarray assay. We compared results from the standard 2-hour hybridization assays performed on the MGX2000 station to those of overnight incubation with rotation. As shown in Figure 14.16, the correlation coefficient of the *4D* (2 hour hybridization) to the overnight (FTC O/N) is 0.88, and is 0.93 to the high-density array. The results demonstrate that microfluidic flow of *4D* arrays allows for markedly shortened hybridization time without loss of performance when compared to a flat surface array or to FTC with much longer hybridization time and without microfluidic flow.

7.   Conclusions

An automated and integrated *4D* Array System based on the FTC and microfluidic technologies has been developed. Data presented here summarize our preliminary findings of the performance characteristics of this novel microarray system for differential gene expression analysis. The uniformity of the *4D* arrays prepared using well-controlled procedures and

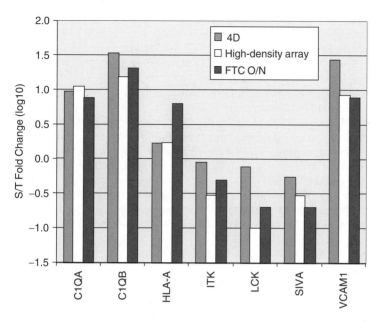

FIGURE 14.16. Fold Change Correlation of Selected Genes on the Inflammation *4D* Arrays. The log fold changes of spleen to thymus from three studies are plotted here. *4D* represents a standard 2-hour hybridization of the Inflammation *4D* Array on the MGX2000 station; FTC O/N denotes that hybridization of the spotted FTC array was performed in a customized chip chamber incubated in a rotisserie oven for overnight rotation. High-density arrays were performed according to the overnight procedures with Cy3 and Cy5 labeled samples.

high reproducibility of the microfluidic delivery and chemiluminescent detection of the *4D* instruments consist of a reliable automated microarray platform. Enhanced hybridization kinetics of the *4D* array can shorten assay times significantly without compromising the accuracy of the results. The cost-effectiveness and user-friendliness of the *4D* Array system are additional attractive features the *4D* system offers. Efforts are ongoing to develop a high throughput *4D* Array system for pharmaceutical screening and biological discovery applications, to continue the improvement of the *4D* assay sensitivity using advanced sample preparation technologies, and to develop proteomic *4D* arrays.

References

1. Afshari, C.A., E.F. Nuwaysir and J.C. Barrett. 1999. Application of complementary DNA microarray technology to carcinogen identification, toxicology, and drug safety evaluation. Cancer Res. 59:4759-60.
2. Beattie, K.L. 1996 Microfabricated, flowthrough porous apparatus for discrete detection of binding reactions. US Patent, 5843767.

3. Benoit, B., A. Steel, M. Torres, Y.Y. Yu, H. Yang and J. Cooper. 2001. Evaluation of three-dimensional microchannel glass biochips for multiplexed nucleic acid fluorescence hybridization assays. Anal. Chem. 73:2412-20.
4. Blanchard, A.P., R.J. Kaiser and L.E. Hood. 1996. High-density oligonucleotide arrays. Biosensors and Bioelectronics. 11:687-90.
5. Brown, P.O. and D. Botstein. 1999. Exploring the new world of the genome with DNA microarrays. Nature Genet. 21:33-37.
6. Cheek, B.J., A.B. Steel, M.P. Torres, Y.Y. Yu and H. Yang. 2001. Chemiluminescence detection for hybridization assays on the flow-thru chip, a three-dimensional microchannel biochip. Anal. Chem. 73:5777-83.
7. Kuschel, M., Muller O., and C. Buhlmann. 2001. Lab-on-a-chip technology. Agilent publication 5988-7995EN.
8. Lehman, V. 1990. Formation mechanism and properties of electrochemically etched trenches in n–type silicon. J. Electrochem. Soc. 137:653-59.
9. Li, X., W. Gu, S. Mohan and D. J. Baylink. 2002. DNA microarrays: Their use and misuse. Microcirculation. 9:13-22.
10. Lockhart, D.J., H. Dong, M.C. Byrne, M.T. Follettie, M.V. Gallo, M.S. Chee, M. Mittmann, C. Wang, M. Kobayashi, H. Horton and E.L. Brown. 1996. Expression monitoring by hybridization to high-density oligonucleotide arrays. Nature Biotechnol. 14:1675-80.
11. Lockhart, D.J. and E.A. Winzeler. 2000. Genomics, gene expression and DNA arrays. Nature. 405:827-836.
12. Marton M.J., J.L. DeRisi, H.A. Bennett, V.R. Iyer, M.R. Meyer, C.J. Roberts, R. Stoughton, J. Burchard, D. Slade, H. Dai, D.E. Jr. Bassett, L.H. Hartwell, P.O. Brown, and S.H. Friend. 1998. Drug target validation and identification of secondary drug target effects using DNA microarrays. Nature Med. 4:1293-301.
13. Schena, M., R.A. Heller, T.P. Theriault, K. Konrad, E. Lachenmeier, and R.W. Davis. 1998. Microarrays: biotechnology's discovery platform for functional genomic. Trends Biotechnol. 16:301-306.
14. Shalon, D., S.J. Smith and P.O. Brown. 1996. A DNA microarray system for analyzing complex DNA samples using two-color fluorescent probe hybridization. Genomic Res. 6:639-45.
15. Steel, A.B., T.M. Herne and M.J. Tarlov. 1998. Electrochemical quantitation of DNA immobilized on gold. Anal. Chem. 70:4670-77.
16. Steel, A.B., M. Torres, J. Hartwell, Y.Y. Yu, N. Ting, G. Hoke and H. Yang. 2000. The Flow-thru chip™: A three-dimensional biochip platform. In Microarray Biochip Technology, Schena, M. Ed., Eaton: Natick, MA, pp 87-117.
17. Steel, A.B., M. Torres, J. Hartwell, Y.Y. Yu, N. Ting, G. Hoke and H. Yang. 2001. The Flow-thru Chip: A miniature, three-dimensional biochip platform. In Integrated Microfabricated Biodevices. Heller M. J. and Gluttman A. Ed., Marcel Dekker: NY-Basel, pp 271-302.
18. Stoss, O., Reuner, B., and T. Henkel. 2002. RNA expression profiling in cardiac tissue-s successful route to new drug targets? Targets. 1:12-19.
19. Wallace, R.W. 1997. DNA on a chip-serving up the genome for diagnostics and research. Mol. Med. Today. 3:384-89.
20. Woolley A.T., and R.A. Mathies. 1994. Ultra-high speed DNA fragment separations using microfabricated capillary array electrophoresis chips. Proc. Natl. Acad. Sci. USA 91:11348-52.

15

Sequencing by Aligning Mutated DNA Fragments (SAM)
A New Approach to Conventional Sequencing and SBH Sequencing

Duncan Cochran[1,2,3], Gita Lala[1,2,3], Jonathan Keith[2,3], Peter Adams[2,3], Darryn Bryant[2,3], and Keith Mitchelson[1,3]
[1]Australian Genome Research Facility, St Lucia 4072, Australia; [2]Department of Mathematics, University of Queensland, St Lucia 4072, Australia; [3]Combinomics Pty Ltd., St Lucia 4072, Australia

Abstract: This paper discusses an original technique, *Sequencing by Aligning Mutants (SAM)*, the purpose of which is to overcome sequencing difficulties caused by problematic genomic regions where local sequence characteristics hinder existing sequencing technologies. It involves forming a number of mutated copies of regions of the target DNA, cloning and sequencing each of these mutated fragments. The effectiveness of SAM technology is demonstrated by sequencing of problematic DNA elements. An application of SAM technology to chip-based sequencing-by-hybridization (SBH) is discussed.

Key words: Unclonable and problematic DNA, sequencing, DNA mutation, SBH.

1. Genome Sequence Assembly

All genome sequencing projects regularly encounter regions that yield no data with current sequencing strategies[1,2,3]. These gaps are present for several reasons including under-representation of sequences in libraries[4,5], the inability to assemble complex sequence regions correctly[6,7] and the presence of DNA motifs that are intractable to cycle sequencing[4]. The human genome project provides many examples of each of these different impediments to sequencing and serves to illustrate the measures that may be taken to overcome them.

1.1 Assembly of the Human Genome

The haploid human genome consists of a total of 3.1 Giga base pairs (Gb). The International Human Genome Project (IHGP) commenced in 1994 as collaboration between medical research agencies, genome institutes and laboratories to sequence the genome. Current achievements and progress[8,9] in this very large sequencing project illustrate the practical difficulty in determining all sequences completely within a genome, despite the enormous resources devoted to it.

The IHGP and other genome programs have established an international quality standard for "finished genomic sequence" to enable the comparison of genomic data from different programs and organisms, and also to define the standard that all genomic sequencing projects should seek to attain. "Finished sequence" is properly defined as the "complete sequence of a clone or genome, with an accuracy of at least 99.99% and no gaps"[10]. A more practical definition is that of "essentially finished sequence", meaning the complete sequence of a clone or genome, with an accuracy of at least 99.99% and no gaps, except those that cannot be closed by any current method. The standards also provide an "end point" for the completeness of genome data with any gap to be smaller than 150 kb and with more than 95% of the *euchromatic* regions as finished sequence. Sequence data that fall short of that benchmark but which can be positioned along the physical map of the chromosomes are termed 'draft'.

Following the publication of the initial drafts of the sequence[1,2] in 2001 in which approximately 80% was sequenced, the project moved deliberately towards the finishing process. Currently, the work is still a mosaic of finished and draft sequence. About 87% of the total genome is finished sequenced[9] and less than 13% is at the draft stage, although the gene-rich *euchromatic* parts of the genome (comprising approximately 2.95 Gb) are about 94% complete "finished" sequence, achieving the IHGP standard. In contrast, *heterochromatic* parts of the genome contain sequences that are more recalcitrant to current sequencing technologies. Additionally, the assembly of many large sequenced repetitive regions is incomplete, as algorithms cannot provide finished sequence at an acceptable standard. Consequently, many *heterochromatic* parts of the human genome must be considered as draft.

1.2   Shotgun Sequencing Strategies

Several strategies for shotgun sequencing, such as "hierarchical" or "map-based shotgun sequencing", "whole-genome shotgun" sequencing and hybrid approaches[11], are still being evaluated for multimegabase- or gigabase-sized genomes of metazoans. The IHGP commenced as a map-based shotgun program, but recently has introduced some hybrid approaches. Perceptively, Waterston et al[5] note that a "great challenge arises in tackling complex genomes with a large proportion of repeat sequences that can give rise to misassembly". They argue that without reference to the IHGP builds, whole genome shotgun assembly methods pioneered by Celera would fail to assemble much larger regions containing repeated DNA motifs and that a whole genome sequencing approach alone would produce fewer assembled regions and more or larger gaps. This assertion highlights a problem common in the assembly of repeated DNA regions, whether using shotgun data or hierarchical shotgun data – that repeated regions possess few "landmarks" that allow definition of the correct order of sub-repeats within the repeated DNA elements.

1.3   Gaps in the IHGP Working Draft Sequence

To illustrate the regions of unfinished sequence of the human genome, Aach et al[4] compared the "gaps" in sequence data and the "gaps" between assembled finished sequence obtained by the IHGP[1] using "hierarchical shotgun sequencing" and Celera's[2] approach using "whole genome shotgun" sequencing. This comparison is shown in Figure 15.1.

Notably, whole genome shotgun sequencing produces many small assembled regions, punctuated by very many gaps frequently several Mb in size.

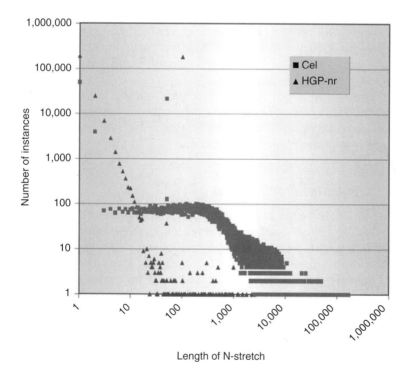

FIGURE 15.1. Size of continuous strings of Ns in the Celera (Cel) and Human Genome Project (HGP-nr) non-redundant genome assemblies. Long strings of Ns are used to represent gaps, but do not always represent gap size. The Celera assembly contained 169,779 stretches of Ns ranging in length from 1 to 168,735. The HGP-nr assembly contained 407,686 stretches of Ns ranging in length from 1 to 2,500. Reprinted with permission from Aach et al (2001) Nature 409, 856.

The hierarchical-shotgun sequencing approach produces much larger assembled regions, punctuated by numerous smaller gaps. These gaps appear in all sequenced regions of the human genome – the *euchromatic* parts, as well as in the *heterochromatic* parts. The gaps that occur in the IHGP working draft sequence fall into several general classes, which are:

i. Gaps within the sequence of ordered clones. Such gaps are mostly small (<200 bp) and represent unclonable regions or unsequenceable motifs within the sequencing libraries.

ii. Gaps between ordered clones and contigs. These are frequently larger gaps up to 0.3 Mb in size. The gaps again represent unclonable or under-represented regions that have not reached quality standards. In addition, the gaps may represent non-reconstructable regions, frequently highly repeated motifs, which despite being able to be sequenced cannot be reconstructed to acceptable quality standards.

iii. Gaps in large repeat regions. Whilst the primary sequence of large repeated regions may be able to be determined, the scale and subtle variant forms of the repeated sub-repeat motifs often prevents accurate assembly and ordering of these larger regions. This problem has been noted particularly for the assembly of megabase long repeated regions, such as telomere and centromere regions[6,12] and their flanking sub-repeated regions[13].

iv. Gaps may also be due to (almost identical) segmental duplications. The detection and correction of errors in assembly of segmental duplications will certainly be of increasing importance for key gene families and regions important for human medicine.

1.4   Segmental Duplications

Eichler[6] and Bailey et al[7] considered large recent duplication events that fell well-below levels of draft sequencing error. Duplications of genomic regions (alignments 90%-98% similar and greater than or equal to 1 kb in length) comprise more than 3.6% of all human sequence. These duplications show clustering within the genome, and have up to 10-fold enrichment within peri-centromeric and sub-telomeric regions – regions of high mobility and sequence rearrangement. Duplicated sequences were found to be over-represented in unordered and unassigned contigs, indicating that duplications are difficult to assign to their correct position, even in recent assembly builds of the human genome. As shown in Figure 15.2, the under-representation or mis-assembly of duplicated sequences are also likely to be a major source of undetected error in current genome assemblies.

Correction of such errors will emerge from re-sequencing projects, from comparison with homologous genomic regions from closely related organisms, and through detailed re-examination of current genomic sequencing data[14].

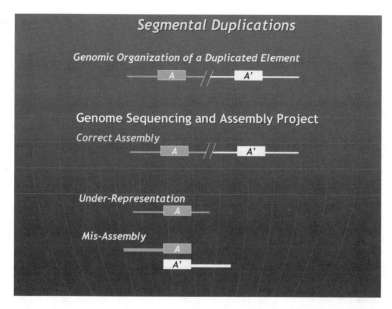

FIGURE 15.2. Errors in the identification and assembly of repeated regions – "segmental duplications" are likely to be a major source of undetected error in current genome assemblies. Correction of such errors will emerge from re-sequencing projects, from comparison with homologous genomic regions from related organisms and through detailed re-examination of current genomic sequencing data and assemblies. Reprinted with permission from Eichler (2001) *Genome Research* **11**, 653-656.

2.   Sequencing by Aligning Mutants

This paper discusses a new and original approach to sequencing of difficult and repeated motifs, *"Sequencing by Aligning Mutants"* (*SAM*)[15-18]. As shown in Figure 15.3, SAM involves forming a number of randomly mutated copies of regions of the target DNA, then several of the mutated fragments are cloned and sequenced. The mutants should ideally be sufficiently altered that they no longer possess the characteristics that caused sequencing difficulties in the target, so existing sequencing techniques are applied more easily and successfully. However, the mutants must also be sufficiently similar to the target such that the original sequence can be inferred by analyzing the mutated sequences.

By aligning the sequenced mutated fragments to each other and to available pieces of the target sequence, mutation sites are identified and corrected for, enabling previously difficult-to-sequence regions of the target to be determined. The information content of the original sequence is not lost, but is merely distributed amongst multiple fragments. The randomness of the mutations is important for another reason: it enables sequence information lost in one mutant to be retained in most of the others.

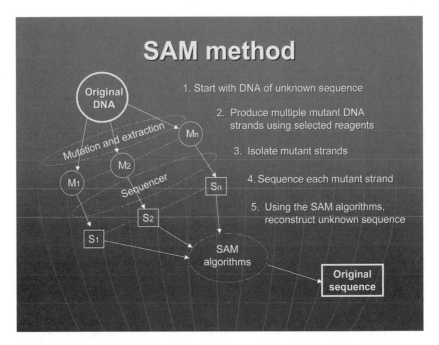

FIGURE 15.3. Sequencing aided by mutation (*SAM*) is a counter-intuitive overall approach to sequencing. Problematic motifs and regions are mutated sufficiently to overcome the obstructive cause to its cloning or sequencing. Random mutations make different altered copies of the region. The sequences determined from a low number of altered copies can be used to reconstruct the original wild-type sequence.

2.1   Local Sequencing Problems

DNA sequence reads are frequently impaired by small motifs that may form secondary structures and other structures that impede the extension of DNA by processive polymerases[19,20]. Such recalcitrant motifs are often AT- or GC-rich repetitive regions that have either high thermal stability or other non-standard characteristics not found in mixed sequence DNA. Yet other small regions that cause sequence gaps may be unclonable or unstable in bacterial cells[3,21]. *SAM techniques* provide a novel method for obtaining sequence data from these intractable regions and could potentially enable genomic researchers to close the gaps present in the genome sequence maps.

Applying *SAM* with a sufficiently high mutation rate can potentially modify inverted repeats and prevent the formation of stem-and-loop structures by disrupting inverted repeat sequences. More generally, mutation can reduce repetition, raise or lower GC content and modify sequences that interfere with cloning host or vector functions or which inhibit DNA manipulations, rendering the recalcitrant fragment amenable to cloning and sequencing. Our calculations[17] indicate that with a substitution rate of 10%, it should be possible to reconstruct the original sequence with an accuracy of less than one error per 10,000 bases using a small number of mutated fragments.

2.2   Base Content

Human chromosomes display wide variation in their regional nucleotide composition that in part reflects the classical *euchromatic* and *heterochromatic* regions. For example, each human chromosome has large swings in GC content:

- one stretch might have as much as 60 percent GC,
- while an adjacent stretch might have only 30 percent GC
- different GC content regions may cause sequencing difficulties for current sequencing technologies and approaches.

Some other organisms have even more extreme nucleotide bias than man. For example, the genome of *Dictylostelium discoideum* is approximately 70% AT, although its chromosomes also display local regions with still higher AT-rich motifs, as well as lower AT content[3]. DNA regions with extreme or high GC or AT composition may also cause sequencing difficulties, causing sequencing enzymes to fall off the DNA template prematurely, or to slip and jump bases[21].

Again, these motifs may interference with either cloning vector or cloning host functions, reducing the representation of the motif within cloned libraries. Indeed, efforts to sequence genomes with extreme AT or CG content are often confounded by a high level of failure to sequence clonable regions, as well as difficulties in gaining library representation of large portions of the genome[3]. Underrepresented regions create gaps, and as they are undefined may in some cases be difficult to isolate even with directed efforts to identify them. Potentially, the introduction of random mutations into whole genomes or into large fractions of genomes could lead to elimination some problem motifs, resulting in increased clonability and larger library representation of variant forms of these recalcitrant sequences.

3.   Dye Terminator Cycle Sequencing

3.1   Repeat Sequences

Many patterns of bases create difficulties for *Taq* DNA polymerase cycle sequencing and DNA amplification. These patterns may be small, extending over some 30-300 bp, or may be larger, extending over 1 kb or up to many kilobases in length. Homopolymer tracts are one example of recalcitrant local motif (see Figure 15.4). Other larger repeated motifs such as microsatellite repeats (~ 3-6 bp repeat motif)[21], LINE and SINE elements[22] may also present barriers to sequencing. In some cases the physical order of the motifs

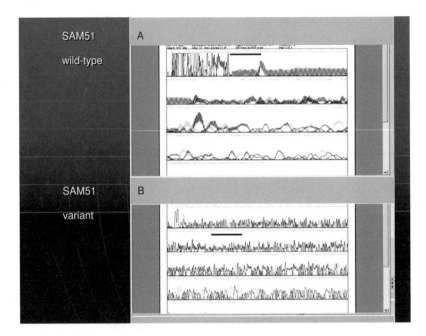

FIGURE 15.4. Comparison of ABI version 2.0 Big-Dye terminator sequencing of wild-type and mutant variant sequences. Bar indicates the problematic motif. A: the wild-type clone (SAM51) contains a homopolymer motif that prevents dye-terminator cycle sequencing. B: Introducing random substitution mutations can reduce the uniformity of the problem motif. One mutated variant of SAM51 is able to be readily sequenced using conventional sequencing technology.

can cause problems for DNA stability and problems for sequencing enzymes. For example, direct repeats may cause problems with PCR, causing primers to mis-prime at multiple sites. Inverted repeats cause hairpins and other complex DNA structures which may be incompatible with plasmid stability or sequencing systems[19-21].

3.2   Improved Sequencing of Simple-Repeats Using SAM Techniques

Figures 15.4 and 15.5 illustrate the use of SAM technologies: homopolymer A tracts that cause difficulties for commercial sequencing kits were used as test molecules.

Figure 15.4A shows the wild-type element in which the sequence could not be determined, with harmonic stutter caused by polymerase slippage and restarts obliterating the usual chromatograph pattern. Following mutation of the region, the sequence of a representative variant shown in Figure 15.4B could be read directly using the same commercial sequencing kit. The

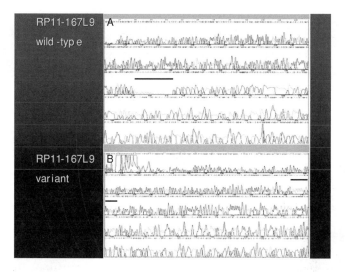

FIGURE 15.5. Comparison of ABI version 2.0 Big-Dye terminator sequencing of wild-type human BAC RP11-167L9, and mutated variants. Bar indicates a problematic polyA motif. A: a region of the wild-type clone contains a problematic motif that prevents dye-terminator cycle sequencing. B: one mutated variant of RP11-167L9 is able to be readily sequenced using conventional technology.

introduction of random substitution mutations reduced the uniformity of the problem motif and allowed *Taq* polymerase to extend through the region. The elimination of stutter bands created a well-defined chromatogram with uniform peaks and good peak separations.

Figure 15.5A displays another wild-type simple repeat region, the sequence trace through the repeat motif is poor, with a weak signal causing the miscalling of some bases and a miscalling of the unit size of the repeat. Although the chromatogram is readable beyond the repeat motif, peaks are broad and potentially miscalling could occur here. Following the mutation of the region, the sequence of a representative variant shown in Figure 15.5B was strong and could be read directly using the same commercial sequencing kit. Again, the introduction of random mutations reduced the uniformity of the motif and allowed *Taq* polymerase to extend through the region and beyond with uniform peaks and good peak separations.

Figure 15.6 shows a Clustal W alignment of sequences from three mutated variant clones of the target region illustrated in *Figure 15.5*, along with the wild-type element read (*polyAwt03_2*) and the published *estimation* of an intractable polyA tract (*target.txt*) of ~36 A residues within the wild-type target. Analysis of fewer than 10 variant sequence reads using *SAM algorithms* recovered the correct sequence and determined the correct size of the repeat as 22 residues.

Sequence alignment of RP11-167L9 and SAM variants
using ClustalW

```
              241       251       261       271       281       291
polyA_P46_    CAGGATTGCGGTGAGCTGATACCGTGCCACTACACTCTAGCCTGGGGGACGGGGTAAGGC
polyA_P47_    CAAGGCCACAGTGAGCTGGTACCGTACCGCTACACTCTAGCCTAGGGGGTGGGGTAAGGC
polyA_P16_    CAAGGCTGCAGTGAGCTGACACCATGCCACTGCACTCTAGCCTGGGGGATGGAGCAAGAC
polAwt03_2    CAAGGCTGCAGTGAGCTGACATCATGCCACTGCACTCTAGCCTGGGGGATGGAGCAAGAC
target.txt    CAAGGCTGCAGTGAGCTGACATCATGCCACTGCACTCTAGCCTGGGGGATGGAGCAAGAC

              301       311       321       331       341       351
polyA_P46_    TTCGCCTCAAAAAA---AGAGGGAAAGAGGA------------GCGGACCTGAACCGGT
polyA_P47_    TCTATCTTAGAAAGGGGAGAGGGAAAAAGG-------------GCGGGCTCGGATTGGT
polyA_P16_    TCTGTCCCAAAAAA---AGAAAAAAGAAAAA-----------GCGGACTTGGATTGGT
polAwt03_2    TCTGTCTCAAAAAAAAAAAAAAAAAAAANNANA-----------GTGGACTTGGATTGGT
target.txt    TCTGTCTCAAAAAAAAAAAAAAAAAAAAAAAAAAAAAAAAAAAAGTGGACTTGGATTGGT

              361       371       381       391       401       411
polyA_P46_    TGCAGATGTGTACTGCAGACTCTGGGG-TGGCCACTACA-TCTCTTTAGAGGGGGGGACA
polyA_P47_    TGCAAACGTGTGATTGTAAACCTTAGGG-TGACCACCGCGCTCTTTTAAGAGGGGAGGT-A
polyA_P16_    TGTAGGTGTATACTGCAAGCTCTAGGG-TGACCACCACA-TCTCTTTGAAAAGAGAAATA
polAwt03_2    TGTAAATGTATATTGCAAACTCTAGGGGTGACCACCACATTTTTTTTAAAAAAAGAAAT-A
target.txt    TGTAAATGTATATTGCAAACTCTAGGG-TGACCACCACATTTTTTTTAAAAAAAGAAAT-A
```

FIGURE 15.6. Clustal W alignment of region of wild-type human BAC RP11-167L9 (polAwt), and mutated variants compared to the published estimation of the polyA tract (target). The wild-type clone contains a problematic motif that prevents accurate dye-terminator cycle sequencing (variable length reads, unreadable N's present), whereas randomly mutated variants of the clone display consistently reduced polyA tract length.

4.   Sequencing by Hybridization (SBH)

4.1   The Limitations of Conventional SBH

"Sequencing by hybridization" (SBH) is a potentially powerful sequencing technology that analyses sequence reads in massively parallel manner by the hybridization of short target fragments to an array of "all" possible oligonucleotide probes. SBH is based on a relatively simple concept, in which a non-redundant array of oligonucleotides (of length p) is arranged on a solid support (typically a silica or glass slide)[23-25]. If a target fragment hybridizes to particular probes in the array, a representative signal is obtained from each of those probes. Signals from each element in the array that hybridizes to each complementary region of the target fragment collectively constitute the "SBH spectrum", which may be analysed by reconstruction algorithms to generate a sequence of the entire target region. Potentially, SBH can provide megabase-scale simultaneous sequencing capacity if practical hybridization and data reconstruction problems can be overcome. Although SBH has been used in detailed local mapping, SNP detection and re-sequencing of relatively small regions[24], doubts remain as to whether SBH can be used for *de novo* sequencing, particularly for megabase scale.

Southern and colleagues[26-28] have demonstrated that steric factors, sequence motifs and differential nucleotide hybridization stabilities each contribute to non-uniformity in hybridization of the target to anchored oligonucleotide probes. Potentially, the breakage of target DNA into small pieces could disrupt the sequence runs and structural barriers that prevent hybridization to probe oligonucleotides. Although some physical intra-strand interactions may be reduced, inter-strand interactions between the smaller target element fragments could be expected to increase.

SBH is a qualitative hybridization technology and many practical barriers to its use for *de novo* sequencing have been identified:

- Hybridization is non-quantitative and non-representative
- Some sequence motifs provide no signals, yet others provide strong signals
- Foldback and other intra-strand hybridizations can limit availability of target regions for hybridization to oligonucleotide probes
- Short repeated regions provide ambiguous reconstruction paths
- AT and CG rich targets have different hybridization efficiency

SBH cannot be used for sequencing simple sequence DNA, such as poly A, (GT)n, repeats, etc. As it cannot quantify the number of repeat copies present in the target, it cannot easily determine the length of the repeat region. SBH also cannot determine the order of repeated sequence elements: these becomes ambiguous if repeat elements longer than length p are present in the target. Practically, SBH is limited to analysis of targets that *lack* 3 or more copies of repeats longer than $(p-1)$. The probability of ambiguity increases exponentially as probe length decreases, suggesting longer oligonucleotide probe lengths as a practical solution. However, as probe length increases the complexity and number of probes located on the SBH array necessarily also increases exponentially. Other informative techniques must be used to supplement SBH data to overcome these ambiguities[25]. These contradictory physical and mathematical problems require novel solutions – SAM technology can provide these solutions.

4.2   The Use of SAM Techniques for SBH Sequencing

SAM techniques can resolve ambiguities in SBH reconstruction. Mutant variants of a target sequence are analysed by SBH, each variant generating a unique SBH spectrum, then each variant spectrum is reconstructed using *SAM algorithms*. Mutant fragments contain *unambiguously reconstructable regions* that span repeated $(p-1)$-mers in the target. These reconstructed regions of mutants are then used in combination as templates to resolve ambiguities in reconstruction of the target sequence spectrum. Computer simulations have shown that analysis using *SAM algorithms* can often get the overall order of maximal sub-strings correct, even when a small number of short sub-strings are misplaced.

Interestingly, Southern and Nguyen[29] and Nguyen et al[30] have recently suggested the introduction of nucleotide analogues into the oligonucleotide probes and use of analogues with chaotropic agents respectively as methods for reducing differential hybridization potential between AT- and CG-rich targets and targets with structural motifs. These methods however do not address the problems of ambiguous reconstruction caused by repeated motifs, they address the differential hybridization stabilities of probes of different base composition.

4.3   Computer Simulations Using SAM for SBH Reconstruction

For simulation studies, random 100 kb pieces of human genome sequence were used as a source of "real" DNA sequence. Randomly mutated variant sequences were generated at different mutation intensities, allowing each base equal probability of variation. Computer simulations of SBH reconstructions of mutant variants of target DNA elements were undertaken using SAM algorithms to direct the build as outlined in *Section 4.2*. The results can be compared to standard SBH reconstructions of the original wild-type human target sequences. Randomly selected 5 kb fragments of human genomic DNA can be *completely reconstructed* in 99.9% of attempts with fewer than 1 error per 1000 bases (97.8% perfectly correct) using 9 mutants and probes of length 13. In contrast, fewer than 1% of reconstructions of 5 kb fragments using standard SBH spectra were correct, even allowing 0.1% error.

The size of DNA fragments that can be reconstructed correctly when SAM techniques are used also increases markedly. Human DNA fragments up to 30 kb long were successfully reconstructed during simulated SAM experiments. In contrast, no fragments of this length could be reconstructed applying conventional builds of the SBH spectrum. There is scope for improving the performance of the current SAM reconstruction algorithms for SBH. For example, the current algorithms assume equal probabilities of mutation of each nucleotide base. Algorithms could be modified to incorporate the probabilities of particular mutation events, which are determined empirically for particular mutation protocols[16,17].

5.   Conclusions

This paper discusses the novel approach to sequencing which we have developed called *Sequencing by Aligning Mutants (SAM)*. It was developed with the purpose of providing a simple and effective method of sequencing DNA motifs that cannot be sequenced by other current techniques. *SAM technologies* include methods to achieve highly controlled levels of mutation in target DNA elements. Preferably these mutations are simple substitution mutations.

The methods also include advanced assembly and reconstruction algorithms to recover original sequence from a small number of altered versions of the target[15-18].

We have shown that improved sequence reads can be obtained using SAM techniques and conventional Dye-terminator cycle sequencing from several model DNA's which contain "difficult to sequence motifs". Although not displayed here for sake of space, the reconstructed sequence recovered from the model targets is accurate even using fewer than 10 variants. The protocols are repeatable and are readily modifiable for different DNA sequence motifs. Although the overall approach is novel, the technologies were developed with the view that they are compatible with use of standard laboratory processes and equipment, and thus available for conventional molecular biological laboratories which may be lacking sophisticated genomic analysis equipment.

The intention of our laboratory is to develop and release portfolios of methods and reagents as well as portfolios of advanced assembly algorithms. Versions of the algorithms may have additional applications for improved sequence comparison. Together these developments are intended to form the basis for several different genomic software tools to be applied along with conventional sequencing kits.

The application of SAM technologies to SBH based sequencing is a new area of development, as the potential for repetitive motifs and other structural motifs that interfere with target::probe hybridization could be diminished within variant target molecules. New array chemistries such as PNA oligonucleotide probes[31] could be used to further alter target::probe interactions and provide a broader spectrum of hybridization signals than conventional DNA::DNA arrays.

References

1. The International Human Genome Sequencing Consortium. Initial sequencing and analysis of the human genome. *Nature* **409**, 860-921 (2001).
2. Venter, C. et al. The sequence of the human genome. *Science* **291**, 1304-1351 (2001).
3. *Dictyostelium* Genome Sequencing Consortium. Sequence and analysis of chromosome 2 of *Dictylostelium discoideum. Nature* **418**, 79-85 (2002).
4. Aach, J., Bulyk, M.L., Church, G.M., Comander, J., Derti, A. & Shendure, J. Computational comparison of two draft sequences of the human genome. Nature **409**, 856-859 (2001).
5. Waterston, R.H., Lander, E.S. & Sulston, J.E. On sequencing the human genome. *Proc. Natl. Acad. Sci. USA* **99**, 3712-3716 (2002).
6. Eichler, E.E. Segmental duplications: what's missing, misassigned, and misassembled—and should we care? *Genome Res.* **11**, 653-656 (2001).
7. Bailey, J.A., Yavor, A.M., Massa, H.F., Trask, B.J. & Eichler, E.E. Segmental duplications: organization and impact within the current human genome project assembly. *Genome Res.* **11**, 1005-1017 (2001).

8. International Human Genome Sequencing Consortium. (2002). Current sequencing status.*http://www.ncbi.nlm.nih.gov/genome/seq*/page.cgi?F=HsProgress. shtml&& ORG=Hs

9. Wolfsberg, T.G., Wetterstrand, K.A., Guyer, M.S., Collins, F.S. & Baxevanis, A.D. Introduction: putting it together. *Nature Genet.* **32** Suppl, 5–8, (2002).

10. Collins, F.S. & McKusick, V.A. Implications of the Human Genome Project for medical science. *J. Am. Med. Assoc.* **285**, 540-544 (2001).

11. Green, E.D. Strategies for the systematic sequencing of complex genomes. *Nature Rev. Genet.* **2**, 573-583 (2001).

12. Horvath, J.E., Schwartz, S. & Eichler, E.E. The mosaic structure of human pericentromeric DNA: a strategy for characterizing complex regions of the human genome. *Genome Res.* **10**, 839-852 (2000).

13. Horvath, J.E., Viggiano, L., Loftus, B.J., et al. Molecular structure and evolution of an alpha satellite/non-alpha satellite junction at 16p11. *Hum. Mol. Genet.* **9**, 113-123 (2000).

14. Bailey, J.A., Gu, Z., Clark, R.A., Reinert, K., et al. Recent segmental duplications in the human genome. *Science* **297**, 1003-1007 (2002).

15. Keith, J.M., Adams, P., Bryant, D., Mitchelson, K.R., Cochran, D.A.E., & Lala, G.H. Inferring an original sequence from erroneous copies: a Bayesian approach. *Proceedings of the 1st Asia-Pacific Bioinformatics Conference* (APBC 2003). (ed. Chen, Y-P.P) **19**, 23-28 (2003).

16. Keith, J.M., Adams, P., Bryant, D., Cochran, D.A.E., Lala, G.H. & Mitchelson, K.R. Inferring an original sequence from erroneous copies: two approaches. *Asia-Pacific BioTech News* **7**, 107-114 (2003).

17. Keith, J.M., Adams, P., Bryant, D., Cochran, D.A.E., Lala, G.H. & Mitchelson, K.R. Algorithms for sequencing aided by mutagenesis. *Bioinformatics* **20**, 2401-2410 (2004).

18. Keith, J.M., Adams, P., Bryant, D., Kroese, D.P., Mitchelson, K.R., Cochran, D.A.E. & Lala, G.L. A simulated annealing algorithm for finding consensus sequence. *Bioinformatics* **18**, 1494-1499 (2002).

19. Razin, S.V., Ioudinkova, E.S., Trifonov, E.N. & Scherrer, K. Non-clonability correlates with genomic instability: a case study of a unique DNA region. *J. Mol. Biol.* **307**, 481-486 (2001).

20. Kang, H.K. & Cox, D.W. Tandem repeats 3' of the IGHA genes in the human immunoglobulin heavy chain gene cluster. *Genomics* **35**, 189-195 (1996).

21. Baran, N, Lapidot, A. & Manor, H. Formation of DNA triplexes accounts for arrests of DNA synthesis at d(TC)n and d(GA)n tracts. *Proc. Natl. Acad. Sci. USA* **88**, 507-511 (1991).

22. Mallon, A.M., Platzer, M., Bate, R., Glöckner, G. et al. Comparative genome sequence analysis of the Bpa/Str region in mouse and man. *Genome Res.* **10**, 758-775 (2000).

23. Southern, E.M. DNA microarrays: History and overview. *Methods Mol. Biol.* **170**, 1-15 (2001).

24. Chechetkin, V.R., Turygin, A.Y., Proudnikov, D.Y., et al. Sequencing by hybridization with the generic 6-mer oligonucleotide microarray: an advanced scheme for data processing. *J. Biomol. Struct. Dyn.* **18**, 83-101 (2000).

25. Drmanac, R., Drmanac, S., Chui, G., et al. Sequencing by hybridization (SBH): advantages, achievements, and opportunities. *Adv. Biochem. Eng. Biotechnol.* **77**, 75-101 (2002).

26. Mir, K.U. & Southern, E.M. Determining the influence of structure on hybridization using oligonucleotide arrays. *Nat. Biotechnol.* **17**, 788-792 (1999).

27. Southern, E., Mir, K. & Shchepinov, M. Molecular interactions on microarrays. *Nat. Genet.* **21** (Suppl.), 5-9 (1999).

28. Shchepinov, M.S., Case-Green, S.C. & Southern, E.M. Steric factors influencing hybridisation of nucleic acids to oligonucleotide arrays. *Nucleic Acids Res.* **25**, 1155-1161 (1997).

29. Nguyen, H.K. & Southern, E.M. Minimising the secondary structure of DNA targets by incorporation of a modified deoxynucleoside: implications for nucleic acid analysis by hybridisation. *Nucleic Acids Res.* **28**, 3904-3909 (2000).

30. Nguyen, H.K., Fournier, O., Asseline, U., Dupret, D. & Thuong, N.T. Smoothing of the thermal stability of DNA duplexes by using modified nucleosides and chaotropic agents. *Nucleic Acids Res.* **27**, 1492-1498 (1999).

31. Weiler, J., Gausepohl, H., Hauser, N., Jensen, O.N. & Hoheisel, J.D. Hybridisation based DNA screening on peptide nucleic acid (PNA) oligomer arrays. *Nucleic Acids Res.* **25**, 2792-2799 (1997).

16

Fabrication of Double-Stranded DNA Microarray on Solid Surface for Studying DNA-Protein Interactions
A High-Throughput Platform for Profiling Bimolecular Interaction

JINKE WANG AND ZUHONG LU
Chien-Shiung Wu Laboratory, Southeast University, Nanjing 210096, China

Abstract: This paper presents two novel methods for fabricating double-stranded DNA (dsDNA) microarray. In the first method, the presynthesized single-stranded DNA (ssDNA) oligonucleotides containing two reverse complementary sequences at their 3′ hydroxyl end were firstly immobilize on the surface of the aldehyde-derivatized glass slides by their 5′ end, and then the two reverse complementary sequences were annealed to form a short dsDNA hairpin structure which provided the primer for later polymerase elongation. Finally, the ssDNA microarrays were converted into the unimolecular dsDNA microarrays by an on-chip polymerase reaction. In the second method, the two kinds of ssDNA oligonucleotides named constant oligonucleotide (CO) and target oligonucleotides (TOs) were synthesized. Then the different TOs harboring the DNA-binding sites were respectively annealed and ligated with the same CO containing an internal aminated dT in tubes. The reaction products were immobilized on the surface of the aldehyde-derivatized glass slides by the aminated dT to fabricate the partial-dsDNA microarrays. Finally, the partial-dsDNA microarrays were converted into the unimolecular dsDNA microarrays by an on-chip polymerase reaction. The excellent efficiency and high accuracy of the enzymatic synthesis in two methods were demonstrated by incorporation of fluorescently labeled dUTPs in Klenow extension and the digestion of dsDNA microarrays with restriction endonuclease. The accessibility and specificity of the DNA-binding proteins binding to dsDNA microarrays were verified by binding Cy3 labeled NF-κB (p50) to dsDNA microarrays. Therefore, the dsDNA microarray containing 66 probes representing 30 all-possible single-nucleotide mutant NF-κB binding targets of Ig-κB and 36 wild-type NF-κB binding targets were fabricated to determine the binding affinities of NF-kB homodimer p50 to all probes on chip. We found the binding results were very consistent with that from x-ray crystallography studies and gel mobility-shift analysis. The unimolecular dsDNA microarray has great potentials to provide a high-throughput platform for investigating the sequence-specific DNA-protein interactions involved in gene expression regulation, restriction and so on.

Key words: Double-stranded DNA microarray, fabrication, DNA-protein interactions.

1.   Introduction

The interactions of DNA-binding proteins and DNA-binding drugs with double-stranded DNA (dsDNA) in genome are involved in many important biological functions, including gene transcription regulation (1), DNA recombination (2), restriction (3), replication (4) and DNA-drugs inter-calation (5,6). Therefore, lots of techniques were used to effectively study DNA-drugs interaction,including nitrocellulose-binding assays (7), gel mobility-shift analysis (8,9), Southwestern blotting (10,11), ELISA (12), reporter constructs in yeast (13), Chromatin immunoprecipitation (ChIP) (14), phage display (15), binding-site signatures (16), in-vitro selection (17), UV crosslinking (18), and methylization interfering assay (19), X-ray crystal-lography (20,21) were developed to effectively examine sequence-specific DNA-protein interactions, and also techniques including UV absorption(22), melting temperature (thermodynamics) (23), NMR (24), X-ray crystallogra-phy (25), free solution capillary electrophoresis (FSCE) (26), scanning force microscopy (SFM) (27,28,29), atomic force microscopy (AFM) (30), surface plasmon resonance (SPR) (31), polymerase chain reaction (PCR) (32) and footprinting (33). However, these techniques using solvent dsDNAs to probe dsDNA interactions with other molecules as proteins, ligands and drugs suf-fered from being laborious, time-consuming and incapable of high-parallel analysis. Therefore, the solid surface-coupled dsDNA had become more and more important for high-throughput examination of sequence-specific DNA/protein (34) and DNA/drug interactions (35,36), it paved the way to new strategies for screening DNA-binding proteins(37), predicting DNA binding sites (38,39), assessing binding affinity (35,36), and screen-ing sequence-specific DNA-binding drugs and finding drugs preferential sequences (35,36).

As the surface-coupled homogenous dsDNAs are employed to examination of the sequence-specific DNA/proteins or DNA/drugs interactions, the num-ber of classes of dsDNAs immobilized on a detectable solid entity determines how much information could be obtained in a single study. The cellulose or agarose-coupled homogenous dsDNAs were traditionally used to isolate the sequence-specific DNA-binding proteins (40,41) by the affinity chromatogra-phy, and the small paramagnetic particles-attached homogenous dsDNA probes were used to identify DNA-binding proteins by matrix-assisted laser desorption/ionization time-of-flight mass spectrometry (MALDI-TOF MS) (37). However, these solid entities carrying homogenous dsDNAs suffered from isolating or identifying only one target molecular every time. Therefore, they were not the high informative strategies for surface-coupled dsDNA applications. Comparatively, the libraries of dsDNA oligonucleotides comprising a plurality of different members immobilized on solid support much improved the limitation (42). These solid-immobilized libraries of dsDNA oligonucleotides provided a useful technique for the screening of numerous biological samples by sequence-specific interactions (42). Thereafter, the fabrication of fast, economical and high informative dsDNA-coupled solid entities became the pivotal problem for extensive surface-coupled dsDNA applications. Nevertheless, the earliest surface-immobilized libraries of different dsDNA oligonucleotides were fabricated by immobilizing very long chemically-synthesized single-stranded DNA (ssDNA) oligonucleotides on solid surface and then forming unimolecular dsDNA oligonucleotides by intrastrand annealing self-complementery elements in immobilized long single-stranded oligonucleotides (42). This kind of surface-coupled dsDNA were high informative, but suffered from economic issues.

The appearance of high density bimolecular dsDNA microarray greatly promoted the application of solid-immobilized duplex nuclei acids (34, 35, 36, 38, 39). It was demonstrated that these high density bimolecular dsDNA microarrays were very effective for high throughput examination of DNA-proteins interaction (38, 39). However, the developments of these bimolecular dsDNA microarrays were impeded by several innate technical and economic drawbacks. The current-developed methods manufacturing bimolecular dsDNA microarrays could be divided into two types, one was by hybridization (35, 36, 43) and the other was enzymatic elongation (34). The former spotted larger number of chemically synthesized ssDNA oligonucleotides onto solid surface to firstly fabricate ssDNA microarray and subsequently converted ssDNA microarray into bimolecular dsDNA microarray by hybridizing the ssDNA microarray with a mixture of complementary ssDNA oligonucleotides. This method suffered from two main problems, one was high costs of the synthesis of the complementary ssDNA oligonucleotides and amino-modification of immobilized ssDNA oligonucleotides, the other more important problem was that the method could not fabricate dsDNA microarrays carrying very sequence-similar probes such as probes with single-nucleotides variation (35, 36). The latter on-chip photo-addressably synthesized high density ssDNA microarray with a constant sequence at far surface-attached 3′ end of each oligonucleotides, and then annealed a general primer to constant sequence and performed enzymatic extension reaction on array to convert ssDNA microarray into bimolecular dsDNA microarray. This method encountered more serious economic and technical issues. In technique, the method replied on currently expensive and proprietary technology of surface photo addressable synthesis of oligonucleotides, but the synthesis of single-stranded oligonucleotides on solid surface was inefficient, with per-nucleotide synthesis efficiencies thought to be only 92-96% (44, 45). For 40-mer oligonucleotides, only 4-20% of the sequences on a chip could be of desired length and sequence (46). In practice, oligonucleotide arrays constructed in this fashion were heavily contaminated with truncated molecules (44, 47). Moreover, the presence of so many competing truncated molecules and single-stranded oligos not accessible to the primers might strongly interfere and mislead binding experiments (46). Finally, considering the possible instability of bimolecular dsDNA oligos in binding or washing reactions, the utility efficiency of this kind of bimolecular dsDNA microarrays might be lowered.

The alternative route to construct dsDNA arrays might resolve the economic and technical problems. Ben-Yoseph and his co-workers fabricated the dsDNA microarray by hybridization and ligation. These investigators attached single-stranded oligonucleotides to gold supports by a thiol linkage, and then to those oligos, hybridized and ligated double-stranded DNAs with the appropriate complementary ends (43). Bulyk and his co-workers fabricated the dsDNA microarray by hybridization and polymerization (38). The alternative methods for dsDNA microarray fabrication promoted the practical application of dsDNA microarray. For example, the dsDNA microarrays, fabricated by microspotting the dsDNA oligonucleotides which were prepared by firstly annealing the set of 64 oligonucleotides representing all possible 3-nt central-finger sites for Zif268 zinc finger with a 5′ amino-tagged universal primer and subsequently polymerizing with Klenow enzyme, were effectively used to explore the DNA-binding specificities of zinc fingers (38). Nevertheless, the utility efficiency of the bimolecular dsDNA micorarray must be lowered by the possible denature of bimolecular dsDNA probes in remove of the bound proteins for more hybridizations.

To overcome the drawbacks existing in the above described dsDNA microarray techniques, we presented two new methods for fabricating unimolecular dsDNA microarray which can be used for many times. First was to immobilize ssDNA oligonucleotides by 5′ end which contained two reverse complementary sequences at 3′ hydroxyl end, and then annealed two reverse complementary sequences to form a short dsDNA hairpin structure which took the role of primer for later polymerization. After an on-chip polymerase elongation reaction, the single-stranded microarrays were converted into unimolecular dsDNA microarrays. Second was to synthesize two kinds of ssDNA oligonucleotides called constant oligonucleotide and target oligonucleotide. The constant oligonucleotide with internal aminated dT was used to capture and immobilize the target oligonucleotides onto solid surface, and also provide primer for later enzymatic extension reaction, while target oligonucleotides took the role of houbouring DNA-binding sites of DNA-binding proteins. The variant target oligonucleotides were annealed and ligated with the constant oligonucleotide to form the new unimolecular oligonucleotides for microspotting. The prepared unimolecular oligonucleotides were microspotted on aldehyde-derivatized glass slides to make partial-dsDNA microarray. At last, the partial-dsDNA microarray was converted into unimolecular complete-dsDNA microarray by a DNA polymerase extension reaction. The excellent efficiency and high accuracy of the enzymatic synthesis in two methods were demonstrated by incorporation of fluorescently labeled dUTPs in Klenow extension and digestion of dsDNA microarrays with restriction endonuclease. The accessibility and specificity of the DNA-binding proteins binding to dsDNA microarrays were verified by binding Cy3 labeled NF-κB to dsDNA microarrays. We fabricated the dsDNA microarray containing 31 probes representing the wild-type and all-possible single-nucleotide mutant NF-κB binding targets of Ig-κB, to determined the binding affinities of NF-kB homodimer p50 to all probes on chip. We found the binding results were very consistent with that from x-ray crystallography studies and gel mobility-shift analysis. The dsDNA microarrays we fabricated have great potentials to provide a high-throughput platform for investigation of sequence-specific DNA-protein interactions involved in gene expression regulation, restriction and so on.

2.   Materials and Methods

2.1   Manufacture of Double-Stranded DNA Microarray

The special ssDNA oligonucleotides listed in tables were chemically synthesized by Shengyou Inc. (Shanghai, China) for manufacturing dsDNA microarrays.

In method 1, the ssDNA oligonucleotides were designed and chemically synthesized as materials to fabricate ssDNA microarray. The prepared single-stranded oligonucleotides contained the seven elements consisting of 5′ NH$_2$—proximal flanking sequence—restriction endonuclease digestion site—DNA-binding protein binding consensus—distal flanking sequence—reverse complementary sequence—hairpin loop bases—reverse complementary sequence 2—3′ OH as listed in Table 16.1. We also synthesized single-stranded oligonucleotides which elements including proximal flanking sequence, restriction endonuclease digestion site and hairpin loop bases were varied as listed in table 16.1 too. The dsDNA microarrays were fabricated according to the scheme displayed in Figure 16.1. Firstly, the 5′-end amino-linked oligonucleotides were printed on glutaraldehyde-derived glass slides to fabricate the single-stranded oligonucleotides microarray. Secondly, the slides were incubated with boiling water to denature the possible partial hairpin structure. Thirdly, the slides were incubated with hybridization buffer to form the hairpin structure primer at the free end of immobilized oligonucleotides. Fourthly, the slides were incubated with Klenow enzyme reaction solution to extent the hairpin primer.

In method 2, the two kinds of ssDNA oligonucleotides named the constant oligonucleotide (CO) and the target oligonucleotides were designed and

TABLE 16.1. Single-stranded oligonucleotides prepared by chemical synthesis

Group	No.	Sequence (5′→3')				
		FS (or SP)	PBS	RS	LB	RS
1	SS	AGTTGAG	GGGACTTTCC	CAGGC	TT	
2	FS0	GAATTC	GGGACTTTCC	CAGGC	TT	GCCTG
	FS1	TGAATTC	GGGACTTTCC	CAGGC	TT	GCCTG
	FS2	TTGAATTC	GGGACTTTCC	CAGGC	TT	GCCTG
	FS3	TTTGAATTC	GGGACTTTCC	CAGGC	TT	GCCTG
	FS4	TTTTGAATTC	GGGACTTTCC	CAGGC	TT	GCCTG
	FS5	TTTTTGAATTC	GGGACTTTCC	CAGGC	TT	GCCTG
3	RS2	TTTTGAATTC	GGGACTTTCC	CA	TT	TG
	RS3	TTTTGAATTC	GGGACTTTCC	CAG	TT	CTG
	RS4	TTTTGAATTC	GGGACTTTCC	CAGG	TT	CCTG
4	LB0	TTTTGAATTC	GGGACTTTCC	CAGGC	T	GCCTG
	LB1	TTTTGAATTC	GGGACTTTCC	CAGGC	TT	GCCTG
	LB2	TTTTGAATTC	GGGACTTTCC	CAGGC	TTT	GCCTG
	LB3	TTTTGAATTC	GGGACTTTCC	CAGGC	TT	GCCTG
5	A3	AAAGGTTCCTT	GGGTTCCTTT	CAGGC	TT	GCCTG
	Lac-rep	AATTGTGAGCGG	AT AACA ATT	CAGGC	GCCTG GCCTG	GCCTG
	CAP	TGTG AGT	TAGCTCACT	CAGGC	GCCTG GCCTG	GCCTG
	GAL4	CGG AGGAC	AGTCCTCCG	CAGGC	GCCTG GCCTG	GCCTG
	TFIID	CTGTGCA	TATAA	CAGGC		GCCTG
	GCN4	AGCGGA	ATGACTCAT	CAGGC	GCCTG GCCTG	
	AP1	CGCTTGA	TGAGTCA	CAGGC	GCCTG GCCTG	
	AP2	AATGTCC	CCGCGGC	CAGGC	GCCTG GCCTG	
	SP1	ACGATCG	GGGCGG	CAGGC		GCCTG

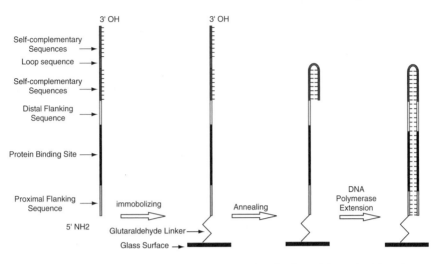

FIGURE 16.1. The scheme for dsDNA microarray fabrication with method 1 and verification with enzymatic digestion.

chemically synthesized as materials to fabricate ssDNA microarray. The CO contained a 7-base single-stranded capture overhang at 3′ end and two reverse self-complementary sequences intervened by an internal dT with a primary NH$_2$ group (amino modifier C6) at 5′ end. The TOs consisted of the 7-base general proximal flanking sequence complementary to capture overhang on CO, target sequence containing 10-base wild or single-nucleotide mutated Ig-κB sites, and the 7-base general distal flanking sequence from 3′ to 5′ end. The CO contained hydroxyl at 3′ end and phosphate at 5′ end, and the TOs had hydroxyl at both 3′ and 5′ ends. The dsDNA microarrays were fabricated according to the scheme displayed in figure 16.7. The different TOs were respectively annealed and ligated with the same pre-self-annealed COs in 1:1 molar ratio in a ligation reaction containing 40 mM Tris-HCl (pH 7.8), 10 mM MgCl$_4$, 10 mM DTT, 0.5 mM ATP and 0.5 U/μl T4 DNA ligase (MBI fermentas). The completed ligation reactions were exchanged into sodium carbonate buffer (0.1M, pH9.0) at a concentration of 50 μM for microspotting, by using CentriSpin-10 spin columns (Princeton Separations, Adelphia, NJ).

The cleaned microscopy glass slides were silanized in 2% aminopropyltri-ethoxysilane (Sigma) dissolved in 95% acetone for 5 min. After rinsing twice with acetone and then baked for 45 min at 75 °C, the silanized slides were activated in glutaraldehyde solution (5% glutaraldehyde, 0.01M PB, pH7.0) for 30 min. The slides were then washed thoroughly with distilled water and blow to dry with N$_2$. The glutaraldehyde-treated glass slides were stored at 4 °C and used in 15 days. A pin-based spotting robot PixSys5500 (Cartesian Technology Inc.) with a CMP3 pin was employed to print the prepared oligonucleotides dissolved in sodium carbonate buffer (0.1M, pH9.0) on slides. After printing, the microarrays were incubated overnight at room temperature, then for 1 h at 37 °C in a humidity chamber containing sodium carbonate buffer (0.1M, pH9.0). The rest of the aldehyde surface was inactivated by a 30-min incubation in 0.28% (w/v) $NaBH_4$/76% (v/v) PBS/24% (v/v) alcohol. After sufficiently washing in sterile ddH_2O, the microarrays were spun dry in a clinical centrifuge.

The dried microarrays were incubated in boiling sterile water for 5min and then treated with hybridization buffer for 1 h at 50 °C. The annealed microarrays were respectively washed with 2×SSC/0.1% SDS and 0.2×SSC/0.1% SDS for 10min at room temperature. After washing sterile ddH_2O and dry in a clinical centrifuge, the microarrays were incubated with DNA polymerase reaction containing 50 mM Tris-HCl (pH 7.2), 10 mM $MgSO_4$, 0.1 mM DTT, 40 µM of each dNTP, 20µg/ml acetylated BSA and 2 U/µl DNA polymerase I large (Klenow) fragment (3′ to 5′ exo⁻; Promega, Madison, WI). After extension reaction, the microarrays were respectively washed with 2×SSC/0.1% SDS, 0.2×SSC/0.1% SDS and sterile ddH_2O for 10min at room temperature. At last, the microarrays were dried in a clinical centrifuge and kept in closed cassette at 4 °C until use.

2.2   NF-κB (p50-p50) Binding to dsDNA Microarray

Transcription factor NF-κB, human recombinant p50 expressed in bacteria from a full-length cDNA encoding 453 amino acids, was purchased from Promega (Madison, WI). The protein provided in glycerol solutions were transferred to sodium carbonate-sodium bicarbonate buffer (pH9.3) and labeled with FluoroLinkTM Cy3 monofunctional dye (Amersham Pharmacia Biotech, Piscataway, NJ) at room temperature for 30min. After labeling, the protein were exchanged into glycerol-free, phosphate-buffer saline (PBS) solution (141 mM NaCl, 7.2 mM Na2HPO4, 2.8 mM NaH2PO4, pH7.4) by BioRad Biospin P6 column. The labeled proteins PBS solutions were kept at 4 °C until use. The dsDNA microarrays were blocked with 10% BSA for 1 h at room temperature, then incubated with DNA-binding buffer (10mM HEPES pH7.9, 50mM KCl, 2.5mM DTT, 0.1mM EDTA, 0.05% NP-40, 10% Glycerol, 5%BSA) containing Cy3 labeled NF-κB (p50) at room temperature for 1 h. After incubation, the microarrays were in turn washed with PBS/0.05% Tween 20 for 15 min, PBS/0.01% Triton 100 for 15 min, and PBS for 15 min at room temperature. After spinning dry in a clinical centrifuge, the microarrays were scanned with ScanArray® Lite of Packard Biochip Technologies in the Cy3 channel at 90% laser power, 80% PMT gain, 5 µm resolution.

2.3   dsDNA Microarray Data Analysis

The signal intensities of the spots on microarray scanned false color images were quantified QuantArray® microarray analysis software (Packard

Biochip Technologies). The signal intensities of the spots referred to the absolute signal intensities arrived from substracting the background fluorescence intensities from detected signal intensities of spots. The relative signal intensities of spots of the single-nucleotide mutated Ig-κB targets were calculated as a fraction of the averaged intensity of spots of the wild-type Ig-κB targets.

3.   Results

3.1   Manufacture of dsDNA Microarray

3.1.1   Manufacture of dsDNA Microarray with Method 1

3.1.1.1   Design of Oligonucleotides for Fabricating the ssDNA Microarray

The prepared ssDNA oligonucleotides were characterized with containing two reverse self-complementary sequences at their 3′ end and an overhang sequence at 5′ end. The two reverse self-complementary sequences linked by loop bases were constituted by the same number of bases, and were used to form a short dsDNA hairpin structure which played the role of primer for enzymatic extension. The overhang sequence contained the DNA-binding site of transcription factors NF-kB and flanking sequence (or spacing sequence) from 3′ end to 5′ end. The overhang sequence was used as template for enzymatic extension. The DNA-binding site of transcription factors NF-kB was used to explore the accessibility of synthesized unimolecular dsDNA oligonucleotide immobilized on chip to DNA-binding protein, NF-kB. The flanking sequence at the far 5′ end served as the arm molecules to avoid the possible steric hindrance from solid surface to DNA/protein interaction. In some oligonucleotides, a HaeIII digestion sites were harbored in flanking sequence to verify the enzymatic synthesis and the accessibility of sequence-specific restriction endonuclease to its target site harbored in immobilized dsDNA. The 5′ end of all synthesized oligonucleotides were modified with one primary amino group, which was used to immobilize the oligonucleotides on aldehyde-derived glass slide surface. The 3′ end of all synthesized oligonucleotides were hydroxyl group, which was used to primer extension. Oligonucleotides with various flanking sequences, reverse complementary sequences and loop bases were synthesized to investigate their influences on enzymatic synthesis, it aims at optimizing the oligonucleotides design for fabricating the most economical and effective dsDNA microarrays with the method provided.

3.1.1.2   Scheme of dsDNA Microarray Fabrication

The scheme of dsDNA microarray fabrication used in this paper is illustrated in Figure 16.1. According to the scheme, dsDNA microarray will be fabricated with three steps. Firstly, the amino-linked oligonucleotides were printed on aldehyde-derived glass slides to fabricate the ssDNA oligonucleotides microarrays. Secondly, the ssDNA oligonucleotides microarrays were denatured with boiling sterile distilled water and then reannealed in hybridization buffer to form partial-ssDNA oligonucleotides microarrays with hairpin primer at the free 3′ end of immobilized oligonucleotides. Thirdly, the partial-ssDNA oligonucleotides microarrays were incubated with Klenow extension reaction to form complete-dsDNA oligonucleotides microarrays.

3.1.1.3   Verification of the Enzymatic Extension of Hairpin Primer

The most important thing for our dsDNA microarray fabrication method is to corroborate DNA polymerase successfully performed nucleotide polymerization from the special hairpin primer on the short template immobilized on solid surface. In order to verify the enzymatic extension of hairpin primer, three ssDNA microarrays were fabricated by spotting oligonucleotide No.A3 in Table 16.1 on glass slides in 4 × 4 format (A3 ssDNA microarray). One A3 ssDNA microarray was incubated with a Klenow extension reaction including Cy3-labled dUTP, unlabeled dGTP, dCTP and dATP, and other two A3 ssDNA microarrays were respectively treated with a reaction containing Klenow plus four unlabeled dNTPs and a reaction including Cy3-labled dUTP, unlabeled dGTP, dCTP and dATP but no Klenow as control fabrications. At the same time, we also fabricated the ssDNA microarray by spotting oligonucleotide No.SS in Table 16.1 on glass slides in 4 × 4 format (SS ssDNA microarray), and incubated it with a Klenow extension reaction including Cy3-labled dUTP, unlabeled dGTP, dCTP and dATP also as control fabrication. As expected, the fluorescent signals of spots were seen over the entire arrays incubated with Klenow extension reaction containing Cy3-dUTP, dGTP, dCTP and dATP (Figure 16.2A). However, no signal intensity

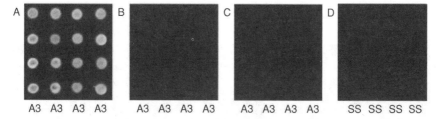

FIGURE 16.2. Fluorescence images of dsDNA microarrays. Image of (A) is dsDNA microarray fabricated with a Klenow extension reaction including Cy3-labled dUTP, unlabeled dGTP, dCTP and dATP. Image of (B) is dsDNA microarray fabricated with a Klenow extension reaction including four unlabeled dNTPs. Image of (C) is dsDNA microarray fabricated with a reaction including Cy3-labled dUTP, unlabeled dGTP, dCTP and dATP but no Klenow enzyme. Image of (D) is ssDNA microarray incubated with a reaction including Klenow, Cy3-labled dUTP, and unlabeled dGTP, dCTP and dATP.

appeared over the three control ssDNA microarrays (Figure 16.2B, 2C and 2D).

As the oligonucleotide No.A3 was designed only containing three adenines at the far 5′ end, therefore, the fluorescent signals of spots appearing on dsDNA microarray of Figure 16.2A demonstrated that not only the hairpin primer has successfully formed, but also the DNA polymerase Klenow fragment enzyme successfully performed nucleotides polymerization reactions to the far 5′ end of immobilized single-stranded oligonucleotides. No signal intensities appearing over the three control ssDNA microarrays verify that the fluorescent signals on dsDNA microarray of Figure 16.2A really depend on the reaction of DNA polymerase Klenow fragment incorporating Cy3-labled dUTP in hairpin primer extension. The results demonstrate that it is feasible to manufacture dsDNA microarrays with the method hairpin primer extension.

The homogeneity of the dsDNA polymerization reactions on larger microarray was also demonstrated by the following 10×30 microarray fabrication (Figure 16.3).

3.1.1.4   Accessibility of Enzymatically Synthesized dsDNA Microarray to DNA-Binding Protein

Because the electrostatic properties at the solid–liquid interface and the local ionic strength of the immobilized dsDNA molecules are greatly different from that in the bulk solution, the interfacial effect can influence the molecular interactions at the solid–liquid interface, especially when the functional motif of molecule becomes increasingly closer to solid surface. Because the dsDNA synthesized on glass slides in our above preliminary experiments was very short (26 bases), as soon as the feasibility of dsDNA microarrays manufacture with hairpin primer extension was confirmed, we expected to find whether the short dsDNA fabricated on array slides was accessible to target

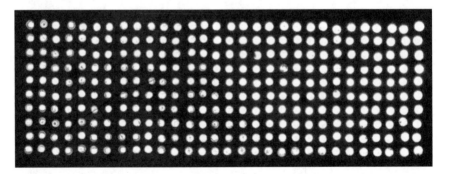

FIGURE 16.3. Large homogeneous dsDNA microarray fabricated with Cy3-labeled dUTP incorporation in polymerization reactions.

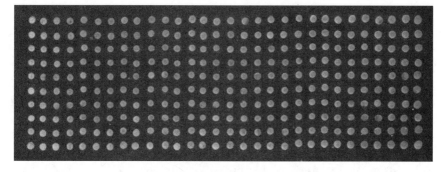

FIGURE 16.4. Large homogeneous dsDNA microarray fabricated with Cy3-labeled dUTP incorporation and dsDNA microarray fabricated with dNTP incorporation which bound by Cy3-labeled NF-kB homodimer p50.

sequence-specific DNA-binding proteins in sample before we do any more successive experiments. Therefore, we synthesized an oligonucleotide which contained a 10-base NF-kB binding site (No. FS5 in Table 16.1) and fabricated a 10×30 dsDNA array by enzymatically incorporating unlabeled dNTP. The fabricated dsDNA microarray was hybridized with Cy3 labeled NF-kB.

The hybridization results revealed that the short dsDNA enzymatically synthesized on glass surface can significantly be bound by Cy3 labeled NF-kB in sample as showed in Figure 16.4. The signal intensities between different spots were very homogeneous. Because only 11 irrelevant base pairs flank the NF-kB binding site in the immobilized dsDNA oligonucleotides, it demonstrates that short spacing distance between glass surface and NF-kB binding site did not prevent the interaction of dsDNA-binding site on glass surface with NF-kB in sample. It also suggested that the spacing distance of 11-base pairs may not be the threshold distance which prevents DNA/protein interaction.

3.1.1.5   Optimizing the ssDNA Oligonucleotides Design

As the feasibility of manufacturing dsDNA microarrays with the method of hairpin primer extension and accessibility of DNA-binding proteins to dsDNA microarrays had confirmed, we wanted to know how long flanking sequence, reverse complementary sequence and loop bases were most economical but effective for dsDNA microarrays fabrication with hairpin primer extension. Therefore, we designed and synthesized three groups of amino-linked single-stranded DNA oligonucleotides displayed in Group 3, 4 and 5 in Table 16.1, which had various flanking sequence, reverse complementary sequence and loop bases in different lengths. We fabricated the ssDNA microarrays by spotting these oligonucleotides on slides in quadruplet format. The fabricated ssDNA microarrays were annealed and then extended with Klenow reaction including Cy3-labeled dUTP, and unlabeled dGTP, dCTP and dATP.

Figure 16.5A displays the fluorescence image of dsDNA microarray fabrication with Cy3-labeled dUTP incorporation. It is clear that the fluorescence intensities of different oligonucleotides are not the same (Figure 16.4B). This demonstrates that the variability of the length of flanking sequence, hairpin loop sequence and reverse complementary sequence can influence the efficiency of Cy3-labeled dUTP incorporation, that is, the efficiency of enzymatic extension reaction. The characteristics of fluorescence intensities of various oligonucleotides revealed that the dsDNA microarray manufacture with enzymatic extension of hairpin primer followed several general laws. First, the efficiency of enzymatic extension reaction enhanced with the increasing length of flanking sequence. The longer of flanking sequence, the easier of enzymatic extension reaction. This may result from the steric hindrance of glass surface to enzymatic extension reaction on surface-near DNA template. Second, the efficiency of enzymatic extension reaction dramatically decreased with the shortening of reverse complementary sequence. It was clear that the longer reverse complementary sequences benefited enzymatic extension reaction, because the longer reverse complementary sequences could more easily form more stable double-stranded hairpin structure which provides the primer for enzymatic extension reaction. Third, the efficiency of

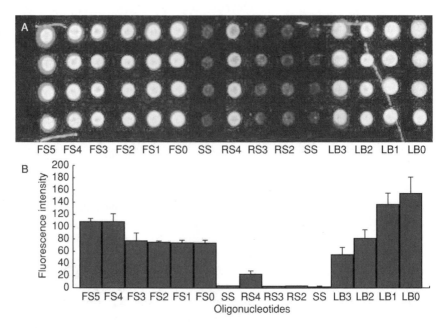

FIGURE 16.5. dsDNA microarray fabricated with Cy3-labeled dUTP incorporation and dsDNA microarray fabricated with dNTP incorporation. The image verified the efficiencies of different single-stranded oligonucleotides for dsDNA microarray fabrication.

enzymatic extension reaction increased with the decrease of loop bases. No loop base resulted in the most effective enzymatic extension reaction. It seems that few loop bases were more beneficial for hairpin structure formation. Therefore, we concluded that the longer flanking sequence, no loop base and the longer reverse complementary sequence are most effective for dsDNA microarray fabrication with hairpin primer extension.

Successively, we confirmed the above characters of dsDNA microarray fabrication with enzymatic extension reaction by binding DNA-binding protein with dsDNA microarray. We fabricated the dsDNA microarrays with same oligonucleotides by Klenow incorporation of unlabeled dNTPs, and hybridized the fabricated dsDNA microarray with Cy3-labeled transcription factor NF-kB. The results showed that all dsDNA oligonucleotides fabricated on microarray were accessible to NF-kB. However, the efficiency of NF-kB binding to various dsDNA probes on microarray was different (Figure 16.5A), and the variation of fluorescence intensities of different dsDNA probes bound by NF-kB (Figure 16.5B) was coincident with that of dsDNA microarray fabrication by enzymatic incorporation of Cy3-labeled dUTP (Figure 16.4A). It suggested that the more dsDNA probes fabricated on spots, the more DNA-binding proteins bound on them. Therefore, we adopted to synthesize the single-stranded oligonucleotides with no loop base and longer reverse complementary sequence (5 bases) to do the latter experiments.

3.1.1.6   Specificity of Interaction of dsDNA Microarray with DNA-Binding Proteins

The sequence-specificity of interactions of dsDNA consensus harbored in immobilized short dsDNA oligonucleotides with their target DNA-binding proteins as transcription factors is critical the most important thing for the applications of dsDNA microarray, because the dsDNA microarray we fabricated mainly focused on screening sequence-specific DNA-binding proteins. To verify the specificity, we synthesized 8 different amino-linked oligonucleotides which respectively contained one DNA-binding site corresponding to one of 8 typical transcription factors from bacteria to human being as listed in Group 8 of Table 16.1. The dsDNA microarray was fabricated with the 8 oligonucleotides and oligonucleotide No. LB0 in Table 16.1 which contains the DNA-binding sites of NF-kB. We hybridized the fabricated dsDNA microarray with Cy3 labeled transcription factors NF-kB.

Figure 16.6 showed the results of the dsDNA microarray hybridized with Cy3-labeled transcription factors NF-kB. It demonstrates that all dsDNA-binding sites of NF-kB were specifically identified and bound by Cy3 labeled transcription factors NF-kB in sample. However, dsDNAs containing other transcription factors were not bound by NF-kB, except little non-specific absorption which could be eliminated by more stringent washing. This primary experiment with DNA-binding proteins NF-kB implies that dsDNA microarray fabricated by hairpin primer extension has great potential to

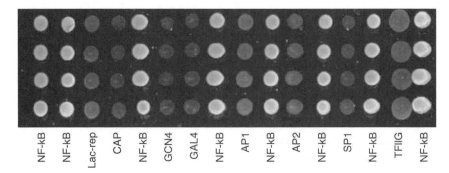

FIGURE 16.6. dsDNA microarray containing 9 different DNA-binding proteins targets which bound by Cy3-labeled NF-kB homodimer p50. The image demonstrated the sequence-specificity of interaction between arrayed dsDNA with protein.

provide a high-throughput technique for DNA-binding proteins screening and DNA/protein interactions.

3.1.2   Manufacture of dsDNA Microarray with Method 2

3.1.2.1   Design of Oligonucleotides for Fabricating the ssDNA Microarray

Two kinds of ssDNA oligonucleotides, constant oligonucleotide (CO) and the target oligonucleotides (TOs), were chemically synthesized to fabricate ssDNA microarray in method 2 (Table 16.2). The CO contained a 7-base single-stranded capture overhang at 3′ end and two reverse self-complementary sequences intervened by an internal dT with a primary NH_2 group (amino modifier C6) at 5′ end. The TOs consisted of the 7-base general proximal flanking sequence complementary to capture overhang on CO, target sequence containing 10-base wild-type or single-nucleotide mutant Ig-κB sites, and the 7-base general distal flanking sequence from 3′ to 5′ end. The

TABLE 16.2. Oligonucleotides synthesized to fabricate dsDNA microarray

	##.	Sequences (5′ →3′)		
		Distal Flanking sequence	Target sequence	Proximal Flanking sequence
Target Oligonucleotide	AP1	CGCTTGA	TGAGTCA	CGTACGC
	AP2	AATGTCC	GCCCGCGGC	CGTACGC
	SP1	ACGATCG	GGGCGG	CGTACGC
	TFIID	CTGTGCA	TATAA	CGTACGC
	NF-kB (NS)	AGTTGAG	GGGACTTTCC	CGTACGC
Constant oligonucleotide	CO	P-GGAATCCCCC T GGGGGATTCC GCGTACG-OH (Aminated dT for immobilization is in bold)		

CO contained hydroxyl at 3′ end and phosphate at 5′ end, and the TOs had hydroxyl at both 3′ and 5′ ends.

3.1.2.2   Scheme of dsDNA Microarray Fabrication

The scheme of dsDNA microarray fabrication is illustrated in Figure 16.7. The main procedures were as followings. Firstly, the CO was denatured and reannealed in tube for forming hairpin-overhang structure. Secondly, the annealed CO was distributed into tubes containing variant denatured TOs, the mixtures were incubated at proper temperature for CO/TO hybridization. After hybridization, bimolecular hairpin oligonucleotides with a long overhang were formed, which, however, contained a nick in hairpin structure. Thirdly, the DNA ligase was added into CO/TO hybridization reaction for eliminating the nick and thus forming unimolecular hairpin oligonucleotides with a long overhang. Fourthly, the CO/TO unimolecular hairpin oligonucleotides were microspotted and immobilized on aldehyde-derivatized glass surface by the primary NH_2 group on the internal dT of CO. After the procedure, the partial-dsDNA microarrays were fabricated. At last, the 3′ end hydroxyls of immobilized unimolecular hairpin oligonucleotides were elongated by DNA polymerase on the template of 5′ end single-stranded overhangs. After the Klenow polymerization, the partial-dsDNA microarrays were converted into complete dsDNA microarrays which harbored the DNA-binding sites of sequence-specific DNA-binding proteins.

3.1.2.3   Verification of CO/TO Annealing and Ligation

To confirm the formation of unimolecular hairpin oligonucleotides by TO/CO annealing and ligating reactions, we synthesized a 5′-end FAM-labeled AP2

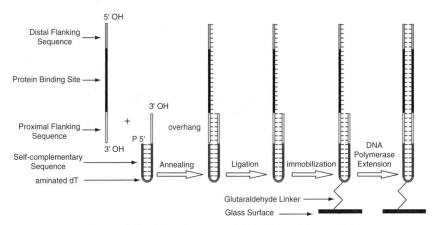

FIGURE 16.7. Scheme of dsDNA microarray fabrication with method 2.

TO and prepared three spotted DNA samples with it. The first was FAM-labeled AP2 TO annealed and ligated with CO, the second was FAM-labeled AP2 TO annealed but not ligated with CO, the third was FAM-labeled AP2 TO itself. The prepared three DNA samples were spotted on aldehyde-derivatized slides in triplet format. After immobilization, the slides were washed with $2 \times$ SSC/0.01% SDS and the fluorescence signals were collected with the standard FAM filter by laser scanning confocal microscope (Leica TCS SP) employing a 488 nm Ar ion laser. The results were displayed in Figure 16.8, which revealed that the first and second DNA samples presented fluorescence signals, while the third DNA sample did not. As without primary NH_2 group, the third DNA sample of FAM-labeled AP2 TO itself could not immobilize on glass slides and shed fluorescence signals, which confirmed the fluorescence signals displayed by the first and second DNA samples were CO-dependent. This implied that the FAM-labeled AP2 TOs in the first and second DNA samples annealed to COs. To further confirm the TO ligation with CO, the above signal-collected slides were denatured in 100 °C sterile ddH$_2$O for 10 min and then washed with $0.2 \times$ SSC/0.01% SDS. After washing, the slides were rescanned with previous laser channel. As expected, the fluorescence signals presented by the second DNA sample in previous scanning disappeared, while that from the first DNA sample still existed. This verified that the FAM-labeled AP2 TO in the first DNA sample was successfully ligated with CO into a unimolecular oligonucleotide. Therefore, the FAM linked at 5′ end of the unimolecular oligonucleotide could not be destroyed by denaturing treatment. However, the FAM-labeled AP2 TO in the second DNA sample broke away from immobilized CO in

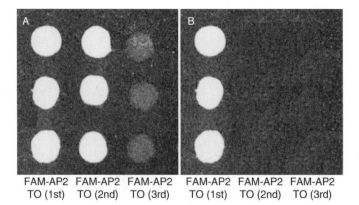

FAM-AP2   FAM-AP2   FAM-AP2   FAM-AP2   FAM-AP2   FAM-AP2
TO (1st)   TO (2nd)   TO (3rd)   TO (1st)   TO (2nd)   TO (3rd)

FIGURE 16.8. Fluorescence images of AP2 array before (A) and after (B) heat denaturation. FAM-AP2 TO (1st), FAM-AP2 TO (2nd) and FAM-AP2 TO (3rd) referred to FAM-labeled AP2 TO hybridized and ligated with CO, FAM-labeled AP2 TO hybridized but not ligated with CO, FAM-labeled AP2 TO itself respectively.

denature because it only annealed with CO to form the bimolecular oligonucleotides which could be melted by heat. In all, the unimolecular oligonucleotides for microspotting can be reliably fabricated by DNA ligase.

3.1.2.4   *Validation of the Enzymatic Extension and Protein Binding to Microarray*

To verify the Klenow polymerization, the Klenow extension reactions with Cy3-labeled dUTP instead of dTTP were done with the unimolecular oligonucleotides microarray fabricated by AP1, AP2, SP1, TFIID and NF-κB TOs listed in Table 16.1. The results were displayed in Figure 16.9A. It demonstrated that the fluorescence signals were seen over the entire arrays incubated with Klenow reaction containing Cy3-dUTP, while no signal appeared over the control arrays incubated with Klenow reaction without Cy3-dUTP (not shown). The Klenow-dependent presentation of fluorescence signals confirmed the occurrence of Klenow polymerization on immobilized short unimolecular hairpin oligonucleotides. We subsequently examined the accuracy of Klenow polymerization by sequence-specific restriction endonuclease digestion. The above signal-presented microarrays were incubated with HaeIII digestion reaction. The HaeIII-digested microarrays were rescanned and fluorescence signals were displayed in Figure 16.9B, which demonstrated the fluorescence intensities of AP2 spots greatly decreased. It agreed with the fact that only AP2 unimolecular dsDNA oligonucleotide harbored HaeIII digestion site (–GGCC-) and only two 5′

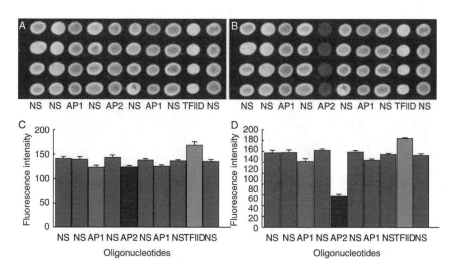

FIGURE 16.9. Fluorescence images and intensity plots of unimolecular dsDNA microarray fabricated by Klenow extension containing Cy3-labeled dUTP before (A and C) and after (B and D) HaeIII digestion in method 2.

end-anchored adenines in AP2 target oligonucleotide allowed Cy3-dUTP incorporation in Klenow polymerization. Therefore, we concluded that the Klenow extension could reach the distal terminals of immobilized oligonucleotides, and the high-fidelity enzymatically synthesized unimolecular dsDNA oligonucleotides on slides were accessible to sequence-specific restriction endonuclease.

The homogeneity of the dsDNA polymerization reactions on larger microarray and the accessibility of DNA-binding proteins to dsDNA microarray probes were demonstrated by the following 6×7 microarray fabrication with Cy3-labeled dUTP incorporation and Cy3-labeled NF-κB p50 binding to microarray fabricated with four label-free dNTP incorporation (Figure 16.10).

3.1.2.5   Specificity of NF-κB p50 Homodimer Binding to DNA Targets on dsDNA Microarray

As we aimed at fabricating unimolecular dsDNA microarray with general application value of study sequence-specific interactions between of DNA-binding proteins and dsDNA targets, we examined the sequence-specific accesiblity of transcription factor NF-κB to its target oligonucleotides immobilized on slides. We fabricated the unimolecular dsDNA microarrays with AP1, AP2, SP1, TFIID and NF-κB target oligonucleotides by Klenow reaction with unlabeled dNTPs. The fabricated microarrays were hybridized with Cy3-labeled NF-κB p50 homodimer and the scanned fluorescence signals were represented by Figure 16.11, which demonstrated that the NF-κB p50

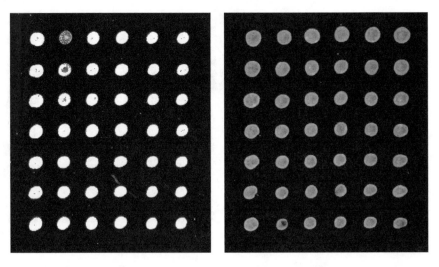

FIGURE 16.10. The homogeneous dsDNA microarray fabrication by Cy3-labeled dUTP DNA polymerase incorporation and the binding of Cy3-labeled NF-kB homodimer p50 to the dsDNA microarray fabricated with dNTP incorporation.

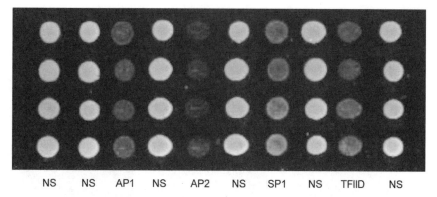

FIGURE 16.11. dsDNA microarray containing 9 different DNA-binding proteins targets which bound by Cy3-labeled NF-kB homodimer p50. The image demonstrated the sequence-specificity of interaction between arrayed dsDNA with protein.

homodimer specifically bound to its target oligonucleotides on slides. To further verify the sequence-specificity, we hybridized 6 similar unimolecular dsDNA microarrays anchoring 6 × 7 NF-κB TOs (NS in Table 16.2) with 6 variant binding reactions containing the same concentrations of Cy3-labeled NF-κB p50 homodimer but 6 different concentrations of cold free NF-κB dsDNA consensus oligonucleotides (Promega) sequencing identical to NS oligonucleotide. The results were displayed in Figure 16.5, which demonstrated that the fluorescence intensities decreased with the increase of cold free NF-κB dsDNA consensus oligonucleotides in binding reactions. It revealed that the binding of NF-κB to immobilized dsDNA targets could be restrained by specific competition. In all, the short unimolecular dsDNA oligonucleotides immobilized on slides could be specifically bound by DNA-binding protein as transcription factor.

3.1.2.6 Sensitivity of Detecting NF-κB p50 Homodimer with dsDNA Microarray

The capability of quantifying DNA-binding protein in detected samples by the unimolecular dsDNA microarray is very useful for the application of the dsDNA microarrays in detecting DNA-binding proteins. To determine the detecting sensitivity of target DNA-binding proteins with unimolecular dsDNA microarray, we hybridized 6 similar dsDNA microarrays anchoring 6 × 7 NF-κB TOs (NS in Table 16.2) with 6 binding reactions containing 6 different concentrations of Cy3-labeled NF-κB p50 homodimer. The results were displayed in Figure 16.12, which revealed that the signal intensities of microarrays coordinately decreased with the concentration of Cy3 labeled NF-κB in samples. The dsDNA microarray allowed detection of Cy3

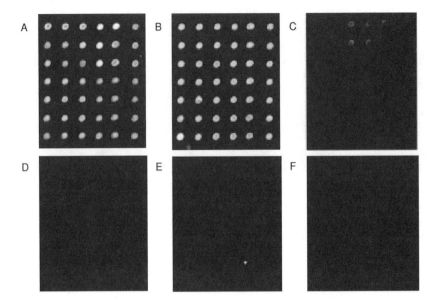

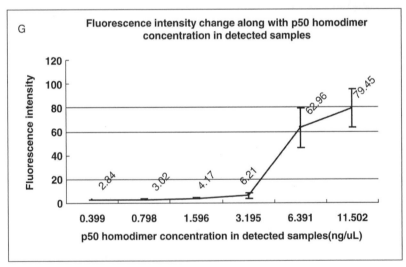

FIGURE 16.12. Fluorescence images and intensity plot of 6 dsDNA microarrays containing 6×7 NF-κB TOs hybridized with 6 variant binding reactions containing 6 different concentrations of Cy3-labeled NF-κB.

labeled NF-kB as few as 0.4ng/μl (0.8mol/ml) with very few protein sample (3μl enough for 6 × 7 spots array, 200μm spot diameter and 300μm spots interval).

3.2   dsDNA Microarray Application: Evaluating the Binding Affinities of NF-κB p50 Homodimer to Wild-Type and Single-Nucleotide Mutations Ig-κB Sites (Method 2)

To high-throughput quantify the binding affinities of NF-κB to variant DNA targets, we fabricated the unimolecular dsDNA microarrays containing the wild-type and all possible single-nucleotide mutant Ig-κB sites and examined the binding affinities of NF-κB (p50 homodimer) to all targets. We hope to define the importance of each nucleotide consisted of 10-base Ig-κB for NF-κB (p50 homodimer) binding with Ig-κB which naturally exists in immunoglobulin light chain κ gene and HIV-LTR. We took the 5′-GGGAC-3′ and 5′-TTCC-3′ as two subsites respectively bound by two p50 monomers and numbered the nucleotides from 5′ to 3′ end as showed in Table 16.3. The microarrays were hybridized with Cy3-labeled NF-κB p50 homodimer.

The results were displayed in Figure 16.13. It was clearly demonstrated that the binding affinities of p50 homodimer to 30 single-nucleotide mutant Ig-κBs were almost all lower than that of p50 homodimer with wild-type Ig-κB site. However, the binding affinities of p50 homodimer to 30 single-nucleotide mutated Ig-κBs were greatly different from each other.

TABLE 16.3. DNA sequences of wild-type NF-κB binding sites

No.	DNA sequence (5′→3′)	No.	DNA sequence (5′→3′)
1CIg-κB	AGTTGAGCGGACTTTCCCAGGC	6G Ig-κB	AGTTGAGGGGACGTTCCCAGGC
1A Ig-κB	AGTTGAGAGGACTTTCCCAGGC	6C Ig-κB	AGTTGAGGGGACCTTCCCAGGC
1T Ig-κB	AGTTGAGTGGACTTTCCCAGGC	6A Ig-κB	AGTTGAGGGGACATTCCCAGGC
2C Ig-κB	AGTTGAGGCGACTTTCCCAGGC	7G Ig-κB	AGTTGAGGGGACTGTCCCAGGC
2A Ig-κB	AGTTGAGGAGACTTTCCCAGGC	7C Ig-κB	AGTTGAGGGGACTCTCCCAGGC
2T Ig-κB	AGTTGAGGTGACTTTCCCAGGC	7A Ig-κB	AGTTGAGGGGACTATCCCAGGC
3C Ig-κB	AGTTGAGGGCACTTTCCCAGGC	8G Ig-κB	AGTTGAGGGGACTTGCCCAGGC
3A Ig-κB	AGTTGAGGGAACTTTCCCAGGC	8C Ig-κB	AGTTGAGGGGACTTCCCCAGGC
3T Ig-κB	AGTTGAGGGTACTTTCCCAGGC	8A Ig-κB	AGTTGAGGGGACTTACCCAGGC
4T Ig-κB	AGTTGAGGGGTCTTTCCCAGGC	9G Ig-κB	AGTTGAGGGGACTTTGCCAGGC
4G Ig-κB	AGTTGAGGGGGCTTTCCCAGGC	9A Ig-κB	AGTTGAGGGGACTTTACCAGGC
4C Ig-κB	AGTTGAGGGGCTTTCCCAGGC	9T Ig-κB	AGTTGAGGGGACTTTCCCAGGC
5G Ig-κB	AGTTGAGGGGAGTTTCCCAGGC	10G Ig-κB	AGTTGAGGGGACTTTCGCAGGC
5A Ig-κB	AGTTGAGGGGAATTTCCCAGGC	10A Ig-κB	AGTTGAGGGGACTTTCACAGGC
5T Ig-κB	AGTTGAGGGGATTTTCCCAGGC	10T Ig-κB	AGTTGAGGGGACTTTCTCAGGC
Wild-type Ig-κB	AGTTGAGGGGACTTTCCCAGGC	Base No.	**GGGACTTTCC** **12345678910**
Constant	P-GGAATCCCCC *T* GGGGGATTCC GCGTACG-OH (*T*: Aminated dT)		
note	Both 5′ and 3′ ends of target oligonucleotides are hydroxyl.		

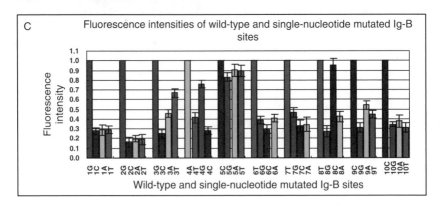

wild	1C	1A	1T	2C	2A	2T
wild	3C	3A	3T	4T	4G	4C
wild	5G	5A	5T	6G	6C	6A
wild	7G	7C	7A	8G	8C	8A
wild	9G	9A	9T	10G	10A	10T
wild	wild	wild	wild	wild	wild	wild

FIGURE 16.13. Fluorescence image and intensity plot of unimolecular dsDNA microarray. (A) One fluorescence image of dsDNA microarray with wild-type and single-nucleotide mutated Ig-κB sites, which was hybridized with Cy3 labeled NF-κB p50 homodimer. (B) DNA targets positions on dsDNA microarray. (C) Plot of statistical fluorescence intensities of 10 dsDNA microarrays.

In 5′-GGGAC-3′ subsite, the guanosines at positions 1 and 2 were most important for high-affinity binding of NF-κB (p50 homodimer) to Ig-κB, as any other three nucleotides replacements of those two guanosines resulted in similar great affinity loses. However, there was still a little difference between the two guanosines. The mutations of guanosine at position 2 presented more affinity loses than that of guanosine at position 1. The guanosine at position 3 is less important for p50p50 /Ig-κB interaction than that at position 2. Its replacements with different nucleotides resulted in distinct binding affinities. The binding affinities of p50 homodimer to the three mutated Ig-κBs became lower as the order of substitution of the 3G with T, A and C. The exchange of the guanosine into cytosine at position 3 resulted in the largest decrease of binding affinity, while the guanosine at position 3 could tolerate A or T substitution. As adenine at position 4 was replaced with G, T and C respectively, the binding affinity loses were different from each other. It was obvious that the nucleotide at position 4 preferred purines other than pyrimidines. All single-nucleotide mutations of cytosine at the position 5 presented the little binding affinity loses. Namely, no matter the cytosine at the position 5 were changed into any other three nucleotides, only a little decrease of the binding affinity happened compared to wild-type Ig-κB site. The adenine and cytosine at position 5 were more beneficial for protecting binding affinity of p50 homodimer to Ig-κB than guanosine.

Compared with the mutation in 5'-GGGAC-3' subsite, most of the single-nucleotide mutations at 5'-TTCC-3'subsite resulted in greater affinity lose in NF-κB (p50 homodimer) binding to Ig-κB. However, the single-nucleotide exchange of thymine into cytosine at position 8 had little effect on high-affinity binding of p50 homodimer to mutant Ig-κB. Moreover, this mutation produced obvious higher binding affinities than wild-type Ig-κB in several experiments. Except the marked affinitiy holding in cytosine replacement at position 8, the exchange of cytosine into adenine at position 9 was more help-ful for keeping binding affinity of p50 homodimer to Ig-κB than other single-nucleotide mutations. Any mutations of cytosine at position 10 could not avoid significant affinity decrease in p50 homodimer binding with Ig-κB. It was noteworthy that the mutations at the axle thymine resulted in significant binding affinity loses.

The results from the unimolecular dsDNA microarray could be steadily repeated on the different microarrays (Fig. 16.14).

4. Discussion

The Church's lab saw the great potentials of dsDNA microarray for studying sequence-specific DNA/protein interaction(34), and fabricated dsDNA microarray for exploring the DNA-binding specificities of zinc fingers with arrayed DNA targets (38, 39). Their creative works verified the feasibility and high effectiveness of dsDNA microarray in studying sequence-specific DNA/protein interaction. Nevertheless, their dsDNA microarray fabrication replied on the Affymetrix proprietary technology of photo-addressable oligonucleotide synthesis which is unaffordable for general laboratories until now. In this paper, we presented two novel methods for fabricating unimolecular dsDNA microarray and verified its reliability. These methods have several significant advantages. Firstly, considering the expensive amino-labeling of oligonucleotides which cost was almost identical to target oligonucleotide synthesis (about 25 base pairs), the free constant oligonucleotide with amino modifier C6 dT was adopted to avoid repeatedly synthesizing long constant oligonucleotide with C6dT on each target oligonucleotides. This strategy can dramatically decrease the cost of unimolecular dsDNA microarray manufacture. Secondly, with free constant C6dT oligonucleotide, we only need to chemically synthesize target oligonucleotides for fabricating unimolecular dsDNA

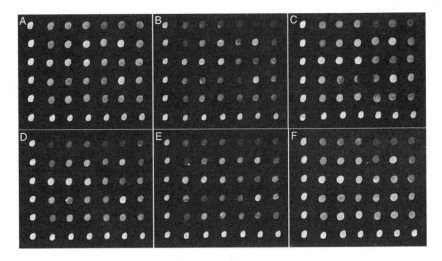

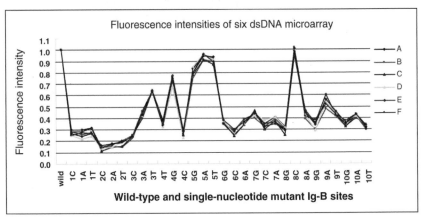

FIGURE 16.14. Fluorescence images and intensity plot of dsDNA microarrays on six different slides. (A-F) Fluorescence image of six dsDNA microarrays which were hybridized with Cy3 labeled p50 homodimer. (G) Fluorescence intensities plot of six dsDNA microarrays (A-F).

microarray. We testified that as short as 23bp (AP2) and 24bp (NF-κB) oligonucleotides harboring protein-binding sites were long enough for proteins binding interaction. To synthesize so short single-stranded oligonucleotides also greatly lowers the cost of unimolecular dsDNA microarray manufacture. It avoids synthesizing long self-complementary single-stranded oligonucleotides for fabricating dsDNA oligonucleotides by annealing (US patent 5556752). Thirdly, as the "second" complementary strands of oligonucleotides were enzymatically synthesized by DNA polymerase, it also greatly lowers the cost of unimolecular dsDNA microarray manufacture. Fourthly, as the exact complements of immobilized target oligonucleotides were synthesized by high-affinity Klenow DNA polymerase I, this method is competent for fabricating dsDNA microarray containing generic or homogenous dsDNA oligonucleotides with similar sequences as single nucleotide polymorphism (SNP). It overcomes the impossibility of fabricating generic or homogenous dsDNA, especially SNP dsDNA microarrays by hybridization (35, 36). This is very important for dsDNA microarray applications to studying sequence-specific DNA/protein interactions. Fifthly, the enzymatically synthesized ultimate dsDNA oligonucleotides immobilized on glass were unimolecular nucleic acids, which can hold their function by reanneating after heat denature. Therefore, the unimolecular dsDNA microarray can be used for many times by removing bound proteins with stringent washing or heat denaturing. This far differs from bimolecular-dsDNA microarray (34-39). Finally, this method of unimolecular dsDNA microarray fabrication accommodates with commercially available microarray spotting robots, it is reachable for extensive application.

For high-throughput investigation of binding affinities of a larger number of DNA targets with NF-κB transcription factors, we fabricated the special unimolecualar dsDNA microarray with a novel scheme to examine the binding affinities of NF-κB p50 homodimer with the wild-type and single-nucleotide mutant Ig-κB sites. We thereby assessed the importance of each nucleotide of Ig-κB to p50-p50/Ig-κB binding interaction by the highly-paralleled microarray pattern. Our studies revealed that each nucleotide of Ig-κB site contributes differently to p50-p50/Ig-κB binding interaction. These findings were useful for not only redefining the role of nucleotides at different positions of Ig-κB site, but also predicting the potential NF-κB-binding targets. The data from unimolecualar dsDNA microarray could be used to pinpoint the base-amino acid contacts in DNA/NF-κB interaction. The crossvalidations of our dsDNA microarray data with that from crystallography, EMSA and statistic model demonstrated that the unimolecular dsDNA microarray provided a valuable general method for exploring the binding affinities of a larger number of DNA targets with DNA-binding proteins or DNA-binding drugs.

Our experiments revealed that the sequence-specific DNA-binding proteins as restriction endonuclease and transcription factor in detected samples could sensitively and specifically bind with dsDNA targets immobilized on microarray. This demonstrates that the unimolecular dsDNA microarray fabricated with our method provides a novel stable high-throughput technique for investigation of DNA/protein interactions. The unimolecular dsDNA microarray should be potentially used to studies including (1) screening sequence-specific DNA-binding proteins, (2) predicting new DNA-binding sites of transcription factors in genome, (3) assessing importance of nucleotides in DNA-binding sites for DNA/protein interactions, (4) monitoring the expression of drug-induced DNA-binding proteins, and (5) screening sequence-specific DNA-binding drugs.

Acknowledgement. This work is supported by the National Natural Science Foundation of China (60201005) and the National Science Fund for Distinguished Young Scholars (60121101).

References

1. Pabo, C.O. and Sauer, R.T. (1992) *Annu. Rev. Biochem.*, 61, 1053–1095.
2. Craig, N.L. (1988) *Annu. Rev. Genet.*, 22, 77–105.
3. Pingoud, A. and Jeltsch, A. (1997) *Eur. J. Biochem.*, 246, 1–22.
4. Margulies, C. and Kaguni, J.M. (1996) *J. Biol. Chem.*, 271, 17035–17040.
5. Brazil, M. (2002) *Nature* 1, 9.
6. Chaires, J.B. (1998) *Current Opinion in structural biology* 8, 314–320
7. Woodbury, C.P. & Hippel, P. H.V. (1983) *Biochem.* 22, 4730–4737.
8. Jansen, C., Gronenborn, A.M. & Clore, G.M. (1987) *Biochem. J.*, 246, 227–232.
9. Ruscher, K., Reuter, M., Kupper, D., Trendelenburg, G., Dirnagl, U., Meisel, A. (2000) *J. Biotech.*, 78, 163–170.
10. Bowen, B., Steinberg, J., Laemmli, U.K. and Weintraub, H. (1980) *Nucleic Acids Res.*, 8, 1–20.
11. Miskimins, W.K., Roberts, M.P., McClelland, A. and Ruddle, F.H. (1985) *Proc. Natl. Acad. Sci. USA*, 82, 6741–6744.
12. Choo, Y. and Klug, A. (1993) *Nucleic Acids Res.*, 21, 3341–3346.
13. Hanes, S.D. and Brent, R. (1991) *Science* 251, 426–430.
14. V. Olando. (2000) *Trands Biochem. Sci.*, 25, 99–104.
15. Rebar, E.J. and Pabo, C. O. (1994) *Sci.*, 263, 671–673.
16. Choo, Y. and Klug, A. (1993) *Proc. Natl. Acad. Sci. USA* 91, 11168–11172.
17. Oliphant, A., Brendl, C. and Struhl, K. (1989) *Mol. Cell Biol.* 9, 2944–2949.
18. Escolano, A.L.G.R., Medina, F., Racaniello, V. R., and Angel, R. M. D. (1997) *Virol.* 227, 505–508.
19. Bilanges, B., Varrault, A., Basyuk, E., Rodriguez, C., Mazumdar, A., Pantaloni, C., Bockaert, J., Theillet, C., Spengler, D., Journot, L. (1999) *Oncogene* 18, 3979–3988,.
20. Müller, C. W., Rey, F. A., Sodeoka, M., Verdine, G. L. & Harrison, S. C. (1995) *Nature* 373, 311–317.
21. Ghosh, G., vanDuyne, G., Ghosh, S. & Sigler, P.B. (1995) *Nature* 373, 303–310.
22. Dougherty, G. & Pigram, W. J. (1982) *CRC Crit. Rev. Biochem.* 12, 103–132.
23. Zimmer, C. & Luck, G. (1992) In Hurley, L. H. (ed.), Advances in DNA Sequence Specific Agents. JAI Press Inc., London, UK, Vol. 1, pp. 51–88.
24. Searle,M.S. (1993) *Prog. NMR Spectrosc.* 25, 403–480.
25. Chaires, J.B. (1992) In Hurley, L. H. (ed.), Advances in DNA Sequence Specific Agents. JAI Press Inc., London, UK, Vol. 1, pp. 3–23.
26. Imad I. Hamdan, Graham G. Skellern and Roger D. Waigh. (1998) *Nucleic Acids Res.*, 26, 3053–3058.
27. Coury, J. E., McFail-lsom, L., Williams, L.D., & Bottomley, L. A. (1996) *Proc. Natl. Acad. Sci. USA*, 93, 12283–12286.
28. Coury, J. E., Anderson, J. R., McFail-lsom, L., Williams, L. D. & Bottomley, L. A. (1997) *J. Am. Chem. Soc.* 119, 3792–3796.
29. Joseph E. Coury, Lori Mcfail-Isom, Loren Dean Williams, and Lawrence A. Bottomley. (1996) *Proc. Natl. Acad. Sci. USA* 93, 12283–12286.

30. Torunn Berge, Nigel S. Jenkins, Richard B. Hopkirk, Michael J. Waring, J. Michael Edwardson, Robert M. Henderson. (2002) Accepted for publication in 'Nucleic Acids Research', 15 May, 2002.
31. Gambari R, Feriotto G, Rutigliano C, Bianchi N, Mischiati C (2000) *J. Pharmacol. Exp. Ther.* 294, 370–377.
32. Passadore M, Feriotto G, Bianchi N, Aguiari G, Mischiati C, Piva R, Gambari R. (1994) *J. Biochem. Biophys. Methods* 29, 307–19.
33. M Broggini, M Ponti, S Ottolenghi, M D'Incalci, N Mongelli and R Mantovani. *Nucleic Acids Res.* 17, 1051–1059.
34. Bulyk, M. L., Gentalen, E., Lockhart, D. J. & Church, G. M. (1999) *Nat. Biotechnol.*, 17, 573–577.
35. Drobyshev, A.L., Zasedatelev, A.S., Yershov, G.M. & Mirzabekov, A. D. (1999) *Nucleic Acids Res.*, 27, 4100–4105.
36. Krylov, A. S., Zasedateleva, O. A., Prokopenko, D. V., Rouviere-Yaniv, J. & Mirzabekov, A. D. (2001) *Nucleic Acids Res.*, 29, 2654–2660.
37. Nordhoff, E., Krogsdam, A.M., Jorgensen, H.F., Kallipolitis, B.H., Clark, B.F.C., Roepstorff, P. & Kristiansen, K. (1999) *Nat. Biotechnol.*, 17, 884–888.
38. Bulyk, M. L., Huang X, Choo Y. & Church, G. M. (2001) *Proc. Natl. Acad. Sci. USA* 98, 7158–7163.
39. Bulyk, M. L., Johnson, P. L. F. & Church, G. M. (2002) *Nucleic Acids Res.*, 30, 1255–1261.
40. Kadonaga, J.T. & Tjian, R. (1986) *Proc. Natl. Acad. Sci. USA* 83, 5889–5893.
41. Kadonaga, J.T. (1991) *Methods Enzymol.* 208, 10–23.
42. Lochart, D.J., Vetter, D. & Diggelmann, M. US patent # 5556752, issue date 9/17/96
43. Braun, E., Eichen, Y., Sivan, U. & Ben-Yoseph, G. *Nature* 391, 775–778 (1998).
44. McGall, G.H., Barone, A.D., Diggelmann, M., Fodor, S.P.A., Gentelen, E. & Ngo. N. (1997) *J. Am. Chem. Soc.* 119, 5081–5090.
45. Southern, E.M. et al. (1999) *Nat. Genet.* (Suppl.) 21, 5–9.
46. Carlson, R. & Brent, R. (1999) *Nat. Biotech.* 17, 536–537.
47. Kwiatkowski, M., Fredriksson, S., Isaksson, A., Nilsson, M. & Landegren, U. (1999) *Nucleic Acids Res.* 27, 4710–4714.
48. Craig, N.L. (1988) *Annu. Rev. Genet.*, 22, 77–105.
49. Pingoud, A. and Jeltsch, A. (1997) *Eur. J. Biochem.*, 246, 1–22.
50. Margulies, C. & Kaguni, J.M. (1996) *J. Biol. Chem.*, 271, 17035–17040.
51. Jansen, C., Gronenborn, A.M. & Clore, G.M. (1987) *Biochem. J.*, 246, 227–232.
52. Bowen, B., Steinberg, J., Laemmli, U.K. & Weintraub, H. (1980) *Nucleic Acids Res.* 8, 1–20.
53. Hanes, S.D. & Brent, R. (1991) *Sci.*, 251, 426–430.
54. Schena, M., Shalon, D., Davis, R.W. & Brown, P.O. (1995) *Sci.*, 270, 467–470.
55. DeRisi, J. L., Iyer, V. R. & Brown, P. O. *Science* 278, 680–686 (1997).

17

Electronic Biosensors Based on DNA Self-Assembled Monolayer on Gold Electrodes

CHEN-ZHONG LI[1,2], YI-TAO LONG[1,2], TODD SUTHERLAND[1,2], JEREMY S. LEE[2], AND HEINZ-BERNHARD KRAATZ[1]
[1]*Department of Chemistry, University of Saskatchewan, 110, Science Place, Saskatoon, SK, Canada* [2]*Department of Biochemistry, University of Saskatchewan, 107, Wiggins Road, Saskatoon, SK, Canada.*
Current Address: Nanobiotechnology/Biosensor Group, Biotechnology Research Institute National Research Council Canada, 6100 Royalmount Avenue, Montreal, Quebec, Canada. e-mail:chenzong.Li@cnrc-nrc.gc.ca

Abstract: Information concerning the immobilization and hybridization of DNA on a surface is paramount to the development of DNA-based electronic biosensors. This study looks at recent investigations of DNA immobilized on gold surfaces using standard electrochemical techniques such as cyclic voltammetry (CV), potential step chronocoulometry and electrochemical impedance electrochemical impedance spectroscopy (EIS). The thiol-gold linkage is exploited for the immobilization of single- and double-stranded DNA onto gold electrodes. Two redox markers of opposite charge, ferricyanide and ruthenium hexaammine, respectively, are used to probe the environment in the vicinity of thiol-derivatized DNA electrodes. M-DNA is a form of DNA which allows the specific incorporation of certain metal ions into its helical structure under stringent conditions (i.e. low ionic strength and pH of 8.5). Single-stranded DNA monolayer and double strands DNA monolayer resistances were evaluated using EIS, respectively, and CV response were compared each other. The addition of Zn^{2+}, under M-DNA formation conditions, led to a dramatic enhancement of electrochemical response compared to B-DNA.

Key words: Electronic, biosensor, M-DNA, SAMs, electrochemistry.

1.   Introduction

Methods for rapid single-nucleotide-polymorphisms (SNPs) detection are critical to the diagnosis of genetic and pathogenic diseases, tissue matching and forensic applications. The driving force behind DNA biosensor and gene chips development lies in the tremendous potential for obtaining sequence-specific information in a faster, cheaper and more reliable manner compared to traditional hybridization assays.[1] Many techniques, including electrochemistry, have been developed or adapted for analyzing nucleic acids. Most detection systems utilize the hybridization of an immobilized target polynucleotide with oligonucleotide probes containing covalently linked reporter groups.[2] The fluorescence-based detection system is commonly used to perform on-chip SNP detection and gene expression. However, the fluorescence-based detection system is not favorable because of its requirement of complicated labeling process, expensive array chips and optical microarray scanners.[3,4]

An electrochemical DNA sensor or biosensor is another means of DNA detection and it has the advantages of sensitivity, rapid screening and easily implemented into conventional solid-state electronic devices. Therefore, the development of electrochemical transducer-based devices for determining nucleotide sequences and measuring DNA damage is an actively researched area. Recently, electrochemical devices have proven very useful for sequence-specific biosensing of DNA. Several groups[5-10] have reported SNP detection based on electrochemical techniques that use surface bound DNA, which do not contain covalently linked reporter molecules. Willner[11] uses a three component system involving tagged liposomes to amplify the Faradaic impedance signal. Barton[5] has reported SNP detection using several diffusible DNA intercalators. Cyclic voltammagrams clearly show an enhanced current response for duplex DNA and a suppressed response for DNA that contains one mismatched base pair. In a similar system, Takenaka[10] has employed ferrocenyl naphthalene diimide (FND) as an intercalator. Their results show an enhanced current for complementary DNA using differential pulse voltammetry (DPV). The basis of this method lies in FND's ability to bind matched DNA. A mismatch in sequence has the effect of creating disorder in the helical structure and as a result, FND does not bind and signal intensity is attenuated.

Information concerning the immobilization of single-stranded DNA (ss-DNA) and duplex DNA (ds-DNA) on a surface is paramount to the development of DNA-based electronic biosensors. Electrochemical detection of DNA immobilization and hybridization usually involves monitoring a current response under controlled potential conditions. Self-assembly of DNA monolayer based on the formation of a gold-thiolate bond is an important method for preparing stable, closely packed monolayers with well-defined structures.[12,13] Although hybridization efficiencies of both ss- and ds-DNA derivatized gold surfaces have been characterized by several groups,[13,14] the electrochemical properties of M-DNA self-assembled monolayers (SAMs) on gold surfaces have never been examined. M-DNA is a form of duplex DNA with divalent metal ions such as Zn^{2+} that forms at basic pH (pH 8.5) and low ionic strength ($\mu = 0.05$). The electrochemical signal can be enhanced due to M-DNA formation on ds-DNA modified electrode surface.[15-17] Until a crystal structure is available, the proposed structure of M-DNA has the metal ions replace the imino protons of guanine and thymine at every base pair (Figure 17-1). Addition of EDTA will sequester the metal ions and convert

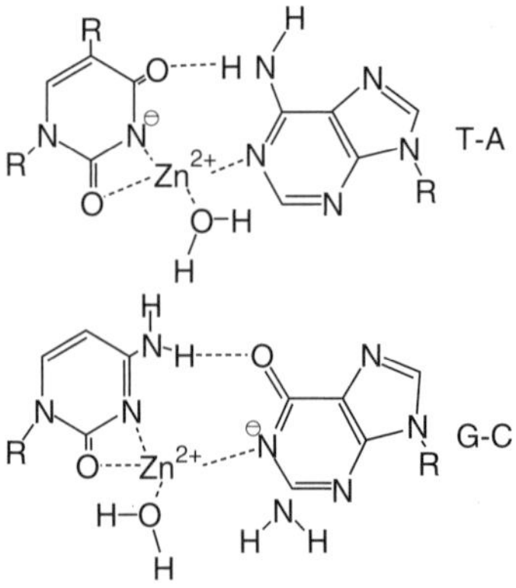

FIGURE 17.1. Possible base-pairing schemes for M-DNA.

the M-DNA back to normal DNA (B-DNA). Alternatively, a decrease in pH will also cause the M-DNA structure to return to B-DNA.

Preliminary evidence demonstrated that M-DNA is an efficient conductor of electrons over distances as long as 500 base-pairs and possibly as long as several microns.[18,19] The enhanced conductivity of M-DNA should allow a greater signal to background current ratio and thus make it more sensitive to perturbations caused by hybridization than B-DNA[15,16] (Figure 17.2).

In this paper a scheme for the voltammetric study of ss-DNA and ds-DNA self assembled monolayers (SAMs) on gold electrodes is presented. Gold surfaces were modified by 20-base-pair 5'-thiol-linked DNA oligonucleotides through the S-Au bond. The electrochemical properties of bare electrode, ss-DNA and ds-DNA modified electrodes were investigated using two redox

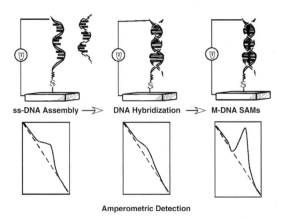

ss-DNA Assembly ⟶ DNA Hybridization ⟶ M-DNA SAMs

Amperometric Detection

FIGURE 17.2. Schematic representation of the electronic DNA sensor for electrochemical detection for DNA hybridization.

markers of opposite charge, ferricyanide and ruthenium hexaammine, respectively. The mediated effect of metal intercalated within ds-DNA on the electron transfer was described to enhance the sensitivity for DNA hybridization detection. The major advantage of these techniques is that the target DNA strand need not be labeled in advance.

2.   Materials and Methods

2.1   General

$Ru(NH_3)_6Cl_3$, $Ru(NH_3)_6Cl_4$, $K_3Fe(CN)_6$ and $K_4Fe(CN)_6$ were purchased from Aldrich and used without further purification. $Zn(ClO_4)_2$, $Mg(ClO_4)_2$ and Tris-ClO_4 were purchased from Fluka. Deionized water (18 MΩzcm resistivity) from a Millipore Milli-Q system was used throughout this work.

2.2   Oligonucleotide Synthesis

Three oligonucleotides were purchased from the Plant Biotechnology Institute-National Research Council (PBI-NRC, Saskatoon) with the following base sequences:

1   SS-5'-GTCACGATGGCCCAGTAGTT-3'
2   5'-AACTACTGGGCCATCGTGAC-3' (complement of **1**)
3   5'-GTCACGATGGCCCAGTAGTT-3' (noncomplementary strand containing the same sequence as **1** but lacks the SS 5' linker)

Note: SS 5' refers to $HO_3PO-(CH_2)_6-SS-(CH_2)_6-OH$

The oligonucleotides were synthesized by standard phosphoramidite chemistry[20] using a fully automated DNA synthesizer then purified by two-step reversed-phase HPLC and characterized by MALDI-TOF mass spectrometry.

2.3   X-Ray Photoelectron Spectroscopy (XPS)

A Leybold MAX200 photoelectron spectrometer equipped with an Al-K$_\alpha$ radiation source (1486.6 eV) was used to collect photo emission spectra. The base pressure during measurements was maintained at less than 10^{-9} mbar in the analysis chamber. The take-off angle was 60°. The routine instrument calibration standard was the Au $4f_{7/2}$ peak (binding energy 84.0 eV).

2.4   Electrochemical Measurements

A potentiostat/galvanostat (EG&G model 283) and Impedance frequency analyzer (EG&G model 1025) connected to a PC running Power Suite (Princeton Applied Research) was used for Impedance spectroscopy measurements. A BAS Model CV-50W potentiostat was used for underpotential deposition (UPD), chronoamperometry and cyclic voltammetry experiments.

2.5   Electrode Characterization and Pretreatment

Gold disk electrodes (Bioanalytical Systems, 1.6 mm diameter, ca. 0.02 cm^2 geometrical area, roughness coefficients between 1.2 and 1.4) were used for the electrochemical measurement. Before modification, the electrode surface was cleaned by electrochemical sweeping in 0.1 M H_2SO_4 from 0 to 1.4 V, then rinsed with water, and ultrasonicated for 5 minutes in fresh piranha solution (30% H_2O_2, 70% H_2SO_4). WARNING: PIRANHA SOLUTION REACTS VIOLENTLY WITH ORGANIC SOLVENTS. The electrode was then sonicated by distilled and degassed ethanol, and finally rinsed with Milli-Q water. A cyclic voltammagram recorded in 0.1 M H_2SO_4 (scan rate 100 mV.s^{-1}) was used to determine the active area of the electrode surface. The real electrode surface area and roughness factor were obtained by integration of the gold oxide reduction peak.[21,22]

2.6   Preparation of DNA SAMs

Duplex DNA modified surfaces were prepared by initially hybridizing the two complementary strands in the absence of a Au surface in a hybridization buffer (100 mM Tris-ClO$_4$, 100 mM NaClO$_4$, pH 7.5) for 24 hours at a DNA concentration of 0.2 mM. The Au electrode was then incubated in the same hybridization buffer for three days at room temperature. Upon completion of monolayer formation, the electrodes were washed repeatedly with (50 mM Tris-ClO$_4$, pH 8.6) for five minutes. The incubation was allowed to continue for one additional day at which time it was rinsed three times with buffer (50 mM Tris-ClO$_4$, pH 8.6). The electrode surface coverage of ds-DNA was quantified to be over 90% by the underpotential deposition (UPD) of Cu.[15,23,24]

The ss-DNA of **1** modified surfaces were formed by dehybridization of ds-DNA from the surface by immersing the ds-DNA modified surface in a water/EtOH solution for 5 minutes at 37 °C. Rehybridization of **1** and **2** was performed at 37 °C for 60 minutes in SSC buffer (300 mM NaCl/30 mM Sodium Citrate, pH 7.0). The concentration of complementary strand **2** or noncomplementary strand **3** was 0.1 mM. Each sample was rinsed thoroughly with an excess volume of 100 mM Tris-ClO$_4$ buffer and dried under a stream of argon prior to characterization. This method of producing a ss-DNA modified surface provide a more reproducible surface compared to straight incubating with ss oligonucleotide **1**. In addition, this methodology will limit the amount of ss-DNA that can form bonds from the exposed base pairs' nitrogen to the gold surface[12].

The ds-DNA monolayer was converted into the M-DNA monolayer by exposure of the monolayer to a solution of 0.3 mM $Zn(ClO_4)_2$ in 20 mM Tris-ClO$_4$ buffer (pH 8.6) for at least two hours.

2.7   Electrochemical Measurements

A normal three-electrode configuration consisting of the modified Au-electrode working electrode, a Ag/AgCl/3M NaCl reference electrode (BAS) and a platinum wire auxiliary electrode. The cell was enclosed in a grounded Faraday cage. A glass-frit salt-bridged reference electrode was used to limiting Cl⁻ ion leakage for the normal Ag/AgCl reference electrode to the measurement system. The open-circuit, or rest potential, of the system was measured prior to all electrochemical experiments in order to prevent sudden potential related changes in the SAM. All electrochemical experiments were started from this rest potential. UPD experiments were carried out in 1 mM $Cu(ClO_4)_2$ in 0.1 M $HClO_4$ at a scan rate of 10 mV⁻s⁻¹, starting at 500 mV (vs. Ag/AgCl), cathodic scanning to 50 mV followed by an anodic sweep to 600 mV. Impedance was measured at the potential of 250 mV vs. Ag/AgCl, to which a sinusoidal potential modulation of ±5 mV was superimposed. The frequencies used for impedance measurements ranged from 100 kHz to 100 mHz. The impedance data for the bare gold electrode, ss-DNA, ds-DNA and M-DNA modified gold electrode were analyzed using the ZSimpWin software (Princeton Applied Research). In all impedance spectra, symbols represent the experimental data, and the solid lines represent the fitted curves.

3.   Results and Discussion

As previously described[12,25], many thiol-derivatized ss-DNA molecules may interact with the gold surface non-specifically. Non-specific interaction is defined as physisorption, such as nitrogen atom or polar side chain interactions, as opposed to chemisorption herein defined as covalent bond formation between Au-S. However, for ds-DNA, because nucleic acid bases are directed toward one another, the non-specific interactions with the Au surface will be very weak and multilayers can simply be removed with buffer rinsing.[26] Thus, in the ds-DNA case, the final structure is most likely to arrive from specific interaction through the covalent Au-S bond formation. A ds-DNA SAM of **1** and **2** results in a mixed monolayer gold, with the ds-DNA-S adjacent to a hydroxylalkyl-S group (Figure 17.3). This arrangement should reduce the efficiency of non-specific interaction of ds-DNA with the gold surface. As previously reported by Tarlov and co-workers, a competitive adsorption step using methylene thiol spacer was deemed necessary to prepare a ss-DNA monolayer with high hybridization ability and few non-specifically adsorbed ss-DNA molecules.[12,13] However, this procedure led to a displacement of some covalently attached DNA-thiolate by the alkylthiol in a well-understood thiol exchange reaction and decreased the surface density of ss-DNA strands.[25,26] In the present study, to prepare ss-DNA modified electrode surface, unlike previous studies, ss-DNA was formed by first adsorbing duplex DNA and then dehybridizing the duplex by immersion in water and

S-CH$_2$-CH$_2$-CH$_2$-CH$_2$-CH$_2$-CH$_2$-OH
|
S-CH$_2$-CH$_2$-CH$_2$-CH$_2$-CH$_2$-CH$_2$-OPO$_3$H-oligonucleotide

di-6-alkyl disulfide modified oligo

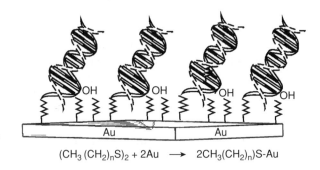

(CH$_3$(CH$_2$)$_n$S)$_2$ + 2Au $\longrightarrow$ 2CH$_3$(CH$_2$)$_n$)S-Au

FIGURE 17.3. Schematic representation of proposed model of the mixed SAMs by dialkyl disulfide ds-DNA on gold surface.

EtOH. The advantages of this methodology are that the loss of covalently attached ss-DNA is low and non-specific binding of DNA is minimized. The ss-DNA, **1**, covered surface will be regenerated and is capable of hybridizing with complementary DNA repeatedly[12], though the chemical or thermally-induced dehybridization of ds-DNA on surface may effect the ability of DNA rehybridization[13] on electrode surface. An important parameter in this methodology is the efficiency[13,27,28] with which the original ds-DNA surface can become dehybridized.

3.1   Analysis of ds-DNA Modified Electrode Surface

Modification of gold surfaces with DNA-hydroxyalkyl disulfide terminated DNA duplexes were confirmed by UPD and XPS experiments. Underpotential deposition of copper has been proved to be a useful tool to evaluate the area of exposed gold remaining after monolayer formation.[24,29] Figure 17.4 shows typical cyclic voltammograms (CVs) of a bare Au electrode and the ds-DNA modified electrode taken in 1.0 mM Cu(ClO$_4$)$_2$, 50 mM HClO$_4$ aqueous solution at a scan rate of 10 mV s^{-1}. As expected, the Cu UPD on the bare electrode produced a pair of well-separated broad peaks.

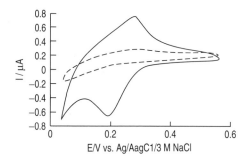

FIGURE 17.4. Under potential deposition of copper on bare (—) and ds-DNA modified gold electrode (-----); 1 mM $Cu(ClO_4)_2$ + 0.05 M $HClO_4$, scan rate was 10 mV$^-$s^{-1}. The gold electrode area was 0.02 cm^2.

During the negative-going scan the Cu is deposited on all accessible Au surface and the the anodic peak corresponds to the stripping wave as Cu is oxidized from the surface.[30,31] In contrast to the bare gold, UPD of Cu was strongly suppressed by a ds-DNA monolayer. Although, a small Cu-UPD stripping peak at 0.310 V, due to Cu/Cu^{2+}, in present results show that the ds-DNA/alkyl SAMs does not completely block the Au electrode even after 4 days of incubation. The integration of the stripping wave for the modified electrode was only 5% of the bare electrode. This indicated that the ds-DNA/6-hydroxylalkyl mixed SAMs can act as an effective barrier for electron transfer even though there are still a few defects in the ds-DNA blocking layer.[15] Note that since the amount of charge calculated from CV often contains contributions from co-adsorbed electrolyte anions to some extent, it is not always correct to determine the actual exposed electrode area from CVs alone.[30,32]

The gold surface was also analyzed by X-ray photoelectron spectroscopy (XPS). As shown in Figure 17.5, the intensity of the Au4f peaks decrease upon attachment of the DNA as expected for a modified surface.[33] The monolayers presence is confirmed by the following new peaks which are absent in bare Au spectra: S2p (162.4 eV), P2p (133 eV) and N1s (400 eV). Furthermore, the value of 162.4 eV for the S2p peak is in good agreement with previous reports for alkylthiol SAMs indicating the specific formation of the Au-S bond of ds-DNA to the gold surface.[34] Film thickness was estimated based on the exponential attenuation of the Au4f signal and calculated to be 45 Å.[35] A 20 base-pair duplex is expected to have a length of about 70 Å so a measured thickness of 45 Å is consistent with the fact that the DNA helices are packed at an angle with respect to the surface.[16] Hence, the morphology of the ds-DNA modified surface appears to involve a densely packed array of duplexes.

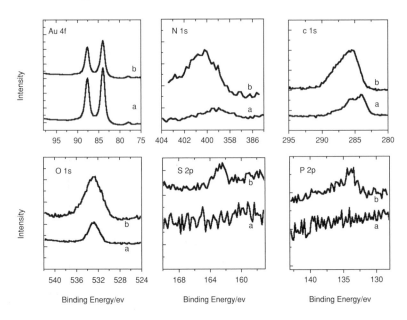

FIGURE 17.5. XPS spectra of (a) bare gold, (b) 20 base-pair thiol-derivatized ds-DNA assembled on gold.

3.2   Electrochemical Impedance Spectroscopy (EIS) for DNA Dehybridization and Rehybridization on Surface

EIS is an effective method to probe the interfacial properties of surface-modified electrodes.[36] EIS data analysis requires modeling the electrode kinetics with an equivalent circuit consisting of electrical components. The general electronic equivalent scheme (Figure 17.6a) for a alkanethiol monolayers-modified electrode is usually described[37] on the basis of the model developed by Randles and Ershler.[38] This equivalent circuit is that of the solution Ohmic resistance R_s in series with a parallel network of the double layer capacitance C_{dl} and the interfacical electron-transfer resistance R_{ct}. Z_w is the

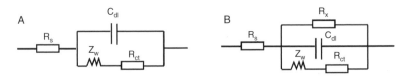

FIGURE 17.6. (a) Standard Randles and Ershler model circuits used to fit the bare electrode; (b) Compartmentalized equivalent circuit used to model DNA modified electrodes. The values of each element in circuits calculated by the computer fitting of the experiment spectra with these circuits are collected in Table 17.1.

Warburg impedance resulting form the diffusion of ions form the bulk electrolyte to the electrode interface. The complex impedance can be presented as the sum of the real, $Z_{re}(\omega)$, and imaginary, $Z_{im}(\omega)$, components originating mainly from the resistance and capacitance of the cell, respectively. The negatively charged $Fe(CN)_6^{3-}/Fe(CN)_6^{4-}$ (1:1 mixture) was used as the redox probe to elucidate the electrical properties of ds-DNA, ss-DNA and rehybridized ds-DNA monolayer by EIS. Figure 17.7a shows a Nyquist plot of the raw data (symbols) for the bare gold electrode and the theoretically best fit curves (solid lines) resulting from the Randles circuit of Figure 17.6.

The semicircle portion, measured at higher frequencies, corresponds to direct electron transfer limited process, whereas the linear portion, observed at lower frequencies, represents the diffusion controlled electron transfer process. In the case of ds-DNA modified electrode, the experimental data in the low frequency region was not adequately fit using Randles circuit model (Figure 17.6a), an additional interfacial resistance, R_x, was added in parallel to the equivalent circuit (Figure 17.6b), termed the modified Randles circuit, that corresponds to electron transfer through the DNA. As shown in Figure 17.6b, the modified circuit gives an excellent fit to the experimental data in all frequency regions for a DNA modified surface. Table 17-1 summarizes the fitting results. ds-DNA shows a larger interfacial electron transfer resistance than that of bare gold indicating that the redox probe is electrostatically repelled by the negatively charged DNA monolayer that is bound to the electrode. After the dehybridization treatment, a significant decrease in R_x and

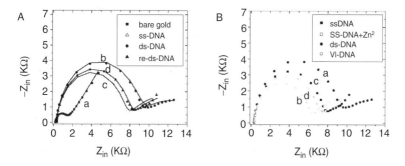

FIGURE 17.7. (A) Nyquist plot (Z_{im} vs. Z_{re}) for the Faradaic impedance measurements at (a) a bare Au electrode, (b) a ds-DNA modified electrode, (c) ss-DNA modified electrode upon de-hybridization treatment of (b), (d) re-hybridized ds-DNA modified electrode. (B) Nyquist plot for the Faradaic impedance measurements at (a) ds-DNA modified electrode, (b) M-DNA formed electrode upon incubating (a) in the buffer solution containing 0.3 mM $Zn(ClO_4)_2$ for two hours. (c) ss-DNA modified electrode, (d) ss-DNA modified electrode (c) treated as the same condition as (b). The measurements were performed in the presence of 5 mM $Fe(CN)_6^{3-/4-}$ in 20 mM Tris-ClO_4 and 20 mM $NaClO_4$ solution, upon application of a biasing potential of 0.25 V vs. Ag/AgCl. Solid Lines correspond to the theoretical fit of the experiment data.

TABLE 17.1. Comparison of the resistance and capacitance values derived from the EIS of the bare electrode and DNA modified electrodes in the presence of 5 mM $Fe(CN)_6^{3-/4-}$ as redox probe upon fitting the experiment data with the equivalent circuits shown in Figure 17.6

Element	Bare Au[†]	ds-DNA[‡]	rehybridized ds-DNA[‡]	M-DNA[‡]	ss-DNA[‡]	ss-DNA with Zn^{2+}[‡]
R_s/Ω	302	320	334	338	337	314
R_x/Ω		16200	15600	12900	15300	14500
$C/\mu F$	2.6	0.29	0.29	0.285	0.28	0.31
R_{ct}/Ω	1230	18800	14900	10000	13500	12100
$W /10^{-5}\Omega s^{-1/2}$	27	3.9	6.6	8.2	7.5	7.9

[†]Values calculated using Randles circuit (Figure 17.6a).
[‡]Values calculated using modified Randles circuit (Figure 17.6b).

R_{ct} was observed. The decrease in the electron-transfer resistance upon dehybridization of ds-DNA is consistent with the idea that the density of the negatively charged phosphates is decreased, which permits penetration of the negatively charged redox marker. Also, an increase in the number or area of defect sites created from the dehybridization cannot be discounted as a plausible explanation for the decrease in R_x and R_{ct}. The rehybridization behavior of surface-immobilized ss-DNA was also determined by impedance spectroscopy. Relatively long hybridization times, 60 minutes, and high salt concentrations, SSC buffer, were used to maximize duplex yield. It was clear form the impedance data that both R_x and R_{ct} increased indicating that complementary strand **2** hybridized with the surface-bound **1**. As a control, rehybridization experiments with noncomplementary strand **3**, showed no increase in the measured electron-transfer resistances. Note that the R_x and R_{ct} for rehybridized surface are still smaller than that of the original ds-DNA modified surface. There are two reasons to be considered. First, the rehybridization efficiency of the ss-DNA attached on gold surface is less than 100%[13], leading to a mixed monolayer that consists of both ss-DNA and ds-DNA. Second, the chemical dehybridization process may result in desorption of ss- or ds-DNA from the surface, resulting in an increase in the amount of surface accessible gold surface.

M-DNA is a novel conformation of duplex DNA in which metal ions, such as Zn^{2+}, are inserted into the helix. M-DNA can also be formed on a surface-immobilized ds-DNA strand under similar condition as solution, except taking much more time. Upon addition of Zn^{2+} to a 20-mer of B-DNA modified gold electrode at pH 8.7 and incubating for 2 hours, the impedance spectrum changed. The differences result in a distinctive pattern with a reduction in Z_{im} and Z_{re} at both high and low frequencies (Figure 17-7b). It is clear that there are significant decreases in R_x and R_{ct} upon addition of Zn^{2+} to ds-DNA modified electrode, which are not found upon addition of Zn^{2+} to ss-DNA modified electrode.[16] The decrease of R_x and R_{ct} following M-DNA formation can be explained by an enhanced rate of electron transfer through the M-DNA monolayer. The fitting results for all modified surfaces are shown in Table 17.1.

3.3 Cyclic Voltammetry and Chronocoulometry at DNA-Modified Electrodes

Generally, two electrochemical systems are commonly used to probe the electrochemical properties of DNA SAMs. One system utilizes an electroactive SAM whereby the redox probe is covalently attached to molecules forming on the DNA monolayer.[17,39,40] The other system, the electron transfer occurs between the gold electrode surface and a redox probe that freely diffuses in solution. The present study focuses on the latter system using an anionic $Fe(CN)_6^{3-/4-}$ or a cationic $Ru(NH_3)_6^{3+/4+}$ redox system.

The voltammagram for these two redox markers at a bare gold electrode is given in Figure 17.8 and Figure 17.9, respectively. Both exhibit a reversible or quasi-reversible, diffusion-limited, one-electron redox process in aqueous buffer solution. A comparison of ss-DNA, ds-DNA and bare gold cyclic voltammagrams with $Fe(CN)_6^{3-}$ is shown in Figure 17.8. Two features of the CV provide evidence that the modified surface is blocked to an anionic redox probe. First, the large peak-to-peak separation (ΔE_p) of both ss- and ds-DNA compared to that of the bare Au and second, the decrease in peak currents (only 5% –15% compared to bare gold). The blocking characteristics of a DNA modified surface is explained by the physical barrier presented to the redox probe. If the redox probe is unable to get close to the electrode then the probability that electron transfer will occur falls off dramatically. In the case of DNA, the monolayer is also negatively charged, due to the phosphate

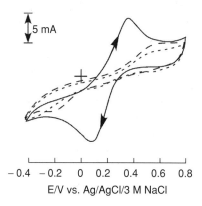

FIGURE 17.8. Cyclic voltammagrams for 2.5 mM $Fe(CN)_6^{3-/4-}$ in 20 mM Tris-ClO_4 buffer (pH 8.6) at a bare electrode (——), a ds-DNA modified electrode (----)and a ss-DNA modified electrode (— — —) upon de-hybridization of duplex DNA modified electrode. The sweep rate was 100 mV·s⁻¹.

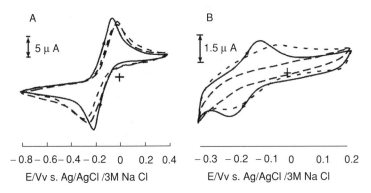

FIGURE 17.9. (A) Cyclic voltammograms for 5.0 mM $Ru(NH_3)^{3+/2+}$ in 20 mM Tris-ClO_4 buffer (pH 8.6) at a bare electrode (—), a ds-DNA modified electrodes (— — —) and a ss-DNA modified electrode (------). (B) Cyclic voltammagrams at a ds-DNA modified electrode (—), a ss-DNA modified electrode (------) in blank buffer (20 mM Tris-ClO_4, pH 8.6) after incubation treatment in 5 mM $Ru(NH_3)^{3+/2+}$ solution, and a ds-DNA modified electrode in blank buffer after incubation treatment (— — —) in 2.5 mM $Fe(CN)_6^{3-/4-}$ solution. The sweep rate was 100 mV⁻s⁻¹.

backbone, and this will electrostatically repel an anionic redox probe. In addition, note the small difference in blocking behaviors for redox probes between ss- and ds-DNA modified gold surfaces. As expected, Figure 17.8 shows the redox peak current for ss-DNA monolayer is slightly larger relative to that of ds-DNA monolayer. This is explained by a decrease in negative charge density on the SAM. The slightly lower negative charge density for ss-DNA manifests in a slightly larger peak current.

In contrast, the positively charged $Ru(NH_3)_6^{3+/4+}$ redox probe is not effectively blocked by the DNA monolayer. Figure 17.9A shows the CVs for ss-DNA modified, ds-DNA modified and bare gold electrodes. The peak separation for bare Au is 0.17 V, whereas, a ds-DNA modified electrode shows only a modest increase to 0.20 V. Assuming the coverage of the Au electrode is the same as in the anionic redox probe scenario, this suggests that cationic species is attracted by the negatively charged DNA backbone. The formation of electrostatic bonds along the phosphate backbone allows the cationic redox probe to approach the electrode surface and give rise to reversible electrochemical behavior similar as that of a bare electrode. Additional support for the electrostatic interaction of DNA SAMs with cationic species is evident from Figure 17.9B. Following extensive rinsing, the $Ru(NH_3)_6^{3+/4+}$ exposed ss- or ds-DNA modified electrodes still indicate a large amount redox probe remains bound to the monolayer. This behavior was not observed for the anionic redox probe.

Conversion of ds-DNA to an M-DNA monolayer was achieved by exposing the electrode to 20 mM Tris buffer (pH 8.5) containing 0.3 mM $Zn(ClO_4)_2$ for 2 hours. Figure 17.10A shows the electrochemical signal due to the

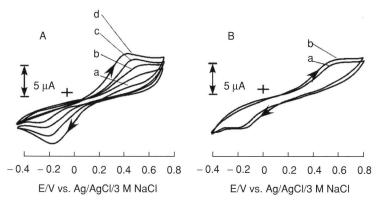

FIGURE 17.10. Cyclic voltammagrams corresponding to the time-dependent formation of M-DNA on a gold surface from (A) ds-DNA modified electrode achieved by re-hybridization of **1** with **2**: (a) 0 minutes, (b) 20 minutes, (c) 50 minutes and (d) 120 minutes. (B) ss-DNA modified electrode treated at the same hybridization condition as (A) except with non-complementary strand **3**: (a) 0 minute, (b) 120 minutes. Data were recorded by the condition outlined in Figure 17-8 with the addition of 0.3 mM $Zn(ClO_4)_2$.

$Fe(CN)_6^{3-/4-}$ is significantly increased during the 2 hour incubation time with Zn^{2+}. The integrated area of the M-DNA peak after 2 hours is at 80% of the bare Au integrated peak area. The peak area remains unchanged for incubation times longer than 2 hours. Conversely, with ss-DNA, the peak area only increases slightly over the same incubation time (Figure 17.10B). The electron transfer kinetics have become faster due to M-DNA formation as is evident from the increase in the peak current and in the decrease of peak separation. Two ss-DNA modified electrode samples were then re-exposed to either the complementary strand **2** or the noncomplementary strand **3**. Hybridization occurred with complementary strand **2**, as expected, and resulted in a slight change in the redox peak current of $Fe(CN)_6^{3-/4-}$. Furthermore, a significant change in the redox peak current was observed following M-DNA formation only when the complementary target **2** was hybridized with the surface-bound ss-DNA (Figure 17.11). Thus, the mediated effect of metal intercalated within M-DNA on the electron transfer provides a powerful tool to enhance differentiation of the electrochemical signal between complementary and non-complementary, and allows electrical detection of the DNA hybridization on a surface.

A charge integration technique, chronoamperometry, has been reported to be a practical charge transport-based technique to electrically characterize SNPs[7,41,42] or to quantitatively determine the surface density of immobilized DNA. This methodology, based on measuring the charge transport through

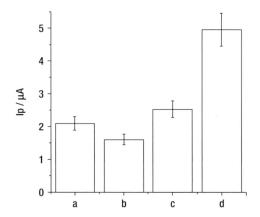

FIGURE 17.11. Sensitivity profile for electrochemical detection of DNA hybridization on an electrode surface. The values represent the redox peak current of $Fe(CN)_6^{3-/4-}$ at different electrodes. (a) **1** modified electrode with non-complement strand **3**. (b) **1** modified electrode with complement strand **2**. (c) **1** modified electrode with non-complement strand **3** under M-DNA forming conditions. (d) **1** modified electrode with complement strand **2** under M-DNA forming conditions.

DNA films, was employed on the same set of redox systems under identical experimental conditions. The initial potential started at 200 mV vs. Ag/AgCl where no electrolysis of ferricyanide occurs. Comparison of the charge passed at ss-DNA, ds-DNA and M-DNA modified electrode is shown in Figure 17.12. During a single step of 12 s to –350 mV, where essentially all the $Fe(CN)_6^{3-}$ is reduced to $Fe(CN)_6^{4-}$. The amount of charge passed on

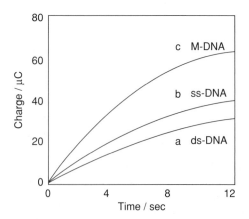

FIGURE 17.12. Chronocoulometric transients at –350 mV of 5 mM ferricyanide in 20 mM Tris-ClO$_4$ buffer (pH 8.6) at (a) ds-DNA modified electrode (b) ss-DNA modified electrode and (c) M-DNA modified electrode.

M-DNA modified electrodes is significantly larger than that of either ss-DNA or ds-DNA modified electrodes. These experiment confirmed the results of CV, that M-DNA is a better electron transfer mediator than both ss- and ds-DNA.

4. Conclusions

The present study has addressed the development of electronic DNA sensors by monitoring changes in the electric properties of ss-, ds- and M-DNA monolayers on gold electrode surface. We have characterized thiol-derivatized DNA attached to gold via a sulfur-gold linkage using XPS, and electrochemical experiments. These results indicate that a ds-DNA monolayer with high surface coverage can be prepared using ds-DNA-hydroxylalkyl disulfide and the SAM is capable of hybridization with complementary DNA after dehybridizaiton treatment. Specifically, our immobilization method avoids the indiscriminate replacement of the DNA probes through competitive alkanethiol replacement. Furthermore, a direct, label-free, electriconic detection of DNA hybridization has been accomplished by monitoring changes in the electrochemical signal at DNA-modified electrodes by EIS, CV and chronoamperometry methods. Our results highlight the sensitivity, based on better conductivity properties of M-DNA, of the hybridization sensing process. Therefore, M-DNA may find widespread applications in nanoelectronics or biosensing since a direct electrical readout of hybridization or DNA binding is now possible.

Acknowledgements. The authors wish to thank CIHR, NSERC and UMDI for financial support, H-B. K. holds a Canadian Research Chair in biomaterials and J. S. L. is supported by a Senior Investigators Award from the Regional Partnership Program of CIHR. The authors also thank Dr. Herrwerth, the University of Heidelberg, Germany, for performing the XPS measurements.

References

1. Razin, S. *Mol. Cell. Probes* **1994**, *8*, 497–511.
2. Hames, B. D.; Higgins, S. J. *Gene Probes 1*; IRL Press: New York, 1995.
3. Fodor, S. P. A.; Rava, R. P.; Huang, X. C.; Pease, A. C.; Holmes, C. P.; Adams, C. L. *Nature* **1993**, *364*, 555.
4. Schena, M.; Shalon, D.; Davis, R. W.; Brown, P. O. *Science* **1995**, *270*, 467.
5. Boon, E. M.; Salas, J. E.; Barton, J. K. *Nat. Biotechnol.* **2002**, *20*, 282–286.
6. Hashimoto, K.; Ishimori, Y. *Lab on a Chip* **2001**, *1*, 61–63.
7. Kelley, S. O.; Boon, E. M.; Barton, J. K.; Jackson, N. M.; Hill, M. G. *Nucleic Acids Res.* **1999**, *27*, 4830–4837.
8. Lee, T. M. H.; Hsing, I. M. *Anal. Chem.* **2002**, *74*, 5057–5062.
9. Lioubashevski, O.; Patolsky, F.; Willner, I. *Langmuir* **2001**, *17*, 5134–5136.

10. Yamashita, K.; Takagi, M.; Kondo, H.; Takenaka, S. *Anal. Biochem.* **2002**, *306*, 188–196.

11. Patolsky, F.; Lichtenstein, A.; Willner, I. *J. Am. Chem. Soc.* **2001**, *123*, 5194–5205.

12. Herne, T. M.; Tarlov, M. *J. J. Am. Chem. Soc.* **1997**, *119*, 8916–8920.

13. Peterlinz, K. A.; Georgiadis, R. M.; Herne, T. M.; Tarlov, M. J. *J. Am. Chem. Soc.* **1997**, *119*, 3401–3402.

14. Georgiadis, R. M.; Peterlinz, K. A.; Peterson, A. W. *J. Am. Chem. Soc.* **2000**, *122*, 3166–3173.

15. Li, C-Z.; Long, Y-T.; Kraatz, H.-B.; Lee, J. S. *J. Phys. Chem. B* **2003**, *107,* 2291–2296.

16. Long, Y-T.; Li, C-Z.; Kraatz, H.-B.; Lee, J. S. *Biophys. J.* **2003**, *84,* 3218-3225.

17. Wetting, S. D.; Li, C-Z.; Long, Y-T.; Kraatz, H.-B.; Lee, J. S. *Anal. Sci.* **2003**, *19,* 23-26.

18. Aich, P.; Labiuk, S. L.; Tari, L. W.; Delbaera, L. J. T.; Roesler, W. J.; Falk, K. J.; Steer, R. P.; Lee, J. S. *J. Mol. Biol.* **1999**, *294*, 477–485.

19. Lee, J. S.; Latimer, L. J. P.; Reid, R. S. *Biochem. Cell Biol.* **1993**, *71 (3-4)*, 162–168.

20. Wincoff, F.; Direnzo, A.; Shaffer, C.; Sweedler, D.; Gonzalez, C.; Scarinje, S.; Usman, N. *Nucleic Acids Res.* **1995**, *25*, 2677–2684.

21. Woods, R.; Bard, A. J.; Dekker, M. *Electroanalytical Chemistry*; 1 ed.: New York, 1980.

22. Hallmark, V. M.; Chiang, S.; Rabolt, J. F.; Swalen, J. D. *Phys. Rev. Lett.* **1987**, *59*, 2879.

23. Eliadis, E. D.; Nuzzo, R. G.; Gewirth, A. A.; Alkire, R. C. *J. Electrochem. Soc.*, **1997**, *144*, 96–105.

24. Sun, L.; Crooks, R. M. *J. Electrochem. Soc.*, **1991**, *138*, L23–L25.

25. Steel, A. B.; Levicky, R. L.; Herne, T. M.; Tarlov, M. J. *Biophys. J.* **2000**, *79*, 975–981.

26. Yang, M.; Yau, H. C. M.; Chan, H. L. *Langmuir* **1998**, *14*, 6121–6129.

27. Kertesz, V.; Whittemore, N. A.; Inanati, G. B.; Manoharan, M.; Cook, P. D.; Baker, D. C.; Chambers, J. Q. *Electroanalysis* **2000**, *12*, 889–894.

28. Kertesz, V.; Whittemore, N. A.; Chambers, J. Q.; McKinney, M. S.; Baker, D. C. *J. Electroanal. Chem.* **2000**, *493*, 28–36.

29. Rubinstein, I.; Steinberg, S.; Tor, Y.; Shanzer, A.; Sagiv, J. *Nature* **1988**, *332*, 426–429.

30. Borges, G.; Kanazawa, K.; Gordon, J. G.; Ashley, K.; Richer, J. *J. Electroanal. Chem.* **1994**, *364*, 281–284.

31. Zei, M. S.; Qiao, G.; Lehmpfuhl, G.; Kolb, D. M. *Ber. Bunsenges, Phys. Chem.* **1987**, *91*, 349–353.

32. Shi, Z.; Lipkowski, J. *J. Electroanal. Chem.* **1994**, *364*, 289–294.

33. Kondo, T.; Yanagida, M.; Shimazu, K.; Uosaki, K. *Langmuir* **1998**, *14*, 5656–5658.

34. Ishida, T.; Choi, N.; Mizutani, W.; Tokumoto, H.; Kojima, I.; Azehara, H.; Hokari, H.; Akiba, U.; Fujihira, M. *Langmuir* **1999**, *15*, 6799–6806.

35. Pressprich, K. A.; Maybury, S. G.; Thomas, R. E.; Linton, R. W.; Irene, E. A.; Murray, R. W. *J. Phys. Chem.* **1989**, *93*, 5568–5575.

36. Bard, A. J.; Faulkner, L. R. *Electrochemical Methods: Fundamentals and Applications*; 2nd edit; John Wiley & Sons, Inc.: New York, 2001.

37. Yamamoto, Y.; Nishihara, H.; Aramaki, K. *J. Electrochem. Soc.* **1993**, *140*, 436–443.

38. Randles, J. E. B. *Discussions Faraday Soc.* **1947**, *1*, 11–19.
39. Ihara, T.; Maruo, Y.; Takenaka, S.; Takagi, M. *Nucleic Acids Res.* **1996**, *24*, 4273–4280.
40. Yamana, K.; Kumamoto, S.; Hasegawa, T.; Nakano, H.; Sugie, Y. *Chem. Lett.* **2002**, 506–507.
41. Boon, E. M.; Pope, M. A.; Williams, S. D.; David, S. S.; Barton, J. K. *Biochem.* **2002**, *41*, 8464–8470.
42. Steel, A. B.; Herne, T. M.; Tarlov, M. J. *Anal. Chem.* **1998**, *70*, 4670–4677.

Part VI

Applications

18

Multi-Parameter Read-Out in Miniaturized Format
Case Studies in Assay Development and High Throughput Screening

THOMAS HESTERKAMP AND ANDREAS SCHEEL
Evotec OAI AG, Schnackenburgallee 114, 22525 Hamburg, Germany

Abstract: The ever increasing need for improved process efficiency in pharmaceutical drug discovery is a major driving force for assay miniaturization and high-density ultra high throughput screening (uHTS). Time- and cost-efficient pharmacophore screening for increasing numbers of molecular targets necessitates generic assay principles, downscaling of assay volumes to the few-microlitre scale or below and versatile uHTS robotic platforms. Fluorescence- and luminescence-based detection technologies, in particular, have superseded conventional radiometric and absorption read-outs. Confocal fluorescence fluctuation techniques are introduced as being ideally suited for miniaturized screening and have been exploited for screening of complex biochemical assays and, more recently, cellular assays as well. Using a confocal fluorescence imaging reader, novel cell-morphological assay formats are amenable to uHTS pharmacophore screening. Improved assay formats in conjunction with robust uHTS hardware contributes to an increased process efficiency in drug discovery and development.

Key words: Confocal – fluorescence fluctuation spectroscopy – microfluidics – generic assay principle – cell imaging – mitogen activated protein kinase – multidrug resistance – receptor internalization.

1. Challenges in Assay Development and UHTS in the Drug Discovery Setting

Pharmaceutical drug discovery and development today is faced with a number of challenges including overall poor process efficiency, strict requirements to drug registration, and an overall shift in paradigm in the preclinical sciences from pharmacologically-driven target discovery to genomically-driven drug discovery. The ever increasing number of validated or potential drug targets poses a formidable task in particular to assay development and ultra high throughput screening (uHTS) in the setting of pharmaceutical drug discovery.

In order to provide for cost- and time-efficient solutions for a growing number of targets, generic assay principles, suited for pharmacophore screening and testing, have to be established for important target classes. Typical assay portfolios include generic assay principles for receptors (G protein coupled receptors, growth factor receptors, nuclear hormone receptors etc.), enzymes (kinases, phosphatases, proteases, oxidoreductases etc), and ion channels. Collectively, these target classes make up 90% of the historic and hence successful pharmaceutical target classes (Drews, 2000). In addition, genomic targets, identified and validated, for instance, by transcriptome/proteome analyses of diseased-versus-control tissues and subsequent genetic manipulations of model organism, enter into drug discovery programs before biochemical and cellular functions of the target have been elaborated. Therefore, additional testing principles are required to identify ligands by mere interaction screening (Wennemers, 2001) or receptor agonists/antagonists by a generic downstream signaling screening (Ungrin et al., 1999; Le Poul et al., 2002). A third layer of complexity is introduced by the front-loading of biochemical and cellular pharmacokinetic test systems into the discovery phase of drug discovery to cope with the increased number of screen-positive compounds and to prevent unsuited compounds from entering into costly development steps (White, 2000). Traditionally, pharmacokinetic profiling had been carried out in the developmental phase of drug candidates.

Two global trends are emerging in order to tackle the above mentioned challenges. First, to miniaturize assay formats to the few-microlitre scale or below for increased throughputs, reduced reagent costs and other reasons (Battersby and Trau, 2002; Wölcke and Ullmann, 2001) and, second, to implement functional cellular assays where applicable (Johnston and Johnston, 2002; Croston, 2002).

These goals are being met by introduction of sophisticated uHTS technologies capable of generating up to 100,000 data points per day. Together with combinatorial chemistry, large and structurally diverse compound collections can thus be screened against many targets in affordable time and at affordable costs for identification of novel pharmacophores. Significant progress in automation has been made along the entire uHTS process chain including compound logistics, microfluidics for compound reformatting and dispensing of assay reagents, detection technologies/read-out devices, data collection and analysis, and subsequent data mining (Sundberg, 2000; Dunn and Feygin, 2000; Battersby and Trau, 2002; Wölcke and Ullmann, 2001). The leading detection technologies rely on radioactivity, absorption, luminescence, fluorescence, and thereof derived detection variants (e.g., Haupts et al., 2000). For instance, the EVOscreen™ Mark II and Mark III uHTS platforms (Evotec OAI) feature a nanolitre compound reformatting module and a nanolitre-to-microlitre reagent dispensing unit in conjunction with confocal fluorescence readers permitting multiple data acquisition and analysis options including high-throughput cell imaging (below). The confocality is particularly suited for fluorescence read-out multiplexing in minimal assay volumes (below). Additionally, the open architecture of the two uHTS platforms allows for implementation of additonal microfluidic devices and conventional macroscopic fluorescence readers.

Before moving on to the specific benefits and aspects of miniaturized uHTS, a number of alternative routes to drug discovery are referred to which in many cases are used to complement approaches by high throughput screening. Depending on the nature of the target and the availability of structural and mechanistic information, rational drug design and focussed screening may be preferred over random screening (e.g., Wlodawer, 2002; Hicks, 2001). Virtual screening of compound collections by docking algorithms can enrich for screen-positives in subsequent wet screens (Langer and Hoffmann, 2001). Evolutionary chemistry (more often referred to as 'chemical genomics') represents an alternative to uHTS in which small tractable chemical entities evolve from poor to good binding affinities for their respective molecular target by random or directed chemical modification (Weber, 2002). Different aspects of the aforementioned approaches are considered to varying degrees in every drug discovery program. It is mostly the order of activities and their relative contribution to the program that makes the difference.

2.   Benefits and Problems of Assay Miniaturization

The benefits of assay miniaturization clearly outweigh the problems encountered. First, a dramatic reduction in reagent consumption can be realized when shifting assay formats from the 100 μl/well (a typical assay volume in 96 well plates) to the 4 μl/well (1536 well) or even 1 μl/well (EVOscreen™ NanoCarrier 2080, Evotec OAI; NanoWell™ Assay Plates, PanVera LLC) format. This benefit holds for precious biological material (receptors, enzymes, primary cells etc.) but also for costly substrates and ligands (e.g., labeled peptides). Second, a manyfold reduction in compound consumption can be similarly achieved. This is relevant to combinatorial chemistry libraries synthesized on the low-milligram scale and to natural product collections, where the resupply of material is extremely costly if possible at all. Further to this, compound collections gain value with the number of screening campaigns perfomed against, because selectivity questions can be directly addressed. A third benefit of assay miniaturization pertains to the reduced storage space and waste loads required and produced in high-density screening compared to macroscale screening. This is particularly relevant to environmentally critical waste components like long-lived radionuclides and scintillation cocktails.

The problems encountered in assay miniaturization fall in two categories. First, principal limitations of miniaturization exist due to increased surface-volume ratios and the associated problems of evaporation, surface-adsorption, surface-tension, and reagent oxidation. These principal problems inevitably restrict the miniaturization scale to the one-microlitre volume range unless novel, for instance chip-based formats are pursued (Khandurina and Guttman, 2002). Miniaturized cellular screening suffers particularly from the above limitations (Johnston and Johnston, 2002). Among others, the inherent heterogeneity of cell samples and overall poor signal-to-background and signal-to-noise factors with limiting cell numbers per well negatively affect assay statistics. The second category of problems deals with the technical realization of the miniaturization process in terms of assay plates, reagent addition in small volumes (microfluidics), positionally precise, fast and sensitive readers, and the problems of data acquisition and analysis. This latter category has been successfully addressed by a number of technology providers (below).

3. Elements of Assay Miniaturization

Current industry standards in uHTS are the 96 well and 384 well plate formats for cellular applications and the 384 and 1536 well plate formats for biochemical applications (Smith, 2002). These three plate formats predominate in routine uHTS, with the 1536 well plate being the future target standard format for biochemical and cellular uHTS (Garyantes, 2002; Dunn and Feygin, 2000). Below the 1536 well few-microlitre scale only two plate formats have found routine application: plates with 2 μl maximum well capacity and 2080 wells per carrier for biochemical uHTS (EVOscreen™ NanoCarrier 2080, Evotec OAI; see Figure 18.1) and 3456 well plates (NanoWell™ Assay Plates, PanVera LLC; Mere et al., 1999).

Future assay formats may rely on even higher density plates (Mander, 2000), virtual well plates (Garyantes, 2002) and chip-based or capillary force-driven micro-compartments (Khandurina and Guttman, 2002; Battersby and Trau, 2002; Sundberg, 2000). After a number of years of investments into infrastructure, however, the current focus of uHTS laboratories apparently has shifted to routine application of assays on the existent technological platforms and plate formats.

Irrespective of the plate format and to cope with the aforementioned time and cost constraints in drug discovery it is always attempted to design homogeneous 'mix-and-measure' assays that do not require time-consuming and error-prone washing, separation or liquid transfer steps. Typical miniaturized uHTS assays are composed of three to five assay components pipetted or dispensed on the nanolitre to low microlitre scale. General requirements to liquid handling systems are precision, low levels of cross-contamination, speed, small dead volumes, and costs of maintenance and consumables. The three leading pipetting/dispensing techniques for the indicated volume range rely on air-/liquid-displacement devices, pintool devices, and drop-on-demand systems (Dunn and Feygin, 2000; Wölcke and Ullmann, 2001).

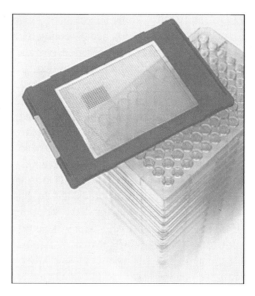

FIGURE 18.1. EVOscreen™ NanoCarrier 2080 with 96 stained wells on a stack of traditional 96 well plates.

In the field of detection systems, fluorescence and luminescence have made up ground at the cost of radiometric and simple absorption systems. This technological transition is mostly driven by the wide appreciation of the exquisite sensitivity and dynamics with which fluorescent and luminescent light can be measured without environmental hazards. Regarding luminescence, the transition from photomultiplyer tube (PMT) based readers to charge-coupled device (CCD) camera-based detection systems offer unprecedented detection speeds while maintaining sensitivity in the field of cellular uHTS (Maffia et al., 1999; CLIPR™ and FLIPR³™, Molecular Devices; ViewLux™, Perkin Elmer Life Sciences). However, due to the problems of liquid handling and assay sensitivity at higher density formats, their implementation in routine high throughput screening has so far been mainly restricted to the 384 well format.

Fluorescent read-out technologies are divided into those that measure a time- and volume-weighted averaged fluorescent signal from a well-macroscopic fluorescence detection – and those that measure fluctuating (stochastic) signals of individual fluorescent molecules within a small confocal reaction compartment – single-molecule approach (e.g., Haupts et al., 2000). The majority of the commercially available fluorescence detection systems (e.g., LJL Acquest, Molecular Devices; Ultra, Tecan; Victor², Perkin Elmer Life Sciences) measure macroscopic fluorescence in a number of facets: fluorescence intensity, time-resolved fluorescence, fluorescence resonance energy transfer, and fluorescence polarisation. Due to the macroscopic nature of these methods, signal quality suffers when moving significantly beyond the 384 well format. Single-molecule fluorescence detection techniques have been pioneered in the form of confocal fluorescence correlation spectroscopy (FCS; Eigen and Rigler, 1994) and meanwhile extended to a suite of versatile and uHTS-compatible read-out options, including the one- and two-dimensional fluorescence intensity distribution analyses (1D-FIDA and 2D-FIDA; Kask et al., 1999; Kask et al., 2000; see Table 18.1). Confocal fluorescence fluctuation technologies are ideally suited for miniaturized assay formats because the femtolitre detection volume is inherently independent of the assay volume (Rüdiger et al., 2001; Haupts et al., 2000). Therefore, these assays can be miniaturized down to the 1 fl level. Along with improvements in the optical setup and signal acquisition times of 1 to 2 seconds per well or below, confocal fluorescence uHTS readers have been realized (FCS+plus detection system, Evotec OAI; ConfoCor2, Analytik Jena) and implemented in uHTS robotic platforms (EVOscreen™ Mark II and Mark III platforms, Evotec OAI). Sophisticated data fitting algorithms allow the extraction of several fluorescence parameters from a single measurement. The following chapter will deal with the benefits of applying different fluorescence read-out options separately or simultaneously to biological questions and elaborate on the set of confocal fluorescence fluctuation techniques developed to date.

4.   Benefits of Read-Out Multiplexing

Read-out multiplexing in the context of this article means the acquisition of multiple read-outs on the same sample probe at the same time. The wealth of physical information contained in fluorescent light (e.g., fluorescence intensity, lifetime, wavelength, molecular polarisation) has been exploited in such a way that different sample analysis methods or algorithms can be applied to the same sample in parallel or close sequential order without major changes in the setup of the detection device. The benefit of this approach is to fully harness, at practically constant time and costs, the information content of biological reactions in terms of read-out statistics, read-out sensitivity to pharmacophore interaction and potential artefact interference, again reducing the overall high attrition rate in follow-up studies. Autofluorescence of compounds, in particular, is a widely held scepticism to fluorescence-based assay read-outs. Macroscopically, autofluorescence is detected by a compound-dependent increase in fluorescence intensity that exceeds the limits of the assay-specific positive or negative controls. Using microscopic fluorescence fluctuation techniques, autofluorescence can firstly be detected by simultaneous acquisition of, for instance, the molecular polarisation and fluorescence intensity values of a sample and secondly be corrected for by applying data analysis routines capable of distinguishing at the single molecule level the overall fluorescence contribution of the fluorescent tracer itself (low particle number, high molecular brightness) and the autofluorescent

TABLE 18.1. Fluorescence fluctuation sample analysis methods

Method	Features	Applied to	Reference
FCS, fluorescence correlation spectroscopy	Resolves and quantifies fluorescent species on the basis of different translational diffusion coefficients	molecular interaction	Eigen and Rigler, 1994
Dual-color FCCS, fluorescence cross-correlation spectroscopy	Uses two dye-labels and two-color detection; resolves and quantifies fluorescent species carrying both dye-labels (coincidence)	molecular interaction	Schwille et al., 1997
Dual-color cFCA, confocal fluorescence coincidence analysis	Uses two dye-labels and two-color detection; resolves and quantifies fluorescent species carrying both dye-labels (coincidence)	molecular interaction; faster data acquisition than FCS	Kettling et al., 1998; Koltermann et al., 1998
FIDA, fluorescence intensity distribution analysis	Resolves and quantifies fluorescent species on the basis of specific molecular brightness	molecular interaction; frequently applied to uHTS of ligand binding to vesicle-/cell-bound receptors	Kask et al., 1999
2D-FIDA, polarization	Uses polarized excitation and detection on two detectors; resolves and quantifies fluorescent species with different specific polarization values via the specific molecular brightness-pair of both detectors	molecular interaction; widely applied to uHTS with ligands of < 10 kDa	Kask et al., 2000
2D-FIDA, dual-color	Uses two dye-labels and two-color detection; resolves and quantifies fluorescent species carrying both dye-labels (coincidence) via the specific molecular brightness-pair of both detectors	molecular interaction; useful if interaction partners are of similar molecular weight	Kask et al., 2000
cFLA, confocal fluorescence lifetime analysis	Resolves and quantifies fluorescent species on the basis of their lifetime of the excited state	molecular interaction; useful in uHTS assays where local environment of fluorophore is affected upon interaction	Palo et al., 2002
FILDA	resolves fluorescent species on the basis of both their specific molecular brightness and the lifetime of the excited state	molecular interaction studies; unifies FIDA and cFLA sample analysis methods	Palo et al., 2002
FIMDA	resolves fluorescent species on the basis of both the specific molecular brightness and the translational diffusion time	molecular interaction studies; unifies FIDA and FCS sample analysis methods	Palo et al., 2000

compound (high particle number, low molecular brightness). The widely applied 2D-FIDA (two-dimensional fluorescence intensity distribution analysis) sample analysis routine (Kask et al., 2000) permits exactly this kind of sample analysis for molecular interaction assays. Another example of read-out multiplexing is offered by the recently introduced FILDA fluorescence fluctuation sample analysis method (Palo et al., 2002). FILDA stands for fluorescence intensity and lifetime distribution analysis and combines the benefits of FIDA and fluorescence lifetime analyses. With very short data acquisition times this method offers further improved assay statistics by sorting receptor-bound and unbound fluorescent tracer according to both, its specific molecular brightness and its excitation-to-detection time interval (Palo et al., 2002). Table 1 provides an overview of the currently available fluorescence fluctuation-based sample analysis methods and their application to biological assay development and uHTS on EVOscreen™ Mark II and Mark III platforms (Evotec OAI).

The availability of a suite of read-out options and their multiplexing in uHTS increases the versatility of the uHTS hardware platform, obviates the need for additional alternative read-outs and contributes to time- and cost-efficient drug discovery with lower overall attrition rates.

5.   Case Studies of Miniaturized Biochemical and Cellular Assay Development and HTS

Irrespective of biochemical versus cellular screening, key statistical figures of uHTS compatible assays and uHTS campaigns are generally defined as follows:

- dynamic range: difference in read-out of the assay negative control (minus inhibitor) and assay positive control (plus inhibitor)
- Z' factor (Zhang et al., 1999): a statistical figure taking into account the dynamic range of the assay system and the standard deviations of the positive and negative controls; generally, assays with Z' factors of >0.5 are considered suitable for uHTS (the theoretically maximal Z' factor would be 1.0)
- Z factor (Zhang et al., 1999): a statistical figure describing the separation compound and positive control values; the Z factor is sensitive to the data variability (hit rate) as well as to the dynamic range of the assay signal
- hit threshold: usually defined by the 3-σ-method, meaning: all compound inhibition data points deviating by more than three standard deviations from the mean of the assay negative control are considered hits or screen-actives; the hit threshold can be calculated in a plate-by-plate manner or cumulatively for an entire screening campaign; the lower the hit threshold, the more sensitively can weak inhibitors be detected

– hit rate: number of hits/screen-actives of all compounds tested; usually, the compound concentration required to achieve a hit rate of 1% is adjusted in a short range-finding study preceeding the uHTS campaign itself

– confirmation rate: number of hits/screen-actives that can be reproduced in follow-up studies employing the same assay setup used for uHTS

In the following, three programs at different process stages and conducted on the one-to-few microlitre scale on EVOscreen™ Mark II and III platforms are detailed. In the first, biochemical uHTS example, a generic kinase assay principle was applied to discovery of novel selective hit series for mitogen-activated protein kinase (MAPK) kinases. The second example is a cellular assay program conducted for library assessment in terms of interference of compounds with an ATP-binding cassette multidrug-resistance pump (P-glycoprotein). The third case study details the use of a confocal fluorescence imaging device for monitoring a complex cellular response to receptor-stimulation.

5.1.   Case Study I: MAPK Kinase Screening

MAPKs play important roles in mediating cellular responses to a variety of extracellular signals, including survival and death/apoptosis signals (Robinson and Cobb, 1997). Three groups of MAPKs are distinguished: extracellular signal-regulated kinases (ERK1/2, p42/44^{MAPK}), c-Jun N-terminal kinases (p46/54^{JNK}), and p38^{MAPK}. MAPKs are activated through distinct upstream so-called dual specificity kinases and hence function in distinct signal transduction pathways. The p46/54^{JNK} and p38^{MAPK} MAPKs are activated by cell stress as exemplified by oxidative agents, UV exposure, or proinflammatory cytokines. Dual specificity kinases activating p46/54^{JNK} are MAPK kinases 4 and 7 (abbreviated MKK4 and 7), whereas p38^{MAPK} is activated by upstream kinases MKK3 and 6. Here, two separate uHTS campaigns for inhibitors of MKK4 on the one hand, and MKK6 plus p38^{MAPK} (pathway screen) on the other hand were conducted. In the latter example, inhibitors of both, MKK6 and activated p38^{MAPK}, can be discovered in one screen. The generic kinase assay principle relies on the selective binding of phosphorylated polypeptide kinase substrates by commercially available antibodies. Rather than direct detection, the polarisation shift upon antibody binding of a pre-phosphorylated and fluorescently labeled tracer peptide is monitored in a binding equilibrium with the enzymatically phosphorylated substrate peptide (Figure 18.2, upper left). Following titration of tracer, substrate, kinase and antibody, an exquisitely sensitive though statistically robust homogeneous assay format with generic character is available. Routinely, assay sensitivity is challenged using well-known kinase standard inhibitors (Figure 18.2, upper right). Figure 18.2, lower left, gives a three-dimensional representation of a MKK6 plus p38^{MAPK} EVOscreen™ NanoCarrier 2080 assay plate, demonstrating the excellent separation of the assay positive controls

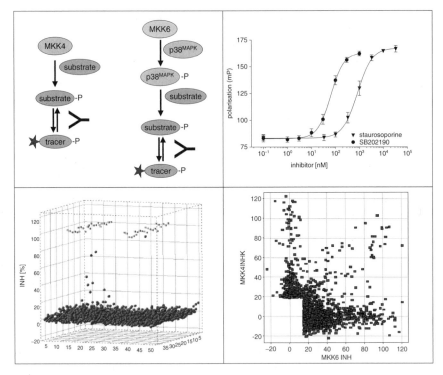

FIGURE 18.2. Miniaturized kinase uHTS examples using a generic assay principle. Upper left, indirect fluorescence polarisation assay principle for MKK4 and MKK6/p38^{MAPK} kinase assays; upper right, inhibition of MKK6/p38^{MAPK} pathway assay by staurosporine and SB202190 standard inhibitors; lower left, EVOscreen™ NanoCarrier 2080 screening plate of MKK6/p38^{MAPK} pathway assay campaign; positive inhibition controls are represented by pale-grey cubes, compound inhibition results are given as dark-grey spheres; lower right, correlation of primary compound inhibition results for MKK4 and MKK6/p38^{MAPK} pathway assay campaign (inhibition results above hit threshold only).

from the non-inhibiting compound wells. A cluster of hits with varying degrees of inhibition can equally well be discerned. A subsequent comparison of primary screening hits obtained for MKK4 and MKK6 plus p38^{MAPK} allows for the selection of common and selective MAPK kinase inhibitors for dose-response-type follow-up studies (Figure 18.2, lower right).

Table 18.2 summarizes a number of key statistical figures pertaining to the two primary uHTS campaigns. The excellent data quality allows the use of low inhibition thresholds and reliable detection of hits that inhibit as little as 15% of the positive controls. This offers a unique opportunity to identify weak but potentially valuable inhibitors. A number of screen-active compounds with selectivity for one or two of the three enzymes MKK4, MKK6,

TABLE 18.2. Key statistical figures for MKK4 and MKK6/p38^{MAPK} uHTS campaigns

Statistical figure	MKK4	MKK6/p38MAPK
number of compounds screened	209,974	109,629
compound screening concentration [microM]	25	25
mean Z'	0.82 (n = 655)	0.84 (n = 334)
mean Z	0.74 (n = 655)	0.72 (n = 334)
hit threshold (3-σ-method) [% inhibition]	19.2	15.1
hit rate [%]	1.33	2.73

and p38^{MAPK} and low-micromolar IC$_{50}$ values have entered follow-up studies to assess their mode of inhibition and potential for optimization by chemical analoging.

Of note, in the MKK6 and p38^{MAPK} pathway assay, the total enyzme protein consumption was 50 micrograms (MKK6) and 225 micrograms (p38^{MAPK}), respectively, for a 110 k uHTS campaign-impressively underlining the benefits of high throughput screening at the 1 µl scale!

5.2.   Case Study II: P-Glycoprotein Screening

P-glycoprotein belongs to the evolutionary conserved family of ATP-binding cassette transporters. It became infamous due to its detrimental role in chemotherapeutic treatment of cancers, where multidrug-resistance is caused by overexpression of the corresponding MDR (multidrug resistance) 1 gene. Meanwhile, a much wider role for P-glycoprotein in many types of malabsorption and poor bioavailability of drugs is recognized. Further to this, some drugs show very strong interactions with P-glycoprotein (e.g., verapamil) such that the normal absorption behavior of co-prescriptions is compromised (drug-drug interactions). Being an integral membrane protein, P-glycoprotein acts as an energy-dependent 'flippase' for all sorts of lipophilic xenobiotics that freely partition into the cell membrane. Unfortunately, lipophilicity of neutral or cationic organic compounds seems to be the only common characteristic of P-glycoprotein substrates and prudent experimentation is required to detect P-glycoprotein interaction of candidate compounds on a case-by-case basis. Here, a miniaturized cellular test system is described that is capable of high-throughput detection of compounds with P-glycoprotein interaction. The miniaturization of this assay to 1536 well format allows for screening of large corporate compound collections for subsequent flagging of suspect molecules and exemplifies the use of early high throughput test systems to support pharmacokinetics.

The P-glycoprotein assay is based upon the passive cellular uptake of lipophilic calceinacetoxymethyl ester (calcein AM; Molecular Probes). A proficient cellular P-glycoprotein pump expels this substrate from the membrane phase. Upon blockage of P-glycoprotein by competing substrates or true blockers (high-affinity interactors), calcein AM gains access to the cytoplasm where it is enzymatically converted to brightly fluorescent calcein. This gives a positive fluorescence read-out for P-glycoprotein interacting compounds. Using the OPERA confocal imaging reader of the EVOscreen™ Mark III screening platform (Evotec OAI) and P-glycoprotein overproducing MES-SA cells, sparsely fluorescent cells (proficient P-glycoprotein; Figure 18.3, upper left) can be easily distinguished from brightly fluorescent cells (saturated P-glycoprotein; Figure 18.3 upper right). Further employing a second spectral detection channel for unspecific cell staining and software algorithms capable of cell identification and counting makes possible the determination of the fluorescence units per cell pixel (Figure 18.3, lower left and right).

Thereby, a background-free specific fluorescence signal is generated. Using this approach, a pilot screen was conducted to assess the overall robustness of the assay under uHTS screening conditions. An average Z' factor of well above 0.5 was found for a set of 30+ 1536 well plates run in an automated

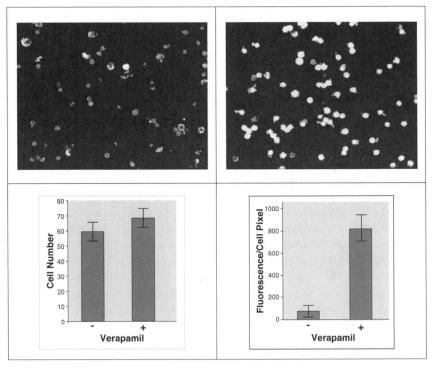

FIGURE 18.3. A cellular assay for P-glycoprotein interacting compounds. Upper left and right, cell stain with fluorescent calcein dye in absence (left) and presence (right) of the P-glycoprotein blocker verapamil; lower left and right, cell number per area unit (left) and calcein fluorescence per cell pixel (right) in absence and presence of verapamil.

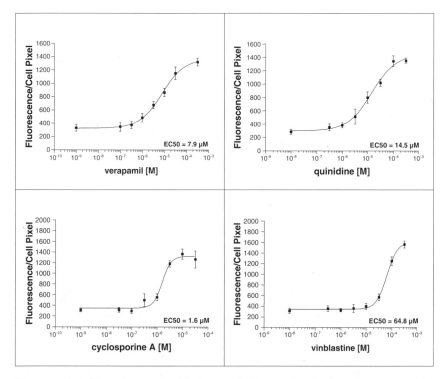

FIGURE 18.4. Dose-dependent blockage of P-glycoprotein by pharmaceuticals, expressed as calcein fluorescence per cell pixel.

mode. Additionally, positive control compounds were tested in an 8-point dose-response mode. Figure 18.4 shows the corresponding EC_{50} plots for verapamil, quinidine, cyclosporine A, and vinblastine, all of which are known as P-glycoprotein interactors. Here, the cell dispensing step in a volume of 1 μl was exerted with the synQUAD (Cartesian Technologies) liquid dispensing device.

5.3.   Case Study III: Receptor Internalization

Another level of complexity is introduced in the third case study. The same OPERA confocal imaging reader, here, was used to quantitatively monitor at subcellular resolution the physiologically relevant translocation of endothelin A receptors from the plasma membrane to an intracellular endosomal compartment following receptor stimulation with endothelin-1. It is a common theme in cell biology that plasma membrane receptors are internalized after stimulation to incorporate the signaling molecule and to desensitize the cells from on-going stimulation. Quantitating this processing step is of great value both as a secondary assay format in functional screens and as a standalone assay format for discovery of inhibitors acting downstream to the ligand binding event (and others types of morphological responses). The assay is based on the simultaneous staining of the endothelin A receptor as a translational fusion to an autofluorescent protein and the cell cytoplasm by typical cell stains (here: Syto59). A differential but simultaneous imaging of the receptor and the cytoplasm then occurs in the two respective spectral channels of the OPERA reader. Figure 18.5 demonstrates the translocation of the endothelin A receptor one and two hours after stimulation by endothelin-1.

The brightly green fluorescent receptor stain on the plasma membrane changes into a punctuate perinuclear staining pattern. Translocation can be quantitated on a per cell basis to generate dose-response plots for the natural agonist endothelin-1 or another model agonist, sarafotoxin 6B (Figure 18.5, lower right). These results were obtained with standard clear bottom 384 well plates and data acquisition times of 2 sec/well. Currently this homogeneous

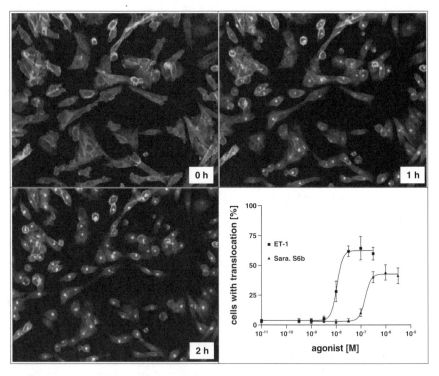

FIGURE 18.5. Double-label cell stain for cytoplasm (red) and the endothelin A receptor (green) before (upper left) and after 1h (upper right) and 2h (lower left) stimulation with endothelin-1. Note the translocation of endothelin A recepetor from the plasma membrane to an intracellular perinuclear compartment following stimulation. Lower right, quantification of cells with translocation following stimulation with increasing doses of agonists endothelin-1 and sarafotoxin 6B.

assay is being reformatted to 1536 well uHTS format for subsequent high throughput screening on an EVOscreen™ Mark III platform. It serves to demonstrate the capability of a sophisticated uHTS platform to quantitatively monitor complex subcellular events of high physiological relevance in a truely high throughput mode. Agonists and antagonists discovered in this way are information-rich and therefore less likely to fail in subsequent biological efficacy studies. Ultimately, complex cellular assay approaches like this will inevitably complement or supersede simpler assay systems in pharmaceutical drug discovery.

Acknowledgements. We thank our colleagues C. Eggeling, K. Herrenknecht, M. Hoffmann-Wecker, J. Jungmann, C. Kirchhoff, J. Krämer, M. Leimkühler, E. Lopez-Calle, D. Ullmann, and J. Wölcke for sharing unpublished results with us, critical reading of the manuscript and continued support.

References

Battersby and Trau (2002) Trends Biotechnol. 20: 167-173, Novel miniaturized systems in high-throughput screening

Croston (2002) Trends Biotechnol. 20: 110-115, Functional cell-based uHTS in chemical genomic drug discovery

Drews (2000) Science 287: 1960-1964, Drug Discovery: a historical perspective

Dunn and Feygin (2000) Drug Discovery Today 5: S84-S91, Challenges and solutions to ultra-high throughput screening assay miniaturization: submicroliter fluid handling

Eigen and Rigler (1994) Proc. Natl. Acad. Sci. U.S.A. 91: 5740-5747, Sorting single molecules: application to diagnostics and evolutionary biotechnology

Garyantes (2002) Drug Discovery Today 7: 489-490, 1536-well assay plates: when do they make sense?

Haupts et al. (2000) Drug Discovery Today (Suppl. S): 3-9, Macroscopic versus microscopic fluorescence techniques in (ultra)-high-throughput screening

Hicks (2001) Curr. Med. Chem. 8: 627-650, Recent advances in NMR: expanding its role in rational drug design

Johnston and Johnston (2002) Drug Discovery Today 7: 353-363, Cellular platforms for HTS: three case studies

Kask et al. (1999) Proc. Natl. Acad. Sci. U.S.A. 96: 13756-13761, Fluorescence-intensity distribution analysis and its application in biomolecular detection technology

Kask et al. (2000) Biophys. J. 78: 1703-1713, Two-dimensional fluorescence intensity distribution analysis: theory and applications

Kettling et al. (1998) Proc. Natl. Acad. Sci. U.S.A. 95: 1416-1420, Real-time enzyme kinetics monitored by dual-color fluorescence cross-correlation spectroscopy

Khandurina and Guttman (2002) Curr. Opin. Chem. Biol. 6: 359-366, Microchip-based high-throughput screening analysis of combinatorial libraries

Koltermann et al. (1998) Proc. Natl. Acad. Sci. U.S.A. 95: 1421-1426, Rapid assay processing by integration of dual-color fluorescence cross-correlation spectroscopy: high throughput screening for enzyme activity

310   Thomas Hesterkamp and Andreas Scheel

Langer and Hoffmann (2001) Curr. Pharm. Design 7: 509-527, Virtual screening: an effective tool for lead structure discovery?

Le Poul et al. (2002) J. Biomol. Screen. 7: 57-65, Adaptation of aequorin functional assay to high throughput screening

Maffia et al. (1999) J. Biomol. Screen. 4: 137-142, Miniaturization of a mammalian cell-based assay: luciferase reporter gene readout in a 3 microliter 1536-well plate

Mander (2000) Drug Discovery Today 5: 223-225, Beyond uHTS: ridiculously HTS?

Mere et al. (1999) Drug Discovery Today 4: 363-369 Miniaturized FRET assays and microfluidics: key components for ultra-high-throughput screening

Palo et al. (2000) Biophys. J. 79: 2858-2866, Flourescence intensity multiple distributions analysis: concurrent determination of diffusion times and molecular brightness

Palo et al. (2002) Biophys. J. 83: 605-618, Fluorescence intensity and lifetime distribution analysis: toward higher accuracy in fluorescence fluctuation spectroscopy

Robinson and Cobb (1997) Curr. Opin. Cell Biol. 9: 180-186 Mitogen-activated protein kinase pathways

Rüdiger et al. (2001) J. Biomol. Screen. 6: 29-37, Single-molecule detection technologies in miniaturized high throughput screening: binding assays for G protein-coupled receptors using fluorescence intensity distribution analysis and fluorescence anisotropy

Schwille et al. (1997) Biophys. J. 72: 1878-1886, Dual-color fluorescence cross-correlation spectroscopy for multicomponent diffusional analysis in solution

Smith (2002) Nature 418: 453-459, Screening for drug discovery -the leading question

Sundberg et al. (2000) Drug Discovery Today 5 (No. 12, Suppl.): S92-S103: Microchip-based systems for target validation and HTS

Ungrin (1999) Anal. Biochem. 272: 34-42, An automated aequorin luminescence-based functional calcium assay for G-protein coupled receptors

Weber (2002) Drug Discovery Today 7: 143-147, Multi-component reactions and evolutionary chemistry

Wennemers (2001) Comb. Chem. High Throughput Screen. 4: 273-285, Combinatorial chemistry: a tool for the discovery of new catalysts

White (2000) Annu. Rev. Pharmacol. Toxicol. 40: 133-157, High-throughput screening in drug metabolism and pharmacokinetic support of drug discovery

Wlodawer (2002) Annu. Rev. Med. 53: 595-614, Rational approach to AIDS drug design through structural biology

Wölcke and Ullmann (2001) Drug Discovery Today 6: 637-646, Miniaturized HTS technologies -uHTS

Zhang et al. (1999) J. Biomol. Screen. 4: 67-73, A simple statistical parameter for use in evaluation and validation of high throughput screening assays

19

Evaluation of the Reliability of cDNA Microarray Technique

Yao Li,[1§] Yao Luo,[1§] Chengzhi Zhang,[1] Minyan Qiu,[2] Zhiyong Han,[2] Qin Wei,[1] Sanzhen Liu,[2] Yi Xie,[1*] and Yumin Mao[1*]

[1] *State Key Laboratory of Genetic Engineering, Institute of Genetics, School of Life Science, Fudan University Shanghai 200433, P. R. China*
[2] *Shanghai BioStar Genechip Inc., Shanghai, 200092, P. R. China*
[§] *These two authors contribute equally to this paper*
[*] *Corresponding authors. E-mail: ymmao@fudan.edu.cn*

1. Introduction

Microarray is a new technological approach developed in early 1990s, which has found wide applications in studies of gene expression patterns in various tissues[1,2]. The technology enables rapid parallel genetic analysis of tens of thousands of genes in one experiment, and makes possible for the genetic researchers to measure the expression of all genes in an organism simultaneously. cDNA microarray technology was established and applied in the research of hepatoma expression pattern in our laboratory[3]. In the early stage, the information about the genes which expressions changed obviously in a specific condition could be generally obtained by ratio analysis[4]. However, there are many variables that will impact on the quality of the data generated by any microarray experiments[5], therefore it is significant to evaluate the reliability of microarray data. Up to now, a large quantity of useful data has been acquired from these experiments. The problems of how to analyze and deal with the data and how to validate the reliability of results have become the key to utilize the approach more effectively.

Currently, there were only a few papers about comprehensive evaluation of cDNA technology. Incyte Ltd. reported the issue of precision, accuracy and reproducibility of microarray data[6]. In order to study the data more efficiently, we evaluated the reproducibility, reliability and variation at several different aspects, and analyzed the advantages and gave an assessment on the whole.

2. Methods and Material

Microarrays of 4096 or 14112 human cDNAs, were manufactured by BioStar Ltd. All clones were verified by being sequenced. The array included spots of plant genes and HCV genes as negative control and spots of preparing solution without cDNA as blank control.

2.1.   Two Sets of Experiments

The experiments for system evaluation were divided into two sets: self-comparison experiments and differential expression experiments. In the self-comparison experiments, Cy5- and Cy3-labeled cDNA were both prepared from the RNA of the same tissue, while in differential expression experiments, the RNA was from two different tissues to measure the differentiation.

2.2.   Preparation of Probes Labeled with Fluorescent Molecules

Donated hepatoma and normal liver tissues were supplied by Changhai Hospital. Two methods for extracting total RNA were used here: Method One[7], as a common method, was used for most total RNA in present article; and Method Two[8] was only used for comparing two methods in 3.1.1. mRNA was purified using Oligotex mRNA Midi Kit (Quagen Company). In the self-comparison experiment, the mRNA (3μ g) from the same tissue or total RNA (50μ g) was labeled with Cy3-dUTP and Cy5-dUTP respectively; While in the differential expression experiments, normal liver tissue was labeled with Cy3-dUTP and hepatoma was labeled with Cy5-dUTP, or vice versa. Labeled cDNA was deposited with ethanol, and then dissolved in hybridizing solution of 20μ L 15×SSC+0.2%SDS.

2.3.   Hybridization and Rinse

The methods were as described[3].

2.4.   Scanning and Analysis

Microarrays were scanned with a Scanarray 4000 laser induced fluorescence scanner from Packard Biochip Technologies Ltd. and signal intensity for each target element was detected with GenePix 3.01 image software from Axon.

2.5.   Data Statistical Study

All data obtained were calibrated on the whole level by Yang's integral correction algorithm[9]. The corrected data were used to calculate the ratio of corresponding signal and determine the cutoff of differential expression by tolerance interval algorithm. CV of each ratio was calculated so as to assess the accuracy and reproducibility of the arrays. According to the cutoff, the distribution of ratio of those differential expression genes was observed and the possible sources of variation were inspected. All data were screened automatically with the image software. Relevant coefficient r, which was usually used to assess the reproducibility of microarray data, was calculated as Pearson relevant coefficient[10]. Therein, x and y represented the corresponding ratio value of two experiments.

$$r = \frac{\sum_i \text{lrb} x_i - \bar{x} \text{ rrblrb} y_i - \bar{y} \text{ rrb}}{\sqrt{\sum_i \text{lrb} x_i - \bar{x} \text{ rrb}^2 \cdot \sum_i \text{lrb} y_i - \bar{y} \text{ rrb}^2}}$$

The concept of "consistence rate" was put forward here as a new parameter to evaluate the reproducibility of microarray data. The consistence rate was the percent of gene number, which showed differential expression in the same direction in both of two experiments, from the total number of all differentially expressed genes, the formula as follows. N_d was the total number of genes showing differential expression, and N_{id} was the number of elements of differential expression in the same direction in both of two experiments therein.

$$CR = \frac{N_d - 2 \times N_{id}}{N_d}$$

3.   Results

3.1.   Self-Comparison Experiments

The same normal liver tissue was labeled with Cy3 and Cy5 to perform self-comparison experiments. Theoretically, the ratio of Cy5/Cy3 should be 1 for all elements arrayed on the slide in self-comparison experiment. However, due to some systematic biases, some deviations from the theoretical value were observed of some genes. The cutoff of the ratio of Cy5/Cy3 to screening differential genes was 2, which was recognized all over the world. Thus, in the self-comparison experiments, any gene of which ratio was higher than 2 or lower than 0.5 was regarded as false positive gene. False positive rate (FPR) refers to the percentage of the number of false positive genes from all genes on the array. The values of false positive rate, relevant coefficient (R) and CV of the ratio were used to evaluate the reliability of microarray data.

3.1.1.   Impact of Different Kinds and Different Processes of RNA on Hybridization

In order to know the impact of different kinds of RNA on reproducibility of hybridization, we performed the following experiments: (1) Performing self-comparison experiments with total RNA, which included three sets of experiments: A) total RNA from the same extraction method—Method One at different time was used; C) total RNA from the same extraction method—Method One at the same time was used; (2) Performing self-comparison experiments with mRNA; (3) Performing self-comparison experiments with total RNA and purified mRNA. The results are shown in Table 1.

According to statistic analysis, the results of Table 1 indicated that there was no obvious differentiation (P>0.05) in the first four sets. False positive rate was usually about 1% when cutoff was defined as 2.0, which was similar to the advanced level in the world[11]. It suggested that the approaches and processes of mRNA extraction would not induce any distinct differentiation, but in the self-comparison experiments of mRNA vs. total RNA, false positive rate was more than 10%. The above results were reliable by several reproducible performances. We could conclude that mRNA and total RNA could be both used in experiments of expression pattern, but only the same kind of RNA can be used in one experiment. For instance, if mRNA was labeled with Cy3, then the kind of RNA labeled with Cy5 should also be mRNA.

3.1.2.   Impact of Probe Labeling Process on False Positive Rate

We demonstrated the reproducibility of two experiments by another means: x and y axis represented ratio in natural log (Ln) scale of the two of replicate experiments respectively. Thus, it was convenient to compare the identity of them (Figure 19.1). Since the false positive rate was usually less than 1% in common condition, the false positive rate after reproducing twice was very low (1%*1%=0.01%) theoretically. In the experiments, when the cutoff was defined as 1.7, Figure 19.1B (The probes of replicated experiments labeled separately) showed that none of genes appeared false positive in both arrays, which accorded with the theoretical value, while Figure 19.1A (the probes of replicate experiments labeled simultaneously) showed that the most false positive genes only appeared in one experiment (the points in diamond), but 5 genes appeared in both experiments (the points in triangle) in the same direction. Analogous results were obtained in the replicated experiments. (Data were not shown here.) Such false positive was due to the bias in the labeling process. It could be concluded that performing two replicated experiments

TABLE 19.1. The result of self-comparison experiments with RNA (mRNA and total RNA) of the same Tissue RNA but different process methods

| | Total RNA/Total RNA | | | mRNA/ mRNA | Total RNA/mRNA (Cy5/Cy3) |
	A Different extraction methods	B The same extraction method	C Extracting at the same time		
FPR (cutoff=1.7)	2.72%	2.83%	2.51%	1.38%	–
FPR (cutoff=2.0)	1.23%	0.97%	0.67%	0.39%	12.40%
correlation coefficient R	0.90	0.97	0.95	0.98	0.76
P value	>0.05				–

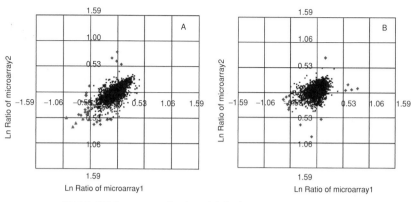

· the spots that show no expression change in both microarrays

▲ the spots that show expression change and have the same direction of change in both microarrays

✦ the spots that show expression change only in one microarray

FIGURE 19.1. The identity comparison between self-comparison experiments. A: Replicated experiments with labeling simultaneously; B: replicated experiments with labeling separately.

with labeling separately can avoid false positive, while performing two replicated experiments with labeling simultaneously will remain at a low false positive rate (0.1-0.2%).

3.2.   Differential Expression Experiments

Analysis of differential expression experiments was using normal liver and hepatoma tissues, which were labeled by Cy3 and Cy5 respectively. We evaluated the system from the following several aspects.

3.2.1.   Impact of Different Concentrations of Target Sequences on Hybridizing Signals on the Multiple-Gene Microarrays

Four genes were chosen to be prepared in 5 concentrations arrayed on a microarray of 14112 genes. The UniGene IDs of the four genes (A-D) were Hs. 181165, Hs. 14376, Hs. 7838, Hs. 148212 respectively. The series of concentrations was 5 ng/µl, 50 ng/µl, 100 ng/µl, 200 ng/µl and 400 ng/µl. Among them, A and B were the genes of high-abundance, which were detected having many copies during sequencing, while C and D were the genes of low-abundance, which were detected having a few copies during sequencing. The results, shown in Figure 19.2 and Figure 19.3, indicated that with the increasing of concentration, the signal intensity enhanced as well, and the intensities of high-abundant genes were obviously stronger than those of low-abundant genes. It also proved that the ratio of these genes were constant in various

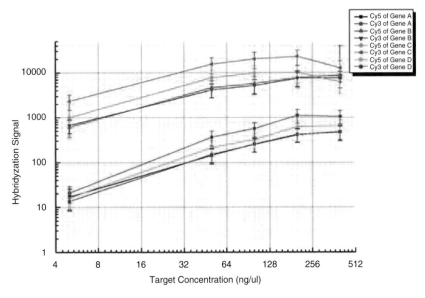

FIGURE 19.2. The value of signal intensity with serial concentrations of target genes (target genes refers to the gene sequences immobilized on the arrays).

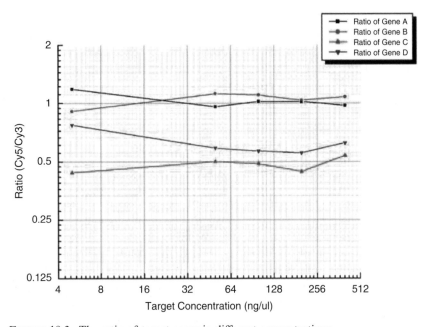

FIGURE 19.3. The ratio of target genes in different concentrations.

concentrations, and the ratio of high-abundant genes were more stable than that of low-abundant genes, which was because deviation increased with intensity decreasing when intensity was lower than 1000.

3.2.2. Impact of Cy5 and Cy3 Reverse Labeling on the Results

In order to investigate the impact of Cy5 and Cy3 labeling on the result, two sets of experiments were performed: in one set, hepatoma RNA was labeled with Cy5, and normal liver RNA was labeled with Cy3; while in the other set, hepatoma RNA was labeled with Cy3, and normal liver RNA was labeled with Cy5. The obtained data were calibrated and then showed in Figure 19.4 and Figure 19.5. Figure 19.4 showed the histogram of corresponding ratio of 50 genes in two sets of experiments. Figure 19.5 showed the scatter plot of corresponding ratio value of all genes in two sets. The relevant coefficient of two sets was −0.909. The histogram and scatter plot demonstrated that labeling with Cy5 or Cy3 has no impact on result. In other words, the ratio of Cy5/Cy3 was not influenced by reverse cross labeling.

3.2.3. Comparison of a Series of Replicated Experiments among Multiple Microarrays

The comparison of microarrays representing different batches was performed with total RNA from normal liver and hepatoma. Meanwhile, the impact of

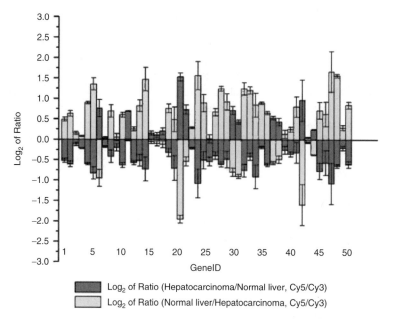

FIGURE 19.4. The histogram of 50 Genes ratio in reverse cross labeling microarrays.

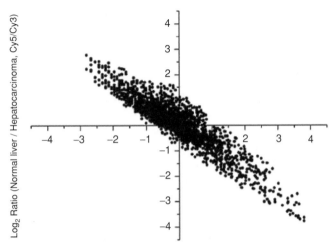

FIGURE 19.5. The scatter plot of ratio in reverse cross labeling microarrays.

fluorescence labeling on the experiments was investigated (Table 19.2). When the cutoff was defined as 0.5 and 2, there were two genes showing contradictory results in the microarray of different batches. Thus, the consistence rate was 99%. There were no more contradictory results in other microarrays and the consistence rate was 100%. But when the cutoff was defined as 0.667 and 1.5, only the same microarray had no genes of contradictory results. Moreover, the microarrays labeling simultaneously had less genes of contradictory results than microarrays labeling separately. The consistence rate commonly ranged from 93.6% to 100%. The consistence rate commonly ranged from 93.6% to 100%. Thus, it was concluded that when the cutoff was defined as 0.5 and 2, microarray had a high reproducibility, and the consistence rate reached 100%.

3.3.   Evaluating the Reproducible Microarray Experiments by Consistence Rate Could Reflect the Reproducibility Better

We put forward a concept of "consistence rate" (CR), defined as the percent of gene number, which showed differential expression in the same direction in both of two experiments, from the total number of all differentially

TABLE 19.2. The comparison with different batches and labeling process

	Replicated gene in the same array	Different batches Labeling simultaneous	The same batch Labeling simultaneous	The same batch Labeling separately
Correlation coefficient (R)	0.942	0.910	0.908	0.878
Cutoff	0.5-2.0	0.5-2.0	0.5-2.0	0.5-2.0
CR	100%	99%	100%	100%
Contradictory genes and proportion	0	0.5% (2 genes)	0	0
Cutoff	0.667-1.5	0.667-1.5	0.667-1.5	0.667-1.5
CR	100%	93.6%	99.92%	99.88%
Contradictory genes and proportion	0	3.2% (12 genes)	0.04% (3 genes)	0.06% (6 genes)
Increasing multiple of co-differential genes	1.40	1.83	2.78	2.33

expressed genes. We considered that consistence rate was better than correlation coefficient (R) and coefficient of variation (CV) to reflect the reproducibility of the results. Table 19.2 showed that compared to correlation coefficient, consistence rate reflected the proportion of contradictory genes to differential genes more exactly. In addition, we found that the number of differential genes had a great impact on the value of correlation coefficient. Three sets of replicated experiments has been analyzed here: self-comparison experiments for replicating twice; low-differential expression replicated experiments for replicating twice (A); high-differential expression replicated experiments for replicating twice (B) (Table 19.3). It was indicated that correlation coefficient was related to the number of differential genes. The lower number of differential genes were, the more the correlation coefficient deviated from 1 and the less it can evaluated the correlation quality of two replicated experiments.

Moreover, the linear working range of scanner was from 800-60,000, but some of the signal intensities of high-abundant or low-abundant genes (especially the weak signals) were out of the linear range, which would make serious impacts on variation coefficient and correlation coefficient. However, such shortcomings could be conquered by consistence rate. In addition, cutoff value could also be determined with evaluating consistence rate. For instance, in terms of the replicated gene in the same array in the Table 2, if cutoff value was determined according to the consistent rate 100%, it is considered that the result was reliable when cutoff was 0.667-1.5. Thus, more differential genes could be obtained.

TABLE 19.3. Comparison of correlation coefficient and consistence rate among different replicated experiments

	Percent of differential genes to the total genes		R	Nd	Nid	CR
	Microarray 1	Microarray 2				
Self-comparison replicated experiments	0.68%	0.70%	0.002	0	0	–
Differential Expression replicated experiments A	2.49%	2.11%	0.882	141	0	100%
Differential Expression replicated experiments B	38.64%	36.35%	0.978	3151	0	100%

4. Discussion

We reported our investigation of the precision, accuracy and reliability of microarray data and the sources of variation here.

In term of cDNA microarrays prepared by arraying cDNA on the slides, the sensibility of the microarrays was related to the concentrations of target genes to some extent. We had already studied the sensibility based on the array of a single gene[12]. The research of multiple-gene hybridization showed that the signal intensity changed with concentrations altering, but the ratio was constant. Thus, the change of ratio could be used to represent the differentiation of gene expression.

In the reverse cross labeling experiments, we found no direct impact of labeling with Cy5 or Cy3 on the results. Two sets of experiments showed a rather good pertinence.

The comparison of the reproducibility of self-calibration experiments indicated that false positive rate of the microarrays was controlled below 3%, similar to the other reports (0.5-3%)[11]. It proved that our system was stable and reliable, and the data were reproducible. Moreover, the genes of differential expression screened from the experiments changed in the same direction in the replicated experiments, so we confirmed that these genes are really differentially expressed, but not a false positive signal.

The variation of expression pattern microarray was divided into biological variation and experiment systematic variation. The biological variation was mainly referred to sample variation, which meant that the samples from different persons perhaps had high differentiation. The expression of some special genes may not be the identical even in the cells from the same tissue. Such difference is difficult to calculate. We focused on discussing experiment systematic variation here. The above data suggested that microarray results were reproducible to some extent, but owing to the existed systematic bias, microarray could only be regarded as a qualitative or half-quantitative approach. We discussed the possible reasons of variation as follows:

1. Linear working range of scanner: Some signal intensities were so weak or so strong that they were out of the linear range, which made the signal values of different arrays fluctuate. Therefore, variation enhanced inevitably.
2. Variation in the process of RNA isolation: The samples of the experiments were from ten different normal liver tissues. The total RNA isolation produced only a little contribution to the variation of the reproducibility (Table 1), but RNA degradation because of improper reservation hasn't been tested.
3. Variation in the process of RNA reverse-transcription: Even to the same gene after normalized, the efficiency of labeling for each time would have a little variation. If we performed labeling separately, we could eliminate the variation by replicating experiments (Figure 19.1).
4. Variation in the process of hybridization: This is the most important source of variation, including the inhomogeneity of solution for hybridization and dilution, different procedure, impurity, background and so on. Thus, the different processes of hybridization brought to different signal intensities.

The sources of variation mentioned in 2., 3. and 4. were always generated randomly. The best method to eliminate the variation was replicating the experiments. Two replicated experiments could avoid most deviations. To biological difference, it was better to replicate more than three times[4]. Comprehensive statistical results could be used to establish a special mathematical model, perform cluster analysis[13] and look for mark genes of diseases and polymorphism.

Another source of variation that could not be ignored was derived from image acquirement and data analysis. This source was related with scanner, analytic software and algorithm[14].

In addition to systematic variation during the experiments, the bias was also owing to the preparation of miraoarray, specially the accuracy of clones. It was reported that accuracy rate of the commercialized clones was about 60%-80%[15]. Thus, every clone should be sequenced to guarantee the reliability of results.

We put forward the "consistence rate" as a new parameter to evaluating the reproducible performance of microarray. Compared to the correlation coefficient and coefficient of variation, consistence rate took advantage in some aspects. Consistence rate, which was put forward based on the microarray as a half-quantitative approach, was used to evaluate the reproducibility by calculating the percent of the genes whose expression changed in the same direction in the replicated experiments from the whole differentially expressed genes. The parameter was not impacted by the number of differentially expressed genes and very high-abundant and low-abundant genes out of the linear range of scanner. The criterion of cutoff was always disputed. Most researchers adopted 2 as cutoff value, and some used 3 or 1.7. We thought that cutoff value could be determined according to the consistent rate of corresponding experiment depended on detailed purpose. For instance, if in some experiments with a little genes of differential expression, the cutoff value could be redefined by reducing the consistence rate, so as to get more useful information from microarray data.

The results presented in this report demonstrated the performance of the cDNA microarray technology platform, and proved that the platform could provide data of high quality to establish a reliable gene expression database. The usefulness of any data acquired from this platform for scientific researchers depended on a strict method how to test the performance of this technology.

References

1. Schena M., Shalon D., Davis R. W., Brown P. O. 1995, Quantitative Monitoring of Gene Expression Patterns with a Complementary DNA Microarray. *Science*, 270(20): 467-470.
2. Schena M., Shalon D., Heller R., Chai A., Brown P. O., Davis R. W. 1996, Parallel human genome analysis: Microarray based expression monitoring of 1000 genes. *PNAS*, 93: 10614-10619.
3. Li Y., Qiu M. Y., Wu C. Q., Cao Y. Q., Tang R., Chen X., Shi X. Y., Hu Z. Q., Xie Y., Mao Y. M. 2000, Detection of Differentially Expressed Genes in Hepatocellular Carcinoma Using DNA Microarray. *Acta Genetica Sinica (Chinese)*, 27(12): 1042-1048.
4. Shalon D., Smith S. J., Brown P. O. 1996, A DNA microarray system for analyzing complex DNA samples using two-color fluorescent probe hybridization. *Genome Res*, 6(7): 639-645.
5. Lee M. L., Kuo F. C., Whitmore G. A., Sklar J. 2000, Importance of replication in microarray gene expression studies: statistical methods and evidence from repetitive cDNA hybridizations. *Proc Natl Acad Sci USA*, 97(18): 9834-9839.
6. LifeArray Chip Validation Study [EB/OL]. http://www.incyte.com
7. Chomczynski P., Sacchi N. 1987, Single-Step Method of RNA Isolation by Acid Guanidinium Thiocyanate-Phenol-Chloroform Extraction. *Anal Biochem*, 162: 156-159.
8. Raha S., Merante F., Proteau G., Reed J. K. 1990, Simultaneous isolation of total cellular RNA and DNA from tissue culture cells using phenol and lithium chloride. *Genet Anal Tech Appl*, 7(7): 173-177.
9. Yang P., Otto C. M., Sheehan F. H. 1997, The effect of normalization in reducing variability in regional wall thickening. *J Am Soc Echocardiogr*, 10(3): 197-204.
10. Eisen M. B., Spellman P. T., Brown P. O., Botstein D. 1998, Cluster Analysis and Display of Genome-Wide Expression Patterns. *Proc Natl Acad Sci*, 95: 14863-14868.
11. Yue H., Eastman P. S., Wang B. B., Minor J., Doctolero M. H., Nuttall R. L., Stack R. Becker J. W., Montgomery J. R., Vainer M., Johnston R. 2001, An

evaluation of the performance of cDNA microarrays for detecting changes in global mRNA expression. *Nucleic Acids Res*, 29: e41.

12. Li Y., Li Y. L., Tang R., Xu H., Qiu M. Y., Chen Q., Chen J. X., Fu Z. R., Ying K., Xie Y., Mao Y. M. 2002, Discovery and analysis of hepatocellular carcinoma genes using cDNA microarrays. *J Cancer Res Clin Oncol*, 128: 369-379.

13. Kuruvilla F. G., Park P. J., Schreiber S. L. 2002, Vector algebra in the analysis of genome-wide expression data. *Genome Biol*, 3: RESEARCH0011

14. Hsiao L. L., Jensen R. V., Yoshida T., Clark K. E., Blumenstock J. E., Gullans S. R. 2002, Correcting for signal saturation errors in the analysis of microarray data. *Biotechniques*, 32(2): 330-332, 334, 336.

15. Kothapalli R., Yoder S. J., Mane S., Loughran T. P. Jr. 2002, Microarray results: how accurate are they? *BMC Bioinformatics*, 3(1): 22

20

High-Throughput Tissue Microarray Technology for the Rapid Clinical Translation of Genomic Discoveries

ZHUOBIN TANG AND YOUYONG LU

Peking University School of Oncology, Beijing Laboratory of Molecular Oncology, Da-Hong-Luo-Chang Street, Western District, Beijing, 100034, P.R.China

Abstract: An unprecedented amount of information about genes has been revealed in genomic surveys by cDNA microarrays and other techniques. The identified potential candidate genes need to be further validated in large-scale studies of well-characterized tissues. Tissue microarrays (TMAs) are a new tool consisting of miniaturized collections of arrayed tissue cores (diameter 0.6 mm) on a microscope glass slide, that allow for high-throughput expression profiling of tissue samples. Different techniques could be employed for identification of specific phenotypic (immunohistochemistry and in situ hybridization) or genotypic (fluorescence in situ hybridization) alterations. This review discusses the validation of TMAs, the technical considerations for construction of TMAs, as well as some applications. The use of paraffin-embedded tissues has some limitations with regard to analysis of RNA or certain proteins. To overcome such limitations, a cryoarray strategy has been developed to allow for the processing of multiple frozen tissue specimens and/or cell lines on a single tissue block. TMAs make it possible to validate potential targets using clinical samples linked to clinicopathological databases. Therefore, TMAs will lead to a significant acceleration of the transition of basic research findings into clinical applications.

Key words: Tissue microarrays; high-throughput; transition; genomic discoveries; clinical applications.

1. Introduction

With several eukaryotic genomes completed and the draft human genome published, we are now entering the postgenomic era. The sequencing of the human genome revealed an unprecedented amount of information about genes, their structure and variation (Lander et al., 2001; Venter et al., 2001). The challenge ahead is to identify, validate, priortize and select the best targets from tens of thousands of candidate genes and proteins. Before any novel gene or protein targets are selected for diagnostic and therapeutic development, they need to be validated in large-scale studies. In order to develop genomics-based medical and biotechnological applications, high-throughput technologies for gene target validation will be essential. Toward this end, a number of new methods have been developed to systematically analyze gene function. One powerful technology that has emerged recently is tissue microarrays. Here we review this new technology that permits high-throughput clinical and functional validation of genes. Specifically, we review how tissue microarrays have provided a high-throughput platform for clinical validation of hundreds or thousands of candidate gene targets at once.

2.   History of Tissue Microarrays

The tissue formalin-fixing, paraffin-embedding, sectioning technique was invented in 1850s. This method only allows one tissue in a single slide. Antibodies can be produced in large numbers in 1975 (Kohler et al, 1975). However, techniques for screening and application of the new antibodies have not kept pace with their production. The identification of the numerous new antibodies can be established by immunohistochemistry method involving numerous normal and tumor tissues. Because these studies are usually done on slides, each of which contains a section from a single specimen, the screening for new antibodies of potential clinical value is costly and time-consuming. To circumvent this problem, Battifora H developed a method by which multiple normal or tumor tissues can be mounted on a single slide, the multi-tissue or "sausage" block technique in 1986 (Battifora, 1986). After that, Battifora H et al. improved the multi-tissue block technique in 1990 (Battifora et al, 1990). This technique permits that the tissues are evenly distributed in a checkerboard arrangement, they can be readily identified by their position in the resulting sections. This device permits rapid and inexpensive screening of new histologic reagents, and facilitates intra- and interlaboratory quality control. As cDNA microarray technology discovers more genes at a faster rate, the need for high-throuhput validation intensifies. Moreover, the number of tissues that can be arranged by the method invented by Battifora H et al. is low. A rapidly growing demand for analyses of thousands of candidate genes in a large sets of well-characterized tissues. Toward this end, one powerful technology that has emerged in1998 is tissue microarrays (Kononen et al., 1998). The use of paraffin-embedded tissues has some limitations with regard to analysis of RNA or certain proteins. To overcome such limitations, a cryoarray strategy have been developed to allow for the processing of multiple frozen tissue specimens and/or cell lines on a single tissue block (Schoenberg et al., 2001). The history of TMAs are summarized in Figure 20.1. This technology permits high-throughput *in situ* analysis of specific molecular targets in hundreds or thousands of tissue specimens at once (Kononen et al., 1998). TMAs are miniaturized collections of arrayed tissue spots on a microscope glass slide that

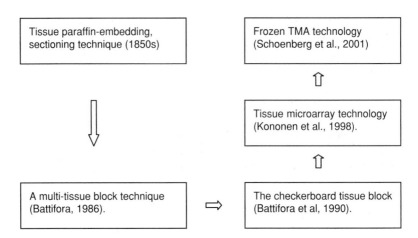

FIGURE 20.1. The history of TMAs.

provide a template for highly parallel localization of molecular targets, either at the DNA, RNA or protein level. TMAs provide a link between molecular targets and tissue or cell morphology, as well as with the clinical data associated with the specimens. TMAs are therefore ideal for the rapid large-scale clinical studies of candidate molecular targets (Kallioniemi et al., 2001). TMAs facilitate the analysis of specific molecular targets in hundreds or thousands of tissue specimens in a massively parallel fasion (Figure 20.2).

3.   Construction and Some Issues of TMAs

3.1.   Construction of TMAs

TMAs are constructed by acquiring cylindrical core from morphologically representative areas of individual tissue samples, followed by insertion of the cores into a new recipient TMA block at defined array locations (Figure 20.3). Selection of tissue specimens and the exact histological area to be sampled are the first steps in the TMA construction. Using a precision instrument, cores from up to 1000 tissue samples are arrayed into a recipient TMA block. Subsequently, up to 300 consecutive sections can be cut from each TMA block, providing slides with the identical configuration of the tissue spots (rows and columns) in each section. TMA slides can be employed for analyses of DNA, RNA and protein targets, using various techniques such as fluorescence *in situ* hybridization (FISH), mRNA *in situ* hybridization, or immunohistochemistry (IHC).

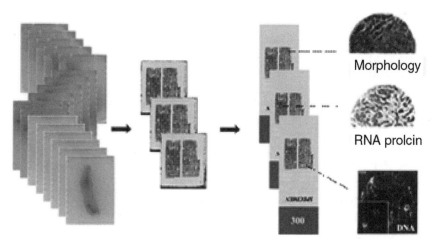

FIGURE 20.2. Principle of TMA analysis. Cyclindrical cores are obtained from a lot of formalin-fixd, paraffin-embedded tissue blocks. These are transferred to a TMA block. Multiple TMA blocks can be generated at the same time. Each TMA block can be sectioned up to 200 times. All the resulting TMA slides have the same tissues in the same coordinate positions. The individual slide can be used for a variety of molecular analysis, such as H&E staining to ascertain tissue morphology, mRNA ISH, DNA ISH or protein immunostaining.

3.2. Advantages of Tissue Microarrays

Tissue microarray (TMA) presents a novel method for high-throughput molecular analysis of thousands of tumors at a time, either at DNA, RNA or protein levels (Kononen et al., 1998; Kallioniemi et al., 2001; Schoenberg et al., 2001). Compared with traditional large section studies, TAMs have many advantages which can be summarized as follows:

The most significant advantage is the fact that a very large number of molecular markers can be analyzed from consecutive TMA sections containing thousands of tissue samples. The tissue microarray containing thousands of tumor samples makes it possible to study a substantially higher number of tumor specimens by FISH and immunohistochemistry than previous methods. A standard histologic section is about 3-5 mm thick. After use for primary diagnosis, the archived block can be cut approximately 200 times for 200 assays depending on the care and skill of the histotechnologist. If this same block is processed for optimal microarray construction, it could routinely be sampled 200-300 times or more, depending on the size of the tumor in the original block. Thus, instead of 200 conventional sections or samples for analysis from one tissue specimen, the microarray technique could produce material for 500,000 assays (assuming 250 core samples per section times two thousand 2.5-μm sections per 5-mm-array block) represented as 0.6-mm disks

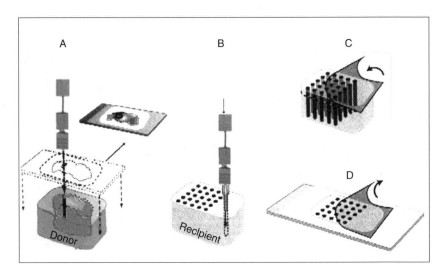

FIGURE 20.3. Tissue microarray construction. (A) A tissue core of 0.6 mm in diameter is punched from a preselected region of a donor block using a thin-wall stainless steel tube. A haematoxylin and eosin (H&E) stained section overlaid on the surface of the donor block guides sampling from representative sites in the tissue. (B) The tissue core is transferred into a premade hole at defined array coordinates in the recipient block. (C, D) An adhesive-coated tape sectioning system assists in cutting the tissue microarray block. This figure was generated and modified according to Kononen J, et al. Nat Med 1998, 4:844-847.

of tissue. Thus this technique essentially amplifies up to 2,500-fold the limited tissue resource. Conventional molecular pathology techniques permit analysis of only approximately 200 molecular targets, which corresponds to a mere 0.5% of all of the estimated 40,000 genes in the human genome. However, TMAs enable whole genome-scale research in tissue blocks.

The second most significant advantage is tissue amplification, especially for scarce materials. This advantage allows the generation of multiple replicate tissue microarray blocks, each having samples from the same tumor specimens at identical coordinates. Depending on the thickness of the original tissue, between 100 and 400 sections can be cut from each array block. Therefore, thousands of replicate tissue microarray slides can be generated, each being stained with different probes or antibodies.

The third advantage is the fact that each tissue sample on a TMA slide containing the same control tissues is treated in an identical manner. This uniformity helps in assuring the specificity and sensitivity of IHC, improving reproducibility of the staining reaction, as well as the speed and reliability of the interpretation. Conventional techniques for gene validation are slow, often highly variable, subjective, and not automated. Using conventional methods, analysis of 1000 cases would require staining and analysis of 1000 individual slides, whereas up to 1000 specimens on a TMA slide can be simultaneously evaluated using the same laboratory procedure, such as anti-body staining or ISH. Thus, a TMA section containing up to 1000 different types of tissues can be used for testing and optimization of probes, anti-bodies and detection systems.

The fourth advantage is its rapid speed. Recently, an automated tissue arrayer has been constructed (Kallioniemi et al., 2001). Allowing the speed of molecular analyses to increase by more than 100-fold. Compared with the TMAs technique, conventional techniques for molecular analysis are labor-intensive and time-consuming. The immunohistochemical interpretation of the microarray containing 2317 bladder cancer samples can be done in 4 hours and the FISH scoring performed in 6 days (Richter et al., 2000). The use of TMA has the potential for allowing validation of new genes at a speed comparable to the rapid rate of gene discovery afforded by cDNA microarrays.

The fifth advantage is that only a very small amount of reagent is required to analyze an entire array in a single experiment. This advantage raises the possibility of using tissue microarrays in screening procedures, a protocol that is impossible using conventional sections. The standardization of stain-ing procedures and reduction of intra-assay variability can also be signifi-cantly improved with this technique.

Finally, the ability to study archival tissue specimens is an important advantage, as paraffin-embedded specimens are usually not suitable for other high-throughput technologies, such as cDNA microarray analysis, serial analysis of gene expression (SAGE) or proteomic screens.

3.3. Limitations of Tissue Microarrays

The major potential limitation of this technique is tissue volume. How can a very small core (0.6mm) acquired from each tumor be representative of the whole tumor specimen, because of tumor heterogeneity? Or alternatively, how many cores would be needed to capture most of the information in an entire section?

To address the influence of tumor heterogeneity and to evaluate the ability of TMAs to yield information on the prognostic value of biomarkers, three studies have directly compared biomarker expression using TMAs and regu-lar sections of the same breast cancers. All studies report >90-95% concor-dance for common breast cancer biomarkers such as the estrogen receptor (ER), the progesterone receptor (PR), and the HER-2 oncoprotein. These studies show that analysis of TMAs produced results representative of the entire tumor, which were appropriate for the analysis of prognostic signifi-cance of such markers (Bucher et al., 1999; Camp et al., 2000; Gillett et al., 2000). Other studies suggest similar results (Sallinen et al., 2000; Schraml et al., 1999; Mucci et al., 2000; Gancberg et al., 2002; Hoos et al., 2001). In addition, the impact of data discrepancies between array and full-sections with regard to patient outcome was also evaluated. Hoos et al. reported that this comparison showed no significant difference in clinicopathological correlations between the two methods, indicating that tissue microarrays may be reliable tools for high-throughput clinicopathological analyses of tumor specimens (Hoos et al., 2001).

The requirement for the use of formalin-fixed paraffin-embedded tissues as a starting material has some limitations with regard to analysis of RNA or certain proteins. One difficulty with paraffin-embedded tissue relates to antigenic changes in proteins and mRNA degradation induced by the fixation and embedding process. Many techniques, especially those using RNA, call for fixations that do not use formalin. To overcome such limitations, this technology has been modified by using frozen tissues embedded in OCT compound as donor samples and arraying the specimens into a recipient OCT block (Schoenberg et al., 2001). This method allows optimal evaluation by each technique and uniform fixation across the array panel. These studies show OCT arrays work well for DNA, RNA, and protein analyses, and may have significant advantages over the original technology for the assessment of some genes and proteins by improving both qualitative and quantitative results (Schoenberg et al., 2001). The use of frozen TMAs has the disadvantage of requiring prospective collection, since most archives contain formalin-fixed paraffin-embedded material. Tissue handling for this technique needs to be optimized for the best possible array quality. These approaches offer the opportunity to conduct pilot and validation studies of potential targets using clinical samples linked to clinicopathological databases. The tissue microarray technology should be viewed as a method for the analysis of molecular alterations at the population level, not as a means of extensively analyzing any single cancer specimen.

3.4.   Technical Considerations for Tissue Microarrays

Some technical issues should be considered before construction of tissue and cell line microarrays. Firstly, sampling a specific area representative for the specimen is critically important for the quality of the array. A fresh H&E-stained slide should be obtained from each donor block and used as a guide to assess morphology and to select an area that represents the specimen. This can vary greatly among tissue types. For example, in intestinal type gastric cancer, it is important that the sampling areas are small and well defined, because stromal areas between the glandular structures of the tumor can be large. A random core sample can easily miss the tumor-cell-rich regions. In other malignancies like esophageal squamocellular carcinoma, tumors consist of densely packed cancer cells that are unlikely to be missed by precise sampling.

Since so many cases can be analyzed on a single slide, the design of the TMA pattern is also important (Figure 20.4). Generally 0.6mm cores can be easily spaced at 0.8mm on center, leaving 0.2 mm between disks. A second consideration is grouping of the cores (Figure 20.5). It is not a good idea to

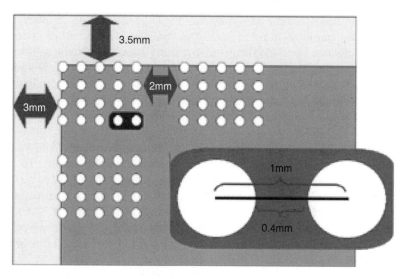

FIGURE 20.4. Diagram of TMA design. Tissue samples are distributed in multiple divisions. Generally 0.6 mm spots is spaced at 1mm on center, leaving 0.4 mm between disks. We recommend leaving 3-3.5 mm margins at the edges of the array block to avoid cracking of the paraffin.

closely space the cores, since occasionally, a small region of the array will be damaged or lost in the cutting or staining process. Analysis of the arrays by the pathologist will be facilitated by leaving a larger space (0.4-0.6 mm) between every 5 or 10 rows. The orientation of the specimens on the array is crucial because confusion about their localization can threaten the evaluation of the experiment. For keeping the orientation of rows simple, we use

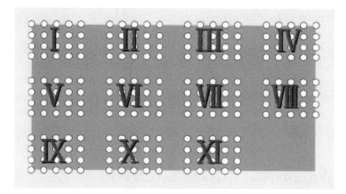

FIGURE 20.5. Diagram of TMA organization. Tissue samples are distributed in 11 divisions. Each division has 20 tissue samples.

different normal tissues or cell lines to allow the identification of every row based on morphology. In addition to this, we place five orientation cores in specific positions outside the geometric margins of the usually square or rectangular arrays to orient the entire microarray section after it has been cut. Finally, controls should be amply placed on each array. Generally normal tissues or cell lines which can be easily obtained are used as controls. The controls will be critical for standardization of automated tissue array readers.

When conventional histologic microtomy technique is used for microarrays, the small disks of tissue are sometimes lost from the section. To circumvent this problem, a tape-based tissue transfer system (Instrumedics, Hackensack, NJ) is recommended (Rimm et al., 2001). This system requires placing an adhesive tape on the face of the section prior to cutting. The tape is removed with the section on it, and placed on special slides with adhesive-coated surfaces. The section is then UV crosslinked to the slide prior to removing the tape with a special "degreasing" reagent.

Another factor that influences tissue loss from the section is the quality of the tissue. If donor blocks were fixed for too long and tissues are dry, they are at higher risk of breaking into fragments than tissue that is in good condition. We recommend that two or three cores per tumor specimen be arrayed to avoid the loss of the specimen.

Another problem of microarray sections is antigenic loss due, presumably, to tissue oxidation. Some investigators have reported loss of antigenicity if sections are stored for as little as a week prior to immunostaining (Jacobs et al., 1996; Bertheau et al., 1998). For tissue microarrays, David L. Rimm and his colleages have found that this loss can be prevented by sectioning without water (using the tape transfer system), removing the degreasing agent after tape release by a short incubation in xylene, and finally recoating the slides in paraffin prior to storage (Rimm et al., 2001).

In addition, staining artifacts at the tissue borders, a well-known phenomenon in immunohistochemistry, occur most frequently in the periphery of the tissue microarray. To minimize this effect, we prefer to frame cores of cancer specimens with one row of normal tissues.

To get as many high-quality sections from one multi-tissue block as possible, it is advisable to array uniformly long tissue cores for each specimen. Unfortunately, this is not always possible because some donor tissues are thin and, therefore, do not allow the punching of long cores. To overcome this problem, more than one thin core from the same tumor area can be punched and stacked on top of another in the same location on the tissue array.

4.   Applications of Tissue Microarrays

The TMA technology permits efficient utilization of the enormous tissue resources available in pathology laboratory archives. A single TMA experiment can assay a molecular target in up to 1000 specimens at once. The molecular data can be correlated with various tumor and patient data, such as clinico-pathological information, survival, and treatment response. TMAs therefore make it possible to perform large-scale clinical studies on a single microscope slide.

Almost all current TMA studies have been published on cancer. Thirty-one TMA studies of cancer have been published (Table 20.1). The size of the materials used in these studies has ranged from 29 to 4700 tumors. These studies are reviewed below.

TABLE 20.1 Tissue microarray studies in cancer research

Tissue type	n	Methods	Molecular targets	Endpoints	Year	Authors
Breast cancer	645	FISH, IHC, MRNA ISH	MYC, HER2, CCND1, 17q23, 20q13, MYBL2, ER, p53	Frequency, methodology comparison, molecular profiles	1998	Kononen et al.
Breast cancer	557	IHC	ER, PR, HER2	Frequency, comparison with whole sections	1999	Bucher et al.
Prostate cancer	371	FISH	NMYC, MYC, HER2, CCND1	Frequency, tumor progression	1999	Bubendorf et al.
Prostate cancer	264	IHC	IGFBP2, HSP-27	Tumor progression, cDNA microarray validation	1999	Bubendorf et al.
Renal cancer	532	IHC	Vimentin	Frequency, cDNA microarray validation	1999	Moch et al.
Multiple(17)	397	FISH	CCND1, CMYC, HER2	Frequency across different tumor types	1999	Schraml et al.
Multiple(135)	4700	FISH	MB-17A	Frequency across different tumor types	2000	Andersen et al.
Breast cancer	612	FISH, IHC	HER2	Frequency, prognosis, automated IHC scoring	2000	Bucher et al.
Breast cancer	668	FISH	S6K, HER2	Frequency, prognosis, cDNA microarray validation	2000	Bärlund et al.

(Continued)

TABLE 20.1. Tissue microarray studies in cancer research—Cont'd

Tissue type	n	Methods	Molecular targets	Endpoints	Year	Authors
Breast cancer	372	FISH	RAD51C, S6K, PAT1, TBX2	Frequency, CGH microarray validation	2000	Barlund et al.
Breast cancer	328	IHC	ER, PR	Methodology, comparison with whole sections	2000	Gillett et al.
Breast cancer	380	IHC	ER, PR, HER2	Methodology comparison	2000	Camp et al.
Prostate cancer	458	IHC	CGA, SYN	Frequency, methodology comparison	2000	Mucci et al.
Prostate cancer	892	IHC	Ki-67	Frequency, ethnic comparison	2000	Perrone et al.
Prostate cancer	632	IHC	NKX3.1	Frequency, tumor progression	2000	Bowen et al.
Bladder cancer	2317	FISH, IHC	Cyclin E	Frequency, tumor progression, prognosis, CGH microarray validation	2000	Richter et al.
Gliomas	418	IHC	IGFBP2, Vimentin, p53	Frequency, tumor progression, methodology cDNA microarray validation	2000	Sallinen et al.
Gliomas	88	IHC	Topo II alpha	Frequency, prognosis	2000	Miettinen et al.
Bladder cancer	2317	FISH	RAF1, FGFR1	Frequency, tumor progression, prognosis	2000	Simon et al.
Breast cancer	113	IHC	CCD1, MEK-1	Frequency, cDNA microarray validation	2001	Hedenfalk et al.
Prostate cancer	738	IHC	Hepsin, pim-1	Frequency, cDNA microarray validation, prognosis	2001	Dhanasek aran et al.

TABLE 20.1. Tissue microarray studies in cancer research—Cont'd

Tissue type	n	Methods	Molecular targets	Endpoints	Year	Authors
Lymphomas/ leukemias	207	IHC	SHP-1	Frequency, cDNA microarray validation	2001	Oka et al.
Breast cancer	750	FISH	FGFR2	Frequency, CGH and cDNA microarray validation	2001	Heiskanen et al.
Prostate cancer	1220	IHC	E-cadherin	Frequency, tumor progression	2001	Rubin et al.
Colorectal cancer	650	IHC	Beta-catenin	Frequency, prognosis	2001	Chung et al.
Breast cancer	553	IHC	ER, PR, p53	Frequency, prognosis, comparison with whole sections	2001	Torhorst et al.
Fibrobastic tumor	59	IHC	Ki-67, p53, pRB	Comparison with whole sections	2001	Hoos et al.
Breast cancer	456	IHC	Smad2, Smad2p, Smad4	Frequency, prognosis	2002	Xie et al
Breast cancer	29	FISH	HER-2	Methodology	2002	Gancberg et al.
Prostate cancer	88	IHC	Ki-67	Methodology, prognosis	2002	Rubin et al.
Lung cancer	193	IHC	Cadherin, Catenin, p120, p27. APC	Frequency, tumor progression, prognosis	2002	Bremnes et al.

APC, adenomatous polyposis coli; AR, androgen receptor; CCND1(CCD1), cyclinD1; CGA, chromograin A; ER, estrogen receptor; FGFR1, fibroblast growth factor receptor 1; FISH, fluorescence *in situ* hybridization; HSP-27, heat shock protein 27; IGFBP2, insulin growth factor binding protein 2; IHC, immunohistochemistry; PR, progesterone receptor; S6K, ribosomal s6 kinase; SHP1, hematopoietic cell specific protein-tyrosine-phosphatase SH-PTP1 (SHP1); SYN, synaptophysin.

TMAs have been extensively used to study gene targets that have been found in genomic surveys by cDNA microarrays and other techniques. For example, Bärlund et al. found overexpression of the ribosomal protein S6 kinase gene in a breast cancer cell line by cDNA microarrays and then applied TMAs to show that this gene is amplified and highly expressed at the protein level in 10–15% of primary breast tumors. This study also indicated that the S6 kinase gene may be a significant prognostic indicator in breast cancer. Similar studies in bladder (Richter et al., 2000), brain (Sallinen et al., 2000), prostate (Bubendorf et al., 1999), renal (Moch et al., 1999) and breast cancer (Barlund et al., 2000; Hedenfalk et al., 2001; Heiskanen et al., 2001), lymphomas/leukemias (Oka et al., 2001) were reported. These studies illustrate how TMA analysis facilitates studies of the clinical significance of new genes discovered in genomic screenings of model systems. Dhanasekaran et al. studied changes in gene expression in different stages of prostate cancer by cDNA microarray and identified several genes with significant expression changes between different groups of tumors (Dhanasekaran et al., 2001). They then assessed the protein expression levels of two of these genes, hepsin and pim-1, using TMAs and showed that their expression was correlated with measures of clinical outcome.

TMAs have so far mainly been used in cancer research. Typical TMAs that have been constructed include multi-tumor, progression, and prognostic arrays.

Multi-tumor TMAs are composed of samples from multiple tumor types. These arrays are utilized to screen different tumor types for molecular alterations of interest. The first example of a multi-tumor TMA contained 397 tumors sampled from 17 different tumor types (Schraml et al., 1999). A larger multi-tumor array containing 4700 tumors representing 135 different tumor types has recently been constructed (Andersen et al., 2000).

Progression TMAs have been used to study molecular alterations in different stages of one type of tumor. For example, amplification of the androgen receptor (AR) gene (Bubendorf et al., 1999) and overexpression of the insulin growth factor binding protein 2 (IGFBP2) protein (Bubendorf et al., 1999) were found to be very common in hormone-refractory end-stage prostate cancers, but infrequent in untreated primary tumors. Bubendorf et al. (Bubendorf et al.,1999), Bowen et al. (Bowen et al., 2000) and Rubin et al.(Rubin et al., 2001) constructed a prostate cancer progression TMA that included all stages of prostate cancer development, starting from normal prostate, benign prostate hyperplasia, prostatic intraepithelial neoplasia, localized clinical cancer, to metastatic and homone-refractory end-stage cancer.

Prognosis TMAs contain samples from tumors with clinical follow-up data and clinical endpoints. With the help of such arrays, novel prognostic parameters can be identified and the value of molecular alterations for prediction of chemotherapy response can be tested. Associations with prognosis were found for cyclin E expression (Richter et al., 2000) and RAF1, fibroblast growth factor receptor 1(FGFR1) (Simon et al., 2001) in urinary bladder cancer, S6 kinase in breast cancer (Bärlund et al., 2000), vimenin expression in kidney cancer (Moch et al., 1999), beta-catenin in colorectal cancer (Chung et al., 2001), and Ki-67 in prostate cancer (Rubin et al., 2002), Topo II alpha in Gliomas (Miettinen et al., 2000), HER2 (Bucher et al., 2000), estrogen receptor (ER), progesterone receptor (PR), p53 (Torhorst et al., 2001) and Smad (Xie et al., 2002) in breast cancer, Cadherin, Catenin, p120, p27, and adenomatous polyposis coli (APC) in lung cancer (Bremnes et al., 2002).

TMAs have also been used to investigate biologic differences between prostate cancer in different racial groups. Perrone et al. reported that racial differences in the tumor cell proliferation (Ki-67-labeling index) are not responsible for the observed more aggressive behavior of prostate cancer in African-American men (Perrone et al., 2000).

We have studied 15 gene alterations in 408 formalin-fixed and paraffin-embedded human tissue blocks, including intestinal metaplasia (IM, n = 72), primary tumors (n = 232), and normal gastric mucosa (n = 104) from patients with gastric cancer. The results show that the number of multiple gene alterations concentrates on 4 to 7 genes. However, the number of gene changes was not significantly different between well differentiated GC and poorly differentiated GC, between lymph node-negative and lymph node-positive, between >5 years survival time and <5 years survival time, and among stage I, II, III, IV ($p>0.05$). Correlation between multiple gene alterations and clinicopathological characteristics in human gastric cancer is shown in Table 20.2.

We can study multiple gene alterations in the same tissue by using TMAs and correlate one gene with other different genes (Figure 20.6). For example, our results demonstrate that metallothionein-II a (MT-II a) expression was significantly associated with inhibitory κB-a (IκB-a) expression, gastrin expression was also significantly associated with c-met expression.

Although most of the applications of the TMA technique described in this review come from cancer research, it is likely that the technology will be equally powerful in other fields of research, for example in inflammatory, cardiovascular or neurological diseases (Ayers, 2000).

TABLE 20.2. Correlation between multiple gene alterations and clinicopathological characteristics in human gastric cancer

Clinical features	Total cases	Number of gene changes										
		0	1	2	3	4	5	6	7	8	9	10	15
Differentiation												
Well	45	0	0	5	3	8	8	8	8	3	1	1	0
Poor	55	0	1	3	5	17	8	9	6	6	0	0	0
Lymph node metastasis												
Negative	29	0	0	4	3	4	6	5	5	2	0	0	0
Positive	71	0	1	4	5	20	11	12	9	7	1	1	0
Clinical stage												
I	7	0	0	0	1	1	4	1	0	0	0	0	0
II	9	0	0	1	2	1	1	0	1	2	0	1	0
III	49	0	0	5	3	12	6	9	7	6	1	0	0
IV	35	0	1	2	2	10	6	7	6	1	0	0	0
Survival time												
>5years	4	0	0	0	1	0	3	0	0	0	0	0	0
<5years	79	0	1	6	4	23	12	14	10	8	0	1	0

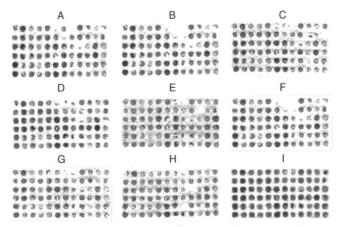

FIGURE 20.6. An example of multiple gene alterations in the same tissue. A. C-met;
B. Gastrin; C. C-erbB2; D. β-catenin; E. MG7; F. p53; G. MT-II a; H. IκB-a; I. p16.

5.   Future Applications of TMAs

Numberous possible applications of TMAs have not been fully explored. It
is likely that this technology will be used in the future for transforming entire
pathology laboratory archives into a "tissue chip" format. TMAs can be used
for cell lines and other experimental tissues such as xenograft tumors or tissues from animal models. TMAs are also an ideal approach to organize large
tissue repositories, such as entire archives from pathology institutes, enabling
tissues having a specific molecular pattern to be found easily.

These published studies illustrate the power of the TMA technology for
rapid translation of cDNA microarray results into the clinical setting
(Richter et al., 2000; Sallinen et al., 2000; Bärlund et al., 2000; Bubendorf
et al., 1999; Moch et al., 1999; Hedenfalk et al., 2001; Heiskanen et al., 2001;
Oka et al., 2001; Dhanasekaran et al., 2001). Each of the studies describes
analysis of hundreds or thousands of tumor samples within the short period
of a few weeks, a task that would have taken years to accomplish using traditional techniques. Virtually all tissues and cell lines are suitable to be placed
into a TMA. Therefore, the range of TMA application is as broad as the
imagination of the users of this technology. Tissue microarrays are useful for
establishing large disease-specific tissue collections for future analysis of new
targets in a particular tumor entity and can be helpful for collaborations
between major institutions. It can be predicted that this powerful research
strategy will be widely applied in the future, as more and more investigators
seek the validation, prioritization and extension of their early cDNA
microarray leads. Continuing development of improved tissue-arraying
instrumentation, automated digital image acquisition, storage, analysis and
standardization will facilitate further expansion of the technology.

6.   Conclusions

High-throughput genomic and proteomic screening technologies have led to a massive increase in the rate of data generation, greatly exceeding the rate at which biological significance and clinical relevance can be determined. The consequence of these new technologies is that the validation of targets has become the rate-limiting step in translating genomic and proteomic information to clinical and therapeutic applications. We have presented here a tissue microarray-based strategy that has the potential of achieving a significant increase in the throughput of clinical and functional validation of molecular targets.

Acknowlegements. We thank Juha Kononen, Guido Sauter, Holger Moch, Lukas Bubendorf, Galen Hostetter, Ghadi Salem, John Kakareka and Tom Pohida for their contribution to the tissue microarray development.

References

1. Andersen CL, Monni OM, Kononen J, barlund M, Bucher C, Hass P, Nocicito A, Bissig H, Sauter G, kallioniemi OP. 2000. High-throughput gene copy number analysis in 4700 tumors: FISH analysis on tissue microarrays identifies multiple tumor types with amplification of the MB-174 gene, a novel amplified gene originally found in breast cancer. Am J Hum Genet. 67: 448.
2. Ayers LW. 2000. Application of the tissue microarray (TMA) method by the midregion AIDS and Cancer Specimen Bank (ACSB) to prepare study sets from HIV infected and control tissues. J Acquir Immune Defic Syndr 23:A18.
3. Barlund M, Forozan F, Kononen J, Bubendorf L, Chen Y, Bittner ML, Torhorst J, Haas P, Bucher C, Sauter G, Kallioniemi OP, Kallioniemi A. 2000. Detecting activation of ribosomal protein S6 kinase by complementary DNA and tissue microarray analysis. J Natl Cancer Inst 92:1252-1259.
4. Barlund M, Monni O, Kononen J, Cornelison R, Torhorst J, Sauter G, Kallioniemi OLLI-P, Kallioniemi A. 2000. Multiple genes at 17q23 undergo amplification and overexpression in breast cancer. Cancer Res 60:5340-5344.
5. Battifora H. 1986. The multitumor (sausage) tissue block: novel method for immunohistochemical antibody testing. Lab Invest 55:244-248.
6. Battifora H, Mehta P. 1990. The checkerboard tissue block. An improved multi-tissue control block. Lab Invest 63:722-724.
7. Bertheau P, Cazals-Hatem D, Meignin V, de Roquancourt A, Verola O, Lesourd A, Sene C, Brocheriou C, Janin A. 1998. Variability of immunohistochemical reactivity on stored paraffin slides. J Clin Pathol. 51:370-374.
8. Bowen C, Bubendorf L, Voeller HJ, Slack R, Willi N, Sauter G, Gasser TC, Koivisto P, Lack EE, Kononen J, Kallioniemi OP, Gelmann EP. 2000. Loss of NKX3.1 expression in human prostate cancers correlates with tumor progression. Cancer Res 60:6111-6115.
9. Bremnes RM, Veve R, Gabrielson E, Hirsch FR, Baron A, Bemis L, Gemmill RM, Drabkin HA, Franklin WA. 2002. High-throughput tissue microarray analysis used to evaluate biology and prognostic significance of the e-cadherin pathway in non-small-cell lung cancer. J Clin Oncol. 20:2417-2428.

10. Bubendorf L, Kolmer M, Kononen J, Koivisto P, Mousses S, Chen Y, Mahlamaki E, Schraml P, Moch H, Willi N, Elkahloun AG, Pretlow TG, Gasser TC, Mihatsch MJ, Sauter G, Kallioniemi OP. 1999. Hormone therapy failure in human prostate cancer: analysis by complementary DNA and tissue microarrays. J Natl Cancer Inst. 91:1758-1764.

11. Bubendorf L, Kononen J, Koivisto P, Schraml P, Moch H, Gasser TC, Willi N, Mihatsch MJ, Sauter G, Kallioniemi OP. 1999. Survey of gene amplifications during prostate cancer progression by high-throughout fluorescence in situ hybridization on tissue microarrays. Cancer Res 59:803-806.

12. Bucher, C., Torhorst, J., Bubendorf, L., Schraml, P., Kononen, J., Moch, H., Mihatsch, M., Kallioniemi, O.P. and Sauter, G. 1999. Tissue microarrays ('tissue chips') for high-throughput cancer genetics: Linking molecular changes to clinical endpoints. Am J Hum Genet 65: 43.

13. Bucher C, Tohorst J, Kononen J, Hass P, Askaa J, Godtfredsen SE, Bauer KD, Seelig S, Kallioniemi OP and Sauter G. 2000. Automated, high-throughput tissue microarray analysis for assessing the significance of Her-2 involvement in breast cancer. J Clin Oncol. Annual Meetng, 2338.

14. Camp RL, Charette LA, Rimm DL. 2000. Validation of tissue microarray technology in breast carcinoma. Lab Invest 80:1943-1949.

15. Chung GG, Provost E, Kielhorn EP, Charette LA, Smith BL, Rimm DL. 2001. Tissue microarray analysis of beta-catenin in colorectal cancer shows nuclear phospho-beta-catenin is associated with a better prognosis. Clin Cancer Res 7:4013-4020.

16. Dhanasekaran SM, Barrette TR, Ghosh D, Shah R, Varambally S, Kurachi K, Pienta KJ, Rubin MA, Chinnaiyan AM. 2001. Delineation of prognostic biomarkers in prostate cancer. Nature 412:822-826.

17. Gancberg D, Di Leo A, Rouas G, Jarvinen T, Verhest A, Isola J, Piccart MJ, Larsimont D. 2002. Reliability of the tissue microarray based FISH for evaluation of the HER-2 oncogene in breast carcinoma. J Clin Pathol 55:315-317.

18. Gillett CE, Springall RJ, Barnes DM, and Hanby AM. 2000. Multiple tissue core arrays in histopathology research: a validation study. J Pathol 192:549-553.

19. Hedenfalk I, Duggan D, Chen Y, Radmacher M, Bittner M, Simon R, Meltzer P, Gusterson B, Esteller M, Kallioniemi OP, Wilfond B, Borg A, Trent J. 2001. Gene-expression profiles in hereditary breast cancer. New Engl J Med 344: 539-548.

20. Heiskanen M, Kononen J, Barlund M, Torhorst J, Sauter G, Kallioniemi A, Kallioniemi O. 2001. CGH, cDNA and tissue microarray analyses implicate FGFR2 amplification in a small subset of breast tumors. Anal Cell Pathol 22: 229-234.

21. Hoos A, Urist MJ, Stojadinovic A, Mastorides S, Dudas ME, Leung DH, Kuo D, Brennan MF, Lewis JJ, Cordon-Cardo C. 2001. Validation of tissue microarrays for immunohistochemical profiling of cancer specimens using the example of human fibroblastic tumors. Am J Pathol 158:1245-1251.

22. Jacobs, TW, Prioleau JE, Stillman IE, and Schnitt SJ. 1996. Loss of tumor maker-immunostaining intensity on stored paraffin slides of breast cancer. J Natl Caancer Inst 88:1054-1059.

23. Kallioniemi OP, Wagner U, Kononen J, Sauter G. 2001. Tissue microarray technology for high-throughput molecular profiling of cancer. Hum Mol Genet 10:657-662

24. Kohler G, Milstein C. 1975. Continuous cultures of fused cells secreting antibody of predefined specificity. Nature (Lond) 256:495-497.

25. Kononen J, Bubendorf L, Kallioniemi A, Barlund M, Schraml P, Leighton S, Torhorst J, Mihatsch MJ, Sauter G, Kallioniemi OP. 1998. Tissue microarrays for high-throughput molecular profiling of tumor specimens. Nat Med 4:844-847.

26. Lander ES, Linton LM, Birren B, Nusbaum C, Zody MC, Baldwin J, Devon K, Dewar K, Doyle M, FitzHugh W, Funke R, Gage D, Harris K, Heaford A, Howland J, Kann L, Lehoczky J, LeVine R, McEwan P, McKernan K, Meldrim J, Mesirov JP, Miranda C, Morris W, Naylor J, Raymond C, Rosetti M, Santos R, Sheridan A, Sougnez C, Stange-Thomann N, Stojanovic N, Subramanian A, Wyman D, Rogers J, Sulston J, Ainscough R, Beck S, Bentley D, Burton J, Clee C, Carter N, Coulson A, Deadman R, Deloukas P, Dunham A, Dunham I, Durbin R, French L, Grafham D, Gregory S, Hubbard T, Humphray S, Hunt A, Jones M, Lloyd C, McMurray A, Matthews L, Mercer S, Milne S, Mullikin JC, Mungall A, Plumb R, Ross M, Shownkeen R, Sims S, Waterston RH, Wilson RK, Hillier LW, McPherson JD, Marra MA, Mardis ER, Fulton LA, Chinwalla AT, Pepin KH, Gish WR, Chissoe SL, Wendl MC, Delehaunty KD, Miner TL, Delehaunty A, Kramer JB, Cook LL, Fulton RS, Johnson DL, Minx PJ, Clifton SW, Hawkins T, Branscomb E, Predki P, Richardson P, Wenning S, Slezak T, Doggett N, Cheng JF, Olsen A, Lucas S, Elkin C, Uberbacher E, Frazier M, Gibbs RA, Muzny DM, Scherer SE, Bouck JB, Sodergren EJ, Worley KC, Rives CM, Gorrell JH, Metzker ML, Naylor SL, Kucherlapati RS, Nelson DL, Weinstock GM, Sakaki Y, Fujiyama A, Hattori M, Yada T, Toyoda A, Itoh T, Kawagoe C, Watanabe H, Totoki Y, Taylor T, Weissenbach J, Heilig R, Saurin W, Artiguenave F, Brottier P, Bruls T, Pelletier E, Robert C, Wincker P, Smith DR, Doucette-Stamm L, Rubenfield M, Weinstock K, Lee HM, Dubois J, Rosenthal A, Platzer M, Nyakatura G, Taudien S, Rump A, Yang H, Yu J, Wang J, Huang G, Gu J, Hood L, Rowen L, Madan A, Qin S, Davis RW, Federspiel NA, Abola AP, Proctor MJ, Myers RM, Schmutz J, Dickson M, Grimwood J, Cox DR, Olson MV, Kaul R, Raymond C, Shimizu N, Kawasaki K, Minoshima S, Evans GA, Athanasiou M, Schultz R, Roe BA, Chen F, Pan H, Ramser J, Lehrach H, Reinhardt R, McCombie WR, de la Bastide M, Dedhia N, Blocker H, Hornischer K, Nordsiek G, Agarwala R, Aravind L, Bailey JA, Bateman A, Batzoglou S, Birney E, Bork P, Brown DG, Burge CB, Cerutti L, Chen HC, Church D, Clamp M, Copley RR, Doerks T, Eddy SR, Eichler EE, Furey TS, Galagan J, Gilbert JG, Harmon C, Hayashizaki Y, Haussler D, Hermjakob H, Hokamp K, Jang W, Johnson LS, Jones TA, Kasif S, Kasprzyk A, Kennedy S, Kent WJ, Kitts P, Koonin EV, Korf I, Kulp D, Lancet D, Lowe TM, McLysaght A, Mikkelsen T, Moran JV, Mulder N, Pollara VJ, Ponting CP, Schuler G, Schultz J, Slater G, Smit AF, Stupka E, Szustakowski J, Thierry-Mieg D, Thierry-Mieg J, Wagner L, Wallis J, Wheeler R, Williams A, Wolf YI, Wolfe KH, Yang SP, Yeh RF, Collins F, Guyer MS, Peterson J, Felsenfeld A, Wetterstrand KA, Patrinos A, Morgan MJ, Szustakowki J, de Jong P, Catanese JJ, Osoegawa K, Shizuya H, Choi S, Chen YJ. 2001. Initial sequencing and analysis of the human genome. Nature. 409:860-921.

27. Miettinen HE, Jarvinen TA, Kellner U, Kauraniemi P, Parwaresch R, Rantala I, Kalimo H, Paljarvi L, Isola J, Haapasalo H. 2000. High topoisomerase IIalpha expression associates with high proliferation rate and and poor prognosis in oligodendrogliomas. Neuropathol Appl Neurobiol. 26:504-512.

28. Moch H, Schraml P, Bubendorf L, Mirlacher M, Kononen J, Gasser T, Mihatsch MJ, Kallioniemi OP, Sauter G. 1999. High-throughput tissue microarray analysis to evaluate genes uncovered by cDNA microarray screening in renal cell carcinoma. Am J Pathol 154:981-986.

29. Mucci NR, Akdas G, Manely S, Rubin MA. 2000. Neuroendocrine expression in metastatic prostate cancer: evaluation of high throughput tissue microarrays to detect heterogeneous protein expression. Hum Pathol 31:406-414.

30. Oka T, Yoshino T, Hayashi K, Ohara N, Nakanishi T, Yamaai Y, Hiraki A, Sogawa CA, Kondo E, Teramoto N, Takahashi K, Tsuchiyama J, Akagi T. 2001. Reduction of hematopoietic cell-specific tyrosine phosphatase SHP-1 gene expression in natural killer cell lymphoma and various types of lymphomas/leukemias : combination analysis with cDNA expression array and tissue microarray. Am J Pathol 159:1495-1505.

31. Perrone EE, Theoharis C, Mucci NR, Hayasaka S, Taylor JM, Cooney KA, Rubin MA. 2000. Tissue microarray assessment of prostate cancer tumor proliferation in African-American and white men. J Natl Cancer Inst. 92:937-939.

32. Richter J, Wagner U, Kononen J, Fijan A, Bruderer J, Schmid U, Ackermann D, Maurer R, Alund G, Knonagel H, Rist M, Wilber K, Anabitarte M, Hering F, Hardmeier T, Schonenberger A, Flury R, Jager P, Fehr JL, Schraml P, Moch H, Mihatsch MJ, Gasser T, Kallioniemi OP, Sauter G. 2000. High-throughput tissue microarray analysis of cyclin E gene amplification and overexpression in urinary bladder cancer. Am J Pathol 157:787-794.

33. Rimm DL, Camp RL, Charette LA, Olsen DA, Provost E. 2001. Amplification of tissue by construction of tissue microarrays. Exp Mol Pathol 70:255-264.

34. Rubin MA, Mucci NR, Figurski J, Fecko A, Pienta KJ, Day ML. 2001. E-cadherin expression in prostate cancer: a broad survey using high-density tissue microarray technology. Hum Pathol 32:690-697.

35. Rubin MA, Dunn R, Strawderman M, Pienta KJ. 2002. Tissue microarray sampling strategy for prostate cancer biomarker analysis. Am J Surg Pathol 26:312-319.

36. Sallinen SL, Sallinen PK, Haapasalo HK, Helin HJ, Helen PT, Schraml P, Kallioniemi OP, Kononen J. 2000. Identification of differentially expressed genes in human gliomas by DNA microarray and tissue chip techniques. Cancer Res 60:6617-6622.

37. Schoenberg Fejzo M, Slamon DJ. 2001. Frozen tumor tissue microarray technology for analysis of tumor RNA, DNA, and proteins. Am J Pathol 159:1645-1650

38. Schraml P, Kononen J, Bubendorf L, Moch H, Bissig H, Nocito A, Mihatsch MJ, Kallioniemi OP and Sauter G. 1999. Tissue microarrays for gene amplification surveys in many different tumor types. Clin Cancer Res 5:1966-1975.

39. Simon R, Richter J, Wagner U, Fijan A, Bruderer J, Schmid U, Ackermann D, Maurer R, Alund G, Knonagel H, Rist M, Wilber K, Anabitarte M, Hering F, Hardmeier T, Schonenberger A, Flury R, Jager P, Fehr JL, Schraml P, Moch H, Mihatsch MJ, Gasser T, Sauter G. 2001. High-throughput tissue microarray analysis of 3p25 (RAF1) and 8p12 (FGFR1) copy number alterations in urinary bladder cancer. Cancer Res 61:4514-4519.

40. Torhorst J, Bucher C, Kononen J, Haas P, Zuber M, Kochli OR, Mross F, Dieterich H, Moch H, Mihatsch M, Kallioniemi OP, Sauter G. 2001. Tissue microarrays for rapid linking of molecular changes to clinical endpoints. Am J Pathol. 159:2249-2256.

41. Xie W, Mertens JC, Reiss DJ, Rimm DL, Camp RL, Haffty BG, Reiss M. 2002. Alterations of Smad signaling in human breast carcinoma are associated with poor outcome: a tissue microarray study. Cancer Res 62:497-505.

42. Venter JC, Adams MD, Myers EW, Li PW, Mural RJ, Sutton GG, Smith HO, Yandell M, Evans CA, Holt RA, Gocayne JD, Amanatides P, Ballew RM, Huson DH, Wortman JR, Zhang Q, Kodira CD, Zheng XH, Chen L, Skupski M, Subramanian G, Thomas PD, Zhang J, Gabor Miklos GL, Nelson C, Broder S, Clark AG, Nadeau J, McKusick VA, Zinder N, Levine AJ, Roberts RJ, Simon M, Slayman C, Hunkapiller M, Bolanos R, Delcher A, Dew I, Fasulo D, Flanigan M, Florea L, Halpern A, Hannenhalli S, Kravitz S, Levy S, Mobarry C, Reinert K, Remington K, Abu-Threideh J, Beasley E, Biddick K, Bonazzi V, Brandon R, Cargill M, Chandramouliswaran I, Charlab R, Chaturvedi K, Deng Z, Di Francesco V, Dunn P, Eilbeck K, Evangelista C, Gabrielian AE, Gan W, Ge W, Gong F, Gu Z, Guan P, Heiman TJ, Higgins ME, Ji RR, Ke Z, Ketchum KA, Lai Z, Lei Y, Li Z, Li J, Liang Y, Lin X, Lu F, Merkulov GV, Milshina N, Moore HM, Naik AK, Narayan VA, Neelam B, Nusskern D, Rusch DB, Salzberg S, Shao W, Shue B, Sun J, Wang Z, Wang A, Wang X, Wang J, Wei M, Wides R, Xiao C, Yan C, Yao A, Ye J, Zhan M, Zhang W, Zhang H, Zhao Q, Zheng L, Zhong F, Zhong W, Zhu S, Zhao S, Gilbert D, Baumhueter S, Spier G, Carter C, Cravchik A, Woodage T, Ali F, An H, Awe A, Baldwin D, Baden H, Barnstead M, Barrow I, Beeson K, Busam D, Carver A, Center A, Cheng ML, Curry L, Danaher S, Davenport L, Desilets R, Dietz S, Dodson K, Doup L, Ferriera S, Garg N, Gluecksmann A, Hart B, Haynes J, Haynes C, Heiner C, Hladun S, Hostin D, Houck J, Howland T, Ibegwam C, Johnson J, Kalush F, Kline L, Koduru S, Love A, Mann F, May D, McCawley S, McIntosh T, McMullen I, Moy M, Moy L, Murphy B, Nelson K, Pfannkoch C, Pratts E, Puri V, Qureshi H, Reardon M, Rodriguez R, Rogers YH, Romblad D, Ruhfel B, Scott R, Sitter C, Smallwood M, Stewart E, Strong R, Suh E, Thomas R, Tint NN, Tse S, Vech C, Wang G, Wetter J, Williams S, Williams M, Windsor S, Winn-Deen E, Wolfe K, Zaveri J, Zaveri K, Abril JF, Guigo R, Campbell MJ, Sjolander KV, Karlak B, Kejariwal A, Mi H, Lazareva B, Hatton T, Narechania A, Diemer K, Muruganujan A, Guo N, Sato S, Bafna V, Istrail S, Lippert R, Schwartz R, Walenz B, Yooseph S, Allen D, Basu A, Baxendale J, Blick L, Caminha M, Carnes-Stine J, Caulk P, Chiang YH, Coyne M, Dahlke C, Mays A, Dombroski M, Donnelly M, Ely D, Esparham S, Fosler C, Gire H, Glanowski S, Glasser K, Glodek A, Gorokhov M, Graham K, Gropman B, Harris M, Heil J, Henderson S, Hoover J, Jennings D, Jordan C, Jordan J, Kasha J, Kagan L, Kraft C, Levitsky A, Lewis M, Liu X, Lopez J, Ma D, Majoros W, McDaniel J, Murphy S, Newman M, Nguyen T, Nguyen N, Nodell M, Pan S, Peck J, Peterson M, Rowe W, Sanders R, Scott J, Simpson M, Smith T, Sprague A, Stockwell T, Turner R, Venter E, Wang M, Wen M, Wu D, Wu M, Xia A, Zandieh A, Zhu X. 2001. The sequence of the human genome. Science 291:1304-1351.

Index